Advanced Technologies in Earth Sciences

Series editors

Ute Münch, Potsdam, Germany
Ludwig Stroink, Potsdam, Germany
Volker Mosbrugger, Frankfurt, Germany
Gerold Wefer, Bremen, Germany

More information about this series at: http://www.springer.com/series/8384

Axel Liebscher · Ute Münch
Editors

Geological Storage of CO_2 – Long Term Security Aspects

GEOTECHNOLOGIEN
Science Report No. 22

Springer

Editors
Axel Liebscher
Telegrafenberg
GFZ German Research Centre
for Geosciences
Centre for Geological Storage CGS
Potsdam
Germany

Ute Münch
Coordination Office
GEOTECHNOLOGIEN
Potsdam
Germany

ISSN 2190-1635 ISSN 2190-1643 (electronic)
Advanced Technologies in Earth Sciences
ISBN 978-3-319-13929-6 ISBN 978-3-319-13930-2 (eBook)
DOI 10.1007/978-3-319-13930-2

Library of Congress Control Number: 2014960205

Springer Cham Heidelberg New York Dordrecht London

Printed on acid-free paper

Springer International Publishing AG Switzerland is part of Springer Science+Business Media (www.springer.com)

Foreword

"Advanced Technologies in Earth Sciences" is based on the German Geoscientific Research and Development Programme "GEOTECHNOLOGIEN" funded by the Federal Ministry of Education and Research (BMBF) and the German Research Foundation (DFG).

This programme comprises a nationwide network of transdisciplinary research projects and incorporates numerous universities, non-university research institutions and companies. The books in this series deal with research results from different innovative geoscientific research areas, interlinking a broad spectrum of disciplines with a view to documenting System Earth as a whole, including its various sub-systems and cycles. The research topics are predefined to meet scientific, socio-political and economic demands for the future.

Ute Münch
Ludwig Stroink
Volker Mosbrugger
Gerold Wefer

Preface

Over the next decades global energy consumption will continue to increase. Despite advances in the transition to renewable energy sources, fossil fuels will nevertheless continue to dominate global primary energy consumption and their combustion will further fuel atmospheric carbon dioxide (CO_2) concentrations and global warming. While energy efficiency and increasing share of non-fossil energy generation will contribute to the de-carbonization of the energy system, *Carbon Capture and Storage (CCS)* is the only available technology that allows reducing CO_2 emissions arising from usage of fossil fuels. CCS may be applied to fossil-fuel-based power generation, industrial processes (steel, cement, refineries) and power generation from biomass; in the latter case, CCS may also result in net negative CO_2 emissions. Central to CCS is the secure and long-term storage of the captured CO_2 in deep geological formations preventing the CO_2 from escaping into the atmosphere. Although geological CO_2 storage builds on experiences from hydrocarbon industry and underground storage of natural and town gas, it poses specific challenges on, amongst others, geochemical and geophysical reservoir understanding, reservoir engineering and storage operation, monitoring techniques, material properties, long-term security and risk assessment and management, as well as communication strategies with the different stakeholders. These challenges have to be addressed and solved to ensure industrial implementation readiness of CCS in the near future.

To address these challenges, the German Federal Ministry of Education and Research (BMBF) has initiated and funded several scientific projects for the development of the necessary scientific and technological knowledge base for geological CO_2 storage in the framework of its Research and Development Programme GEOTECHNOLOGIEN. Since 2005, a total of 33 research projects on the different aspects of geological CO_2 storage have been and are still funded within three successive funding periods. Due to this effort, Germany has gained substantial scientific and technological know-how in the field of geological CO_2 storage. From a scientific and technological perspective, the results and outcomes of the different projects lay the fundamentals for secure and reliable implementation of demo-scale projects on CO_2 storage.

In eleven individual chapters, this volume compiles and reviews the main results from the most recent research and development projects on geological storage of CO_2 funded by the third funding period under the GEOTECHNOLOGIEN Programme. The projects concentrate on the development of innovative technologies and processes for the reliable assessment of the operational/long-term safety of potential and existing CO_2 storages. The main goals were the development of methods for the identification and comprehension of security relevant processes of possible weak points (e.g. faults) of the cap rock—from the reservoir to the surface as well as the prognosis of the influence of industrially sequestered CO_2 on reservoir rock and cap rock. Furthermore, experimental analysis of thermodynamic and kinetic data were carried out to model complex reactions between injected CO_2, associate material, natural formation water and mineral phases. In addition, further developments on reactive multiphase transport models were realized. In this context, the analysis of coupled hydraulic and geo-mechanical processes in the reservoir and cap rock with regard to the deformation and mechanical reactions of the cap rock and alterations at the surface are explained in the volume. Furthermore, hydrodynamical reservoir modelling with regard to the replacement of formation fluids to avoid contaminations (e.g. of the groundwater) were developed to ensure the long-term security of a CO_2 storage site. Therefore, the development and realization of effective and efficient technologies to monitor the lateral CO_2 and pressure extension and footprint and the replacement of brines were investigated.

Geophysical methods, numerical modelling procedures, microbiological investigations as well as the development of new optical sensor technologies and their combination will be explained in detail in this volume. Laboratory as well as field studies in Germany, e.g. at the Ketzin pilot site, but also with international partners in Canada and Australia were carried out to merge the already existing expertise in the field of Carbon Capture and Storage (CCS).

In addition to the technical natural science-dominated methods and techniques, social scientists were involved in several projects to evaluate the opportunities and limits for the acceptance of CCS in Germany. On the basis of data from completed projects, using new research approaches to close the existing gaps in acceptance research were investigated. Furthermore, analyses of factors influencing acceptance as well as an assessment of participatory methods were analysed.

Axel Liebscher
Head of Centre for Geological Storage CGS,
GFZ German Research Centre for Geosciences

Ute Münch
Head of the GEOTECHNOLOGIEN coordination office

Contents

Joint Research Project CO_2MAN (CO_2MAN Reservoir Management): Continuation of Research and Development Work for CO_2 Storage at the Ketzin Pilot Site

Sonja Martens, Ronald Conze, Marco De Lucia, Jan Henninges, Thomas Kempka, Axel Liebscher, Stefan Lüth, Fabian Möller, Ben Norden, Bernhard Prevedel, Cornelia Schmidt-Hattenberger, Alexandra Szizybalski, Andrea Vieth-Hillebrand, Hilke Würdemann, Kornelia Zemke and Martin Zimmer

Abstract The joint project CO_2MAN (CO_2 Reservoir Management) was a scientific programme accompanying geological CO_2 storage at the Ketzin pilot site in the German Federal State of Brandenburg. The project which was funded by the German Federal Ministry of Education and Research (BMBF) from 1 September 2010 to 31 December 2013 enclosed six scientific institutions and seven industry partners. The Ketzin pilot site is the longest-operating on-shore CO_2 storage site in Europe. In advance of the CO_2MAN project, CO_2 injection had already started in June 2008 and storage operation had been accompanied by one of the world's most extensive scientific research and development programmes. The CO_2MAN project took advantage of this unique potential of the site in order to answer further technical and scientific questions on CO_2 storage and to inform about this highly debated technology. The CO_2MAN project demonstrates safe geological CO_2 storage at the Ketzin site on a pilot scale.

1 Introduction

CO_2MAN which stands for CO_2 Reservoir Management was a scientific programme accompanying geological CO_2 storage at the Ketzin pilot site. Within the framework of CO_2MAN a total of six German scientific institutions and seven

S. Martens (✉) · R. Conze · M. De Lucia · J. Henninges · T. Kempka · A. Liebscher · S. Lüth
F. Möller · B. Norden · B. Prevedel · C. Schmidt-Hattenberger · A. Szizybalski
A. Vieth-Hillebrand · H. Würdemann · K. Zemke · M. Zimmer
Helmholtz Centre Potsdam, GFZ German Research Centre for Geosciences,
Telegrafenberg, 14473 Potsdam, Germany
e-mail: martens@gfz-potsdam.de

A. Liebscher and U. Münch (eds.), *Geological Storage of CO_2 – Long Term Security Aspects*, Advanced Technologies in Earth Sciences,
DOI 10.1007/978-3-319-13930-2_1

industry partners from Norway, Austria and Germany participated (Fig. 1). The joint project was coordinated by the GFZ German Research Centre for Geosciences in close cooperation with all partners.

The Ketzin pilot site for geological storage of CO_2 is located about 25 km west of Berlin in the German Federal State of Brandenburg (Fig. 2). It is the longest operating European on-shore CO_2 storage pilot site and provides an in situ laboratory for CO_2 storage in a saline aquifer of the Northeast German Basin. Injection of CO_2 at Ketzin started on 30 June 2008 and was accompanied by one of the most comprehensive scientific research and development (R&D) programmes worldwide with key objectives being R&D on injection operation, monitoring and modelling. The first results that had been achieved at the pilot site Ketzin were promising (Schilling et al. 2009a; Würdemann et al. 2010). However, for a final assessment of the technology further investigations were necessary (Stroink et al. 2009). Hence the aim of the CO_2MAN project was to take advantage of the unique potential and the infrastructure of the Ketzin pilot site and to continue CO_2 injection and the R&D activities with a dedicated public outreach programme. The key objectives of the joint research project were:

- to monitor the migration of the injected CO_2, to determine the sensitivity of individual monitoring methods and to test and develop geophysical monitoring concepts for CO_2 storage sites,
- to characterize and quantify CO_2-induced interactions between fluid, rock and microbial community in the storage system,
- to validate tools for static modelling and dynamic simulations for the Ketzin pilot site, and
- to inform the public, stakeholders, decision makers and regulatory authorities about CO_2 storage.

Fig. 1 CO_2MAN partners from academia and industry. The consortium includes GFZ German Centre Research for Geosciences (coordinator) in Potsdam, Friedrich-Alexander University of Erlangen-Nürnberg, University of Stuttgart (Department of Hydromechanics and Modelling of Hydrosystems (LH2)), University of Leipzig and the Helmholtz Centre for Environmental Research -UFZ in Leipzig. Industry partners are Dillinger Hüttenwerke, OMV, RWE, Saarstahl, Statoil, Vattenfall and VGS

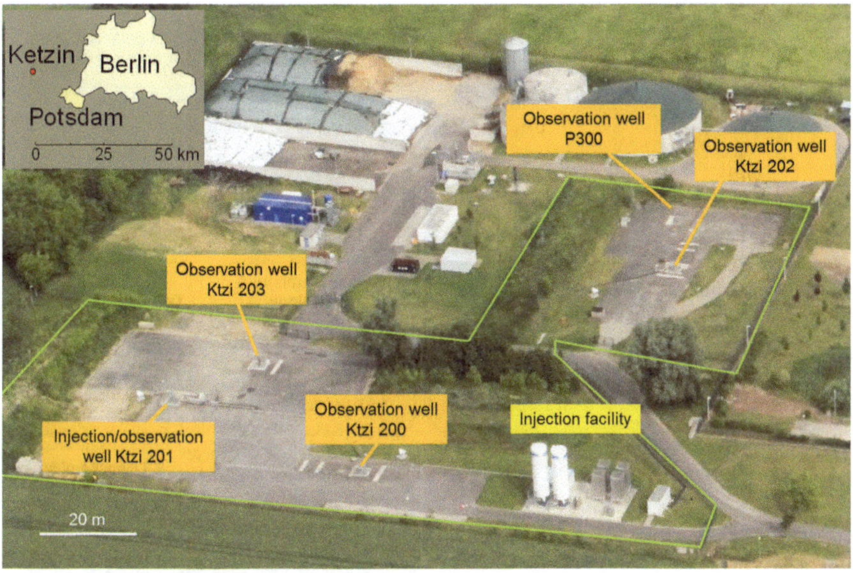

Fig. 2 Geographic location of the Ketzin pilot site in the Federal State of Brandenburg and aerial picture of the site with its research infrastructure in June 2013

The core of the project consisted of the four work packages "Research Infrastructure", "Geophysical Monitoring", "Reservoir Processes" and "Modelling and Simulations". Each work package comprised three to five sub-projects (Fig. 3). The work packages were complemented by the activities "Data Management", "Information Centre Ketzin" and "Project Management". This contribution comprises an overview of the work carried out and the results obtained under the joint research project CO_2MAN.

2 Research Infrastructure at the Ketzin Pilot Site

The Ketzin project on CO_2 storage was initiated by the GFZ German Research Centre for Geosciences in 2004 (Würdemann et al. 2010). Hence, there was already a certain research infrastructure available at the pilot site and the CO_2 injection was on-going since June 2008 when the CO_2MAN joint project started in September 2010. The infrastructure consisted of an injection facility (Fig. 4) with two intermediate CO_2 storage tanks and four large ambient air heaters as well as the injection/observation well Ktzi 201 and the pure observation wells Ktzi 200 and 202. In the course of the CO_2MAN project the CO_2 injection was continued until the end of August 2013 and two additional monitoring wells were drilled.

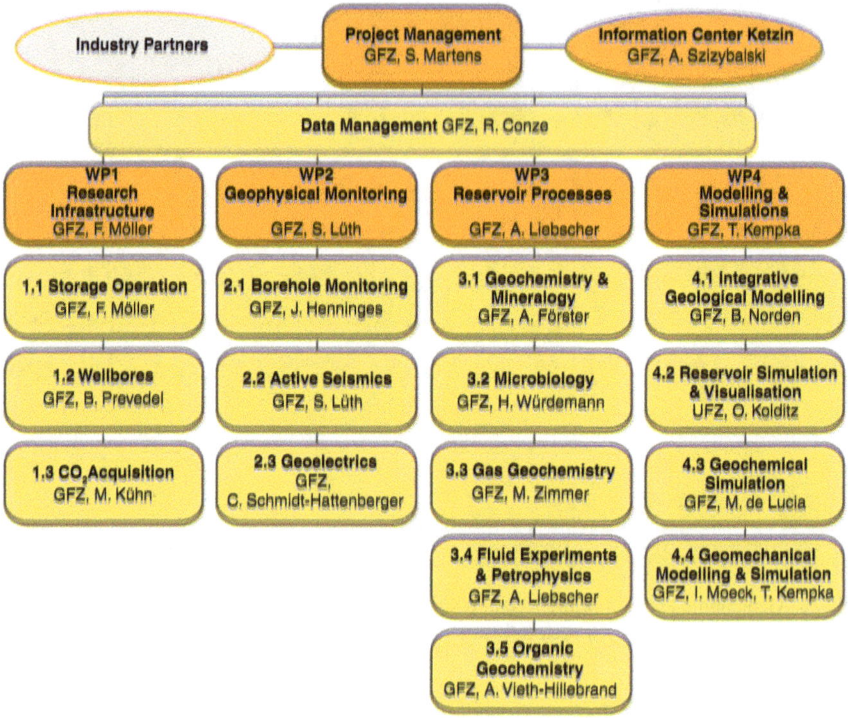

Fig. 3 Organization chart of the joint project CO_2MAN, showing the different work packages, sub-projects and respective lead scientists

2.1 Storage Operation

Injection of CO_2 at the Ketzin pilot site started already prior to the CO_2MAN project on 30 June 2008 and ended on 29 August 2013. The CO_2 was delivered by road tankers in a liquid state and stored at about −18 °C and 21 bars in the two intermediate storage tanks on site (Fig. 4). Prior to injection the pressure was raised by plunger pumps to the necessary injection pressure and the CO_2 was heated by ambient air heaters and an electrical heater. Then the CO_2 was transported via a pipeline of about 100 m length to the injection well Ktzi 201. A total amount of 67 kt of CO_2 was safely injected over the more than 5 years period. Thereof, 29,337 t were injected within the CO_2MAN project (Fig. 5). During most of the time food-grade CO_2 (purity > 99.9 vol%) was used with monthly injection rates between 1,000 and 2,300 t. From May to June 2011, 1,515 t of CO_2 captured from the Vattenfall Schwarze Pumpe oxyfuel pilot plant (purity > 99.7 vol%) were used.

CO_2 injection was accompanied by a comprehensive operational pressure-temperature monitoring programme (Liebscher et al. 2013). Due to CO_2 injection the reservoir pressure increased to about 76–79 bars already after eighth months of injection (Fig. 5). After this initial increase the reservoir pressure slightly decreased

Fig. 4 Injection facility at Ketzin with two intermediate CO_2 storage tanks (*left*) and ambient air heaters (*middle*) before decommissioning in December 2013

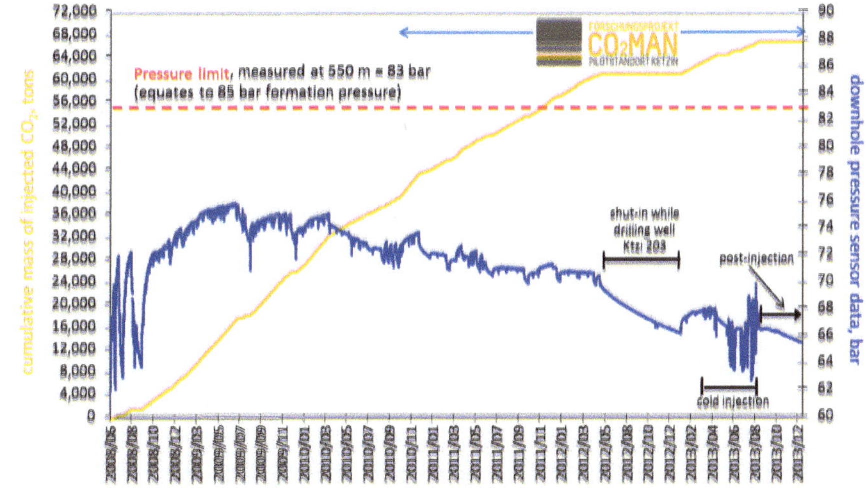

Fig. 5 Overall injection history at the Ketzin pilot site showing cumulative mass of injected CO_2 (*orange*) and measured pressure at 550 m depth in well Ktzi 201 (*blue*) from June 2008 to December 2013. The reservoir pressure at 630 m is about 2 bars higher than the measured pressure at 550 m. The *red line* refers to the maximum permitted pressure of 85 bars at reservoir depth (=83 bars at 550 m) given by the Mining Authority

and stabilized between about 72 and 75 bars reflecting a stable injection regime (Liebscher et al. 2013). Between March and July 2013, the injection temperature was lowered stepwise down to 10 °C to demonstrate the feasibility of a "cold injection" process, i.e. without pre-heating the CO_2. Despite high dynamics of the measured pressure within the injection well Ktzi 201 (Fig. 5), the entire injection process ran smoothly and the experiment could be carried out successfully.

Longer shut-in phases, e.g. during drilling of the Ktzi 203 well, and the beginning of the post-injection phase at the end of August 2013 were characterized by a continuous decrease of the reservoir pressure. The maximum approved reservoir pressure as defined by the Mining Authority is 85 bars at 630 m depth, which transforms into 83 bars at 550 m depth, i.e. installation depth of the pressure sensor in well Ktzi 201. During the entire storage operation the reservoir pressure was always well below this maximum approved value. The injection facility was dismantled in December 2013 whereas post-injection monitoring continues.

During the CO_2MAN project the four deep wells at Ketzin were inspected by comprehensive wellbore logging campaigns on an annual basis. Logging included magnetic measurements, saturation measurements using pulsed neutron gamma (PNG) logging and borehole inspections with a video camera to enable visual inspection. Based on the results of the logging campaign and the video material the good condition of the wells could be repeatedly confirmed. Fluid samples from the wells were gained during all logging campaigns and used for further investigations on reservoir processes (compare Sect. 4).

The storage operation at Ketzin is carried out under the framework of the German Mining Law. The injection phase was realized with well trained personnel from the gas storage industry and monitored, in addition to the scientific investigations, by two consulting engineering companies. The consultants reported to the mining authority independently from the GFZ on a regular basis. This constellation proved to be successful in the way that over more than 5 years of injection no health, safety or environmental issues occurred. The safe storage of CO_2 at Ketzin was only possible due to a close and trustful cooperation between the storage operator, the consultants, the mining authority and of course all scientific personnel. This also ensured that all scientific experiments within CO_2MAN could be carried out while at the same time all legal requirements were met.

2.2 *Drilling and Well Abandonment*

Prior to the start of the CO_2MAN project three wells (Ktzi 200, Ktzi 202, Ktzi 202) had already been drilled at Ketzin in 2007 and were completed with a smart casing concept (Prevedel et al. 2009). To meet the scientific needs of the CO_2MAN project two additional monitoring wells (P300, Ktzi 203) were drilled in 2011 and 2012.

The well P300 was drilled in summer 2011 to 446 m depth into the lowermost aquifer (Exter Formation) above the cap rock of the CO_2 storage reservoir (Fig. 6) to allow for above-zone monitoring in the indicator horizon. The geological succession

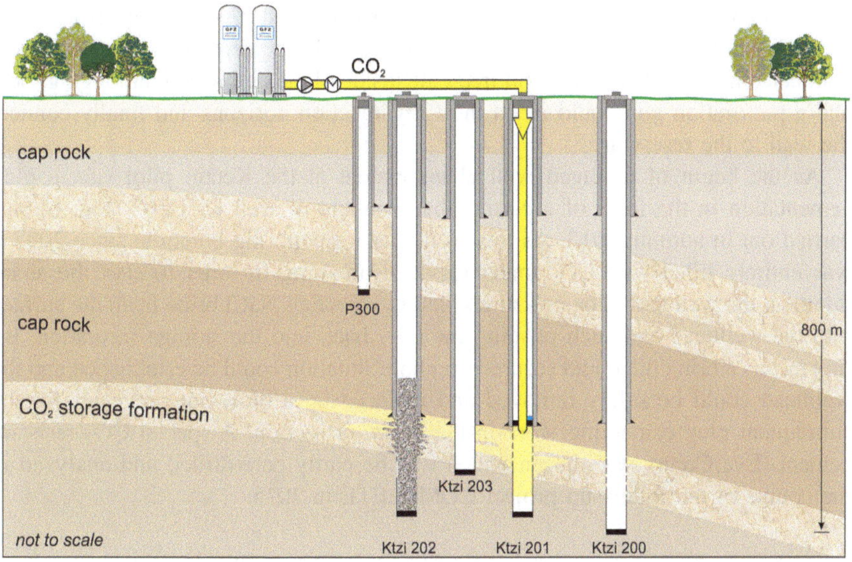

Fig. 6 Schematic vertical profile of the Ketzin pilot site showing its four deep wells (Ktzi 200 to Ktzi 203) and one shallow observation well (P300). Both wells P300 and Ktzi 203 were drilled in the course of the CO_2MAN project. A plug cementation was carried out in the lower part of the observation well Ktzi 202 in autumn 2013

encountered at well P300 (Martens et al. 2013) was very similar to the geology observed at the other Ketzin boreholes (Förster et al. 2009). 40.9 m of best quality cores could be retrieved, allowing a detailed analysis of the mineralogy and geochemical properties of the Exter Formation. The well was completed with a combined high resolution pressure and temperature sensor at 418 m depth, a level sensor at 21 m and a non-cemented monitoring string with a U-tube fluid sampling system to allow for pressure and fluid monitoring of the indicator horizon (compare Sect. 3.4).

In August and September 2012, the well Ktzi 203 was drilled about 25 m apart from the injection well Ktzi 201 (Fig. 2) to 701 m depth to gain cores of the cap rock and reservoir sandstones that were in contact with the injected CO_2 for more than 4 years. This well was planned and drilled in accordance with a slim-hole concept and recovered 90 m of cores. The well showed a geological profile very similar to the Ktzi 201 borehole. At the top of the storage (Stuttgart) formation the approximately 18-m thick reservoir sandstone is present. At final depth the reservoir section was cased and cemented with fiber glass casing pipe for testing the applicability of corrosion resistant composite pipe material for future storage operation and abandonment steps. The well Ktzi 203 was completed with a combined 3.5″ steel/fiber glass production casing with two distributed temperature sensing (DTS) cables and two pressure-temperature sensors on the outside at 305 and 610 m depth. Subsequent wireline logging runs in the Ktzi 203 well revealed an obstruction inside the 3.5″ production casing at 557 m that required a work-over operation in order to remove this section which was blocked by cement. This operation was conducted in April/May 2013 and freed the inside of the casing so that logging tools and a perforation gun could reach the CO_2 injection horizons and finally connect the well to the reservoir.

As the begin of a staged well abandonment at the Ketzin pilot site, a plug cementation in the form of a partial abandonment of well Ktzi 202 (Fig. 6) was carried out in autumn 2013. As this well was a monitoring borehole since 2007 it was entirely filled with CO_2 under elevated pressure. In order to start the abandonment the well had to be pressure killed by injecting NaCl brine from the surface into the wellhead and such pushing the CO_2 back into the storage formation. By that means a brine filled and secured borehole situation could be established and the wellhead could be safely removed and replaced by a blow-out preventer for the subsequent plug cementing work in the reservoir section. A special CO_2 resistant cement (EverCrete) was chosen which will be partly core-drilled and analyzed in the course of the follow-up project COMPLETE in 2015.

3 Monitoring of the Ketzin Pilot Site

R&D on monitoring of the Ketzin pilot site was one of the key objectives of the CO_2MAN project. Already in advance of this joint research project, a comprehensive monitoring concept which combined operational, geophysical, geochemical and microbiological monitoring techniques had been tested and established at the pilot site (Würdemann et al. 2010).

In the framework of CO_2MAN, geophysical techniques including borehole monitoring, active seismic and geoelectric methods were continued in order to further monitor the migration of the injected CO_2 on different scales. Gas geochemical monitoring focused on CO_2 soil flux measurements at the surface and fluid sampling from the wells via permanently installed capillary riser tubes and a U-tube system.

3.1 Borehole Monitoring

The migration of the injected CO_2 close to the boreholes was monitored using a combination of different well logging techniques. The temperature distribution along the deep boreholes was continuously recorded with permanently installed distributed temperature sensing (DTS) cables. The saturation conditions within the CO_2 storage horizon and the cap rock were investigated using pulsed neutron-gamma (PNG) logging. Furthermore, the heat-pulse method was tested for monitoring of saturation changes. Here the formation thermal conductivity is determined based on temperature changes under the influence of a controlled heat source. For this purpose, a combined opto-electric sensor-heater cable was installed behind casing in the new observation well Ktzi 203.

After filling up with CO_2 the temperature conditions in the deep observation wells are controlled by a heat-pipe process (Henninges et al. 2011). Temperature changes are caused by phase transitions during evaporation and condensation of CO_2, and characteristic temperature gradients are established in the two-phase zone. The pressure evolution along the borehole depends on the composition and the phase distribution within the fluid column (Loizzo et al. 2013).

For the Ktzi 203 well, in situ thermal conductivities were calculated by numerical inversion of the data acquired during heat-pulse measurements. The thermal conductivity profiles show a good correlation with lithological changes, with values ranging between 1 and 4.5 $Wm^{-1}K^{-1}$. They display similar characteristics as the profiles previously determined for the other wells (Freifeld et al. 2009). Within the storage horizon, changes of thermal conductivity in the order of 20–30 % could be observed between different measurements, but the results show a high sensitivity against external thermal influences.

PNG logs were acquired within all deep wells in about annual intervals and saturation conditions were calculated by comparison of the data with previously recorded baseline and repeat measurements (Fig. 7). The highest CO_2 saturations

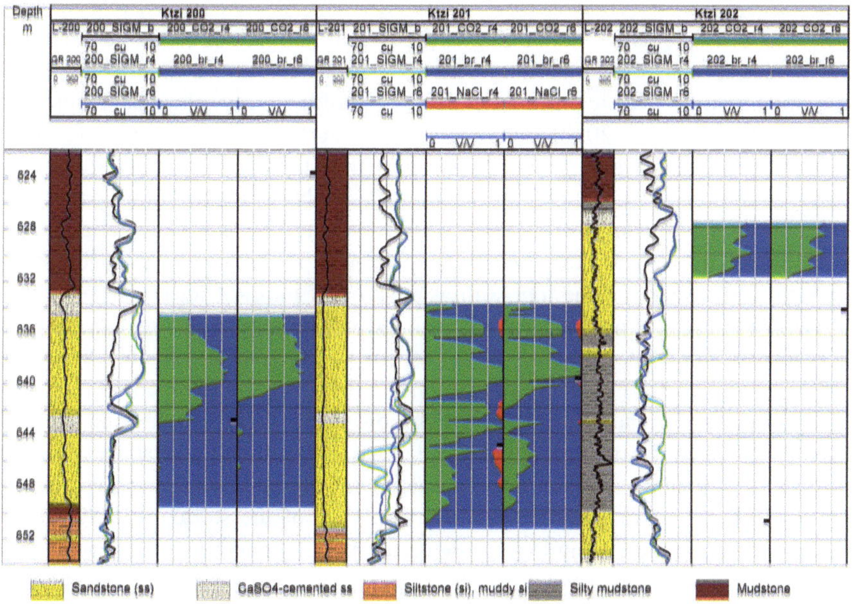

Fig. 7 PNG logging data and calculated CO_2, brine (br) and halite (NaCl) saturations in the wells Ktzi 200 (*left*), Ktzi 201 (*middle*) and Ktzi 202 (*right*). SIGM: measured macroscopic formation capture cross-section; b: baseline (June 2008), r4: repeat 4 (March 2011), r6: repeat 6 (October 2012). The positions of the brine levels within the wells are indicated with *black* markers. Lithology after Förster et al. (2010)

occur at the injection well Ktzi 201, with average values of 68 %, and up to 100 % locally. At the observation wells CO_2 saturations are lower, with average values of > 60 % at well Ktzi 200, and a further decrease towards well Ktzi 202 (averages < 60 %). A new PNG saturation model for CO_2 and NaCl brine was developed which besides displacement also accounts for evaporation and precipitation processes (Baumann 2013; Baumann et al. 2014). Based on this model and the PNG measurements salt precipitation at the CO_2-brine contact in the Ktzi 201 near-well area could be shown for the first time.

Within the cap rock, no indications for accumulation of CO_2 in shallower aquifers were observed. This is important evidence that no significant migration of CO_2 along the boreholes is occurring. The calculated saturations were used as input parameters for estimates of the CO_2 mass contained within the storage horizon based on seismic data (Ivanova et al. 2012; see below) and for evaluation of electrical resistivity tomography data (Bergmann et al. 2012; see below). The established fiber-optic sensor cable network also exhibits favorable properties for seismic surveys using the newly emerging method of distributed acoustic sensing DAS (Daley et al. 2013).

3.2 Seismics

The main task of seismic monitoring at the Ketzin pilot site is to image the lateral and vertical propagation of the injected CO_2 in the reservoir. To this end, high-resolution vertical seismic profiling (VSP), star-profile surveys close to the injection location (Ivandic et al. 2012) and large scale 3D surface seismic surveys (Ivanova et al. 2012, 2013) were repeated providing time-lapse observations at various scales. Additionally, the emerging technology applying a fiber optic cable as a seismic (acoustic) receiver array DAS was investigated on site (Daley et al. 2013). As all four deep wells on the site are equipped with a fiber optic cable, conditions are ideal for a simultaneous four-well acquisition of multi-offset DAS-VSP data.

The VSP and star-profile surveys, acquired in February 2011, revealed a clear CO_2 related amplitude signature at the top of the storage formation. Due to the limited spatial aperture of these measurements the CO_2 signature was restricted to the close vicinity of the injection well and showed only a part of the complete CO_2 signature imaged by the previous and subsequent full 3D repeat surveys. An analysis of time-lapse amplitude variations in the vertical direction showed clearly that no CO_2 signature was detected above the top of the storage (Stuttgart) formation indicating there is no leakage detected by high resolution reflection seismic surveys. It could also be shown that the sparse acquisition geometry, concentrating on seven profiles (“star”) in the area close to the injection site, is able to detect the CO_2 in the reservoir. However, the time-lapse data are characterized by a smaller degree of repeatability than are the time-lapse data of the full 3D repeat measurements (Ivandic et al. 2012).

The second 3D repeat survey was acquired in autumn 2012 after 61 kt CO_2 injected in the storage formation. The data were time-lapse processed and amplitude variations were extracted for the top of the Stuttgart Formation. The lateral distribution of time-lapse amplitudes at the top of the Stuttgart Formation shows a high degree of anisotropic propagation and confirms several features of propagation detected by the first repeat survey (2009) which is shown in Fig. 8. After the injection of 61 kt the CO_2 could be imaged with a west-east extension of about 700 m in autumn 2012 (Fig. 8, right).

For seismic reservoir monitoring, the use of fiber optic cables is currently discussed as an emerging technology with considerable potential of replacing conventional wireline-based seismic acquisition in boreholes and also in surface applications (Parker et al. 2014). In May 2013, a simultaneous DAS-VSP survey was acquired using 23 vibro points and acquiring the seismic wave field along the fiber optic cable deployed in four wells and with a spatial sampling of 1 m. The survey was performed within 4 days. The acquired DAS-VSP shot gathers show clear onsets of the downgoing compressional wave and of reflected upgoing waves. The data acquired in this survey are the basis for a high-resolution 3D imaging of the reservoir layer between the injection and monitoring wells of the Ketzin pilot site.

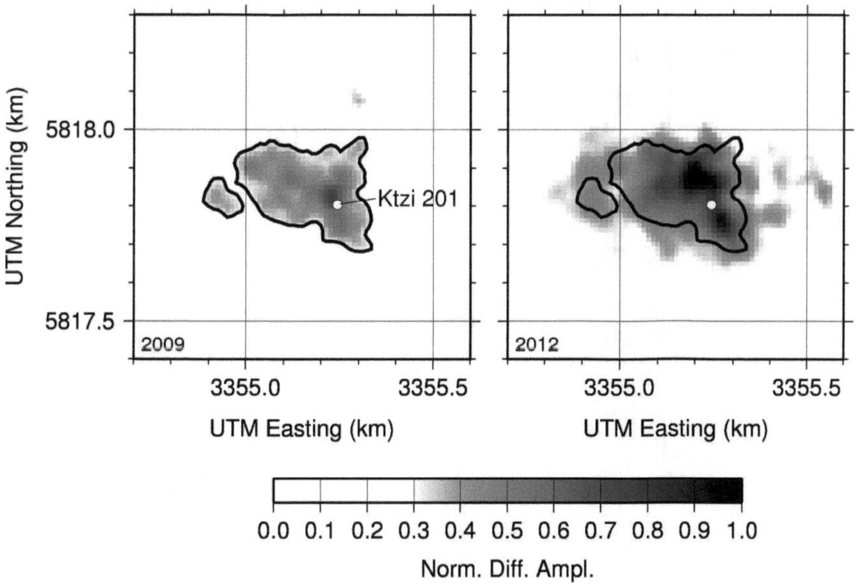

Fig. 8 Map displaying the normalized time-lapse amplitudes at the top of the Stuttgart Formation for the first 3D repeat survey in 2009 (*left*) and the second repeat survey in 2012 (*right*). Time-lapse amplitudes exceeding a background noise level of 0.3 are displayed in *grey-scales*. The location of the injection well (Ktzi 201) is indicated by a *white circle*. For the comparison of the lateral extents of the CO_2 in 2009 and 2012, the contour of the 2009 image (*black line*) has been projected onto the 2012 image

3.3 Geoelectrics

The geoelectrical monitoring programme at Ketzin comprised direct current (DC) geoelectric measurements in three different setups: (1) surface-to-surface, (2) surface-to-downhole and (3) crosshole data acquisition. The first two setups were performed as dipole-dipole measurements on two crossed profiles (length ~4.8 km each of them) and two sparsely settled rings with radii of 0.8 km and 1.5 km, respectively. In addition, the setups (2) and (3) made use of 45 electrodes permanently installed in the wells Ktzi 200, Ktzi 201 and Ktzi 202 (Fig. 9).

At a detection level of about ~600 t of injected CO_2 the geoelectric measurements indicated relevant subsurface resistivity changes associated with the

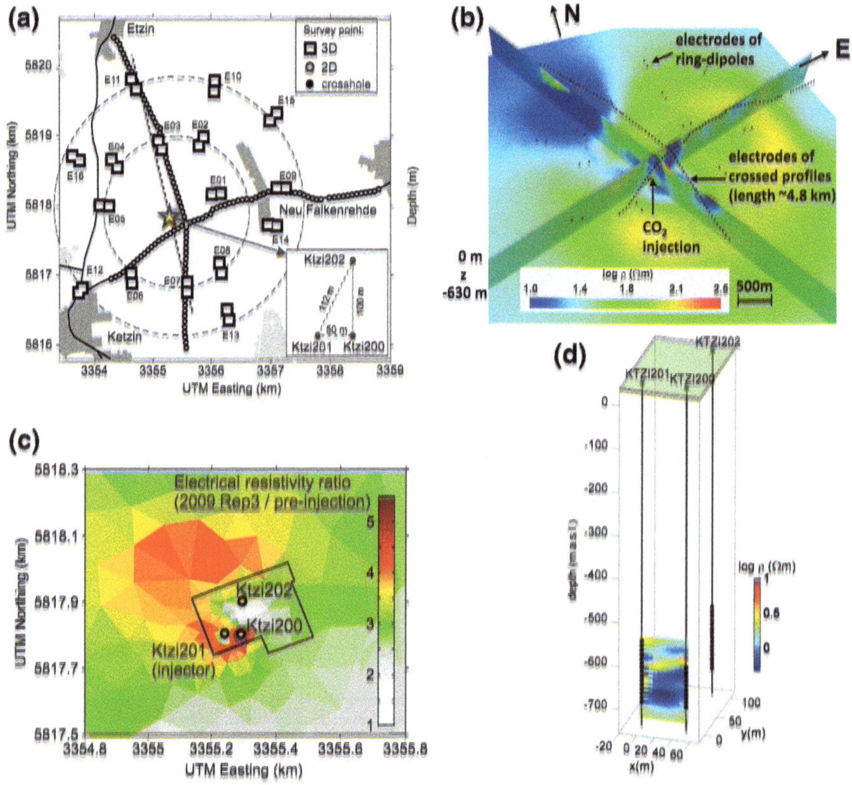

Fig. 9 **a** Schematic of the DC geoelectric measurement concept at the Ketzin pilot site. Large-scale surveys were acquired in 2008, 2009, 2011 and 2012. Weekly-measured crosshole surveys were conducted in the wells Ktzi 200, Ktzi 201 and Ktzi 202 (see inlay for the borehole distances). The major results are: **b** 3D resistivity distribution from inversion of the crossed profiles and the surface-downhole survey in 2011, **c** Resistivity change (repeat 3/2009 vs. baseline) from constrained inversion of surface-downhole data (modified after Bergmann et al. 2014), and **d** Corresponding crosshole results in the major observation plane Ktzi 200–Ktzi 201

migration of CO_2 in a depth of about 630 m. The measured resistivity contrast was in consistency with laboratory measurements on Ketzin sandstone core samples, where Archie fluid substitution revealed a resistivity increase of a factor of about 3 (Kiessling et al. 2010). The permanent downhole electrode array was the subject of engineering developments, such as modular system components, automated optimization of data pre-processing and remote-controlled data acquisition in order to achieve the operational requirements of CO_2 storage sites. From the weekly-measured crosshole data (Schmidt-Hattenberger et al. 2012) and the periodically measured large-scale surface-downhole surveys (Bergmann et al. 2012) consistent time-lapse images of the CO_2 plume migration were derived which correlated fairly well with other monitoring results obtained from seismic surveys, borehole logging and geochemical data.

As a promising tool of geophysical data integration, a structurally constrained inversion approach was applied that incorporates seismic structural information as a priori information into the resistivity inversion (Bergmann et al. 2014). The resulting time-lapse resistivity signature of the constrained inversion was found to collocate clearer with the time-lapse signature from the repeated 3D seismic investigations. The asymmetrical extension of this signature indicates preferential CO_2 migration towards the northwest direction which was also in good agreement with the results from the seismic interpretation.

3.4 Gas Geochemistry

Long-term background data on the natural spatial and timely CO_2 distribution and variability are indispensable for a reliable monitoring and the detection of a potential leakage. In order to obtain this information for the Ketzin pilot site, gas-chemical and isotope investigations have been performed since 2005. Up to now, no indication of any CO_2 leakage has been detected with this comprehensive gas monitoring network system.

The monitoring network comprises 20 sampling locations for soil gas flux, soil moisture and temperature measurements distributed across an area of approximately 2 km × 2 km around the pilot site. In March 2011, eight permanent automated soil gas samplers were added in the direct vicinity of the boreholes together with a meteorological station. Since the start of injection in 2008, no change in soil CO_2 gas flux could be detected as compared to the pre-injection baseline (Zimmer et al. 2011a; Martens et al. 2013). Mean CO_2 flux as averaged over all sampling locations ranged from 2.4 to 3.5 $\mu mol/m^2s$ for the pre-injection period and from 2.2 to 2.5 $\mu mol/m^2s$ after the start of injection (Zimmer et al. 2011a). The spatial variability of soil CO_2 gas flux is 1.0–4.5 $\mu mol/m^2s$ for all sampling locations reflecting the different organic carbon and nitrate contents, both serving as nutrients for bacterial life in the soil. The data show that soil temperature is the key factor controlling the biogenic CO_2 production and subsequently the CO_2 flux rate.

A U-tube system in the shallow observation well P300 enables above-zone monitoring and the possible detection of a potential leakage through the first cap rock at an earliest possible stage. Formation water from well P300 was permanently sampled from a depth of 417 m (Exter Formation) and analyzed for dissolved cations, anions, gases and $^{12}C/^{13}C$ isotope ratio of CO_2 and revealed no impact of the injected CO_2 on the Exter Formation.

From March 2010 to October 2011 a riser tube was installed in well Ktzi 200 which allowed for continuous sampling and analyses of gas from 600 m depth. The measured gas composition was relatively constant with about 99 % CO_2 and traces of nitrogen, helium and methane. Two tracer tests were performed where both krypton and sulfur hexafluoride were added in the injection well Ktzi 201 in May and June 2011. Both gaseous tracers were detected in well Ktzi 200 after the injection of 608 and 701 t of CO_2, respectively, since the tracer test started. In October 2011 the riser tube was transferred to well Ktzi 202. The analyzed gas composition from 600 m depth showed constant values until the end of the measurements in October 2013 and consists of 99.5 % CO_2 with traces of nitrogen, helium and methane. Following the partial closure of the observation well Ktzi 202 in autumn 2013, a gas membrane sensor (Zimmer et al. 2011b) for real time observation of gas at depth was installed at 500 m (21 m above the cement head) to monitor the tightness of the cementation.

Since the beginning of the CO_2 injection in 2008, stable isotope measurements have been conducted for a detailed geochemical characterization of the reservoir and overlying formations, comprising $\delta^{13}C$ and $\delta^{18}O$ data of brine dissolved inorganic carbon (DIC) and H_2O (Nowak et al. 2013). Isotope measurements in connection with gas tracer tests were also carried out when CO_2 from the Schwarze Pumpe oxyfuel pilot plant was used for injection in May and June 2011 (Martens et al. 2012). The $\delta^{13}C$ of DIC proved to effectively trace the migration of the injected CO_2 at Ketzin (Myrttinen et al. 2010). When the $\delta^{13}C$ CO_2 isotopic composition of gas samples from the wellhead of Ktzi 201 and well Ktzi 200 were analyzed, a change in the $^{13}C/^{12}C$ composition of the CO_2 was detected during the temporary use of CO_2 from Schwarze Pumpe (Martens et al. 2012).

4 Fluid Experiments and Processes in the Storage Reservoir

In order to examine the potential interactions between injected CO_2, formation fluid, the storage system and its microbial community at the Ketzin pilot site, laboratory experiments and investigations of samples from the site were carried out. On the one hand, the processes occurring in the reservoir should be characterized and quantified. On the other hand measurements were taken to monitor these processes. The main focus was on the study of fluid, gas and rock samples from

Ketzin, especially from the newly drilled observation well Ktzi 203. The work on the natural samples was supplemented by experimental studies under defined laboratory conditions.

4.1 Fluid Experiments and Petrophysics

Geochemical experiments on the fractionation of Fe, Cu and Zn between CO_2 and formation fluid showed that CO_2 acts as a solvent for trace elements and mobilizes small but measurable amounts of Fe, Cu, Zn regardless of the CO_2 density. By CO_2 dissolved organic compounds act as potential complexing agents and increased the extraction ability for trace elements. In terms of long-term CO_2 storage, the potential consequences of these results include the precipitation of carbonate minerals in shallower, more distal regions of the aquifer and the transferal of metals to adjacent aquifer systems (Rempel et al. 2011).

Several batch experiments under in situ conditions have been performed using CO_2, formation fluid and rock samples from Ketzin wells (Ktzi 201, 202). Results neither show clear changes nor uniform trends over time (up to 40 months) with regard to porosity, pore-size distribution, capillary pressure or geomechanical parameters. Nuclear magnetic resonance (NMR) spectroscopy and mercury injection porosimetry (MIP) were used to characterize samples before and after experiments. For two core sets geomechanical parameters were determined before and after CO_2 treatment. Baseline measurements are consistent with porosity logging data of corresponding wells and show high variability due to the heterogeneous lithology of the Stuttgart Formation. This natural variability of the reservoir hampers the comparability of whole rock samples before and after the experiments. Nonetheless, several lines of evidence indicate that mineralogical-geochemical changes occur due to CO_2 exposure; but these are quantitatively subordinate. In conclusion, effects of injected CO_2 on the reservoir system integrity are minor (Fischer 2013; Fischer et al. 2013).

Porosity measurements on core samples recovered before the injection and the corresponding NMR log (Ktzi 201) compared with porosity measurements of approximately 100 cores of the newly drilled well Ktzi 203 (gained after about 4 years of CO_2 exposure) show that the impact of CO_2 injection on pore size related properties of reservoir and cap rocks is minor and within the natural variability (Fig. 10). The variation of the porosity estimated by different methods is generally low for the corresponding depths of the lithological sections. The mineralogical investigations on the samples show also no significant dissolution or precipitation of minerals and find only minor quantities (usually < 2 vol%) of various species of newly precipitated carbonates with three potential CO_3^{2-} sources: dolomite dissolution, reactions of injected CO_2 with the formation fluid, and the drill mud (Bock et al. 2013). The influence of the potash-containing drilling fluid of the well Ktzi 203 could be determined by porosity investigation on twin samples from inner and outer parts of the cores. Especially for porous outer core samples NMR core

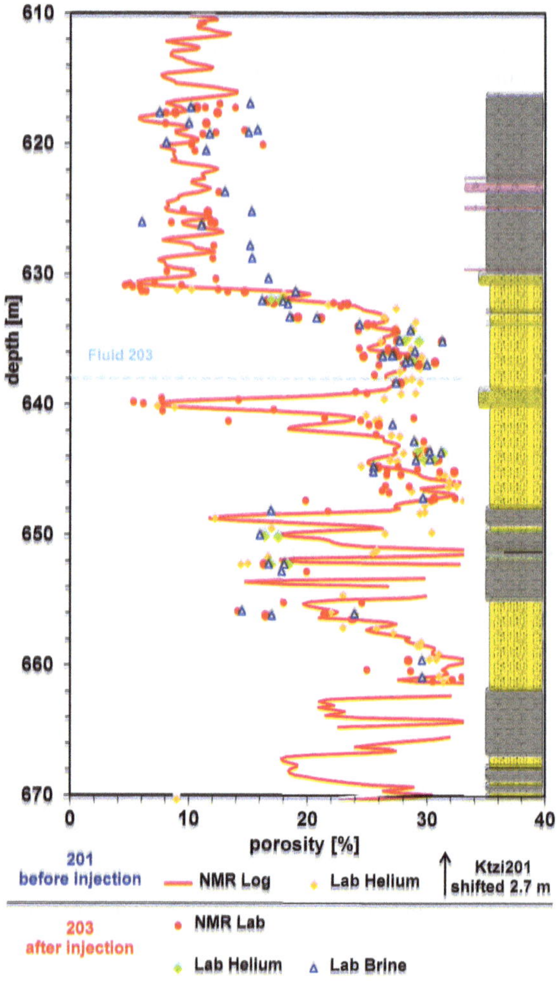

Fig. 10 Porosity—cross plot of neighboring wells with similar lithology: Ktzi 201 before the injection of CO_2 and Ktzi 203 after four years of injection with schematic lithology. Ktzi 201: porosity from NMR log and from Helium pycnometry reported by Norden et al. (2010). Depth was shifted by 2.7 m to correlate the cemented sandstone at 640 m. Ktzi 203: porosity from NMR, Helium pycnometry and brine saturation after Archimedes from laboratory measurements of cores

measurements show a reduced fraction of larger pores together with lower porosities and so only inner core samples were used for further investigation.

Comparison with data from the open hole logging (e.g. gamma ray, neutron-neutron, resistivity, sonic, PNG) of the neighboring wells show clear difference for corresponding units before and after the injection for the reservoir sandstone and no difference in the overlaying cap rock units, because methods are more sensitive to the CO_2 in the pore fluid and do not exhibit changes in the rock matrix, which affect the capacity, injectivity or integrity of the storage system.

Due to the heterogeneous character of the Stuttgart Formation it is difficult to estimate definite CO_2-induced changes from petrophysical measurements. However, given the only minor differences between rock samples from pre- and post-injection, it is reasonable to assume that the potential dissolution-precipitation processes appear to have no severe consequences on reservoir and cap rock integrity or on the injection behavior. This is also in line with the continuously recorded injection parameters which do not point to any changes in reservoir injectivity.

4.2 Microbiology

The microbial biocenosis was monitored in fluids taken from the deep wells of the Ketzin pilot site before and during the CO_2 injection. Drilling fluids, cleaning procedures and the CO_2 injection itself affected the microbial community of the fluids in its composition and abundance as detected by genetic fingerprinting (SSCP) and fluorescence in situ Hybridisation (FISH). Before CO_2 injection, up to 10^7 cells ml^{-1} with up to 40 % sulphate reducing bacteria (SRB) were enumerated in well fluids with DAPI staining (Morozova et al. 2010). The clean-up of wells prior to CO_2 injection reduced temporarily the total cell numbers by one order of magnitude, the numbers of SRB were below detection limit and the TOC declined from 600 to 20 mg l^{-1}. However, after 5 months of exposure to CO_2 cell numbers and the microbial activity increased again (Morozova et al. 2010). SRBs were quantified to 20 % of total cell numbers and remained at a high level until the third year of storage. Subsequent to the CO_2 exposure and during 5 years of monitoring the TOC concentration showed high variations between 1 and 160 mg l^{-1} in the injection and observation wells. Going along with decreasing TOC contents over the years also the cell numbers declined slowly. After 5 years, the SRB numbers were again below the detection limit in all wells.

The results indicate that the microbial community was able to adapt to the changes of the deep biosphere environmental conditions caused by the CO_2 injection and was mainly influenced by the availability of TOC strongly affected by technical procedures. Thus, a decrease in total cell numbers and the SRB numbers was first observed after 3 years of CO_2 exposure and corresponded to the decreasing concentrations of organic carbon in the well fluids. This long-term effect of organic carbon to the deep subsurface on the microbial biocenosis clearly emphasizes the need to limit the use of organic substances for drilling and maintenance procedures. Especially the high abundance of SRB in the first 3 years of CO_2 storage might pose a risk as SRB are known to be involved in microbial induced corrosion processes. The increase of iron concentrations after start of the CO_2 injection and arrival of the CO_2 at the observations wells was suspected to be a result of mobilization effects due to the injected CO_2 and/or of corrosion processes. However, in the Ketzin case, well inspections gave no indications for technical relevant corrosion effects to the casings.

The quantity and diversity of subsurface microorganisms differed significantly between fluid- and sediment-associated populations within the same formation, which is in accordance with other studies in deep aquifers (Fry et al. 1997; Hazen et al. 1991). Cell numbers enumerated in sandstone cores of the Stuttgart Formation

(Wandrey et al. 2010) were at least four orders of magnitude lower than in the well fluids collected via hydraulic testing and downhole sampling. In the reservoir sandstones, the concentration of low molecular weight organic acids like acetate and formate were below 0.1 % (Scherf et al. 2011). However, activity and numbers of microorganisms could be significantly increased if organic carbon is introduced by drilling procedures or mobilized and enriched by injected CO_2.

4.3 Organic Geochemistry

At typical CO_2 storage conditions, CO_2 is known as an excellent solvent for low to medium polar organic compounds (Hawthorne 1990) and extraction and mobilization of the organic matter present in the reservoir by the injected CO_2 is most likely. When studying storage of CO_2 in coal seams, the release of hydrocarbons from the coals has been investigated in more detail (Kolak and Burruss 2006). During and after CO_2 injection into a deep saline aquifer (Frio Formation, USA) a strong increase in DOC (dissolved organic carbon) as well as an increase in concentrations of formate, acetate and toluene has been observed in fluid samples (Kharaka et al. 2009). At the Ketzin pilot site, rock samples from the reservoir formation have been tested for the mobilization of different organic compounds during extraction with CO_2 at in situ reservoir conditions in lab experiments (Scherf et al. 2011).

Within the CO_2MAN project, the effects of injected CO_2 on the natural organic matter in reservoir and cap rock have been characterized using three different approaches: (i) characterization of natural organic matter in reservoir and cap rock samples prior and after the injection of CO_2; (ii) monitoring of changes in DOC in fluid samples from the wells Ktzi 200, 201 and 202 prior and during the injection of CO_2; (iii) monitoring of organic compounds being transported with the injected CO_2.

To characterize the quality and quantity of organic compounds that are transported with the injected CO_2, one riser tube (stainless steel, inner diameter 5 mm, total volume 12 l) installed in well Ktzi 202 was used for the continuous sampling of fluids from the reservoir horizon (compare Sect. 3.4). Fluids were produced from a depth of 600 m with a flow rate of 5 l/min. The produced gas was percolating through a gas washing bottle filled with distilled water or dichloromethane (DCM) over different times. In general, the amount of organic compounds increases with increasing washing times. This clearly shows that solubility is not limiting the extraction and that these compounds are not strongly degraded in the gas washing bottle. Variable yields of total DOC and of different aliphatic acids have been detected in the water samples over the monitoring period (Fig. 11). In all samples, yields of propionate, butyrate and valerate are very low whereas yields of acetate and formate are much higher. It seems as if the yields of individual organic acids are quite variable within our monitoring but that yields are not related to the duration of

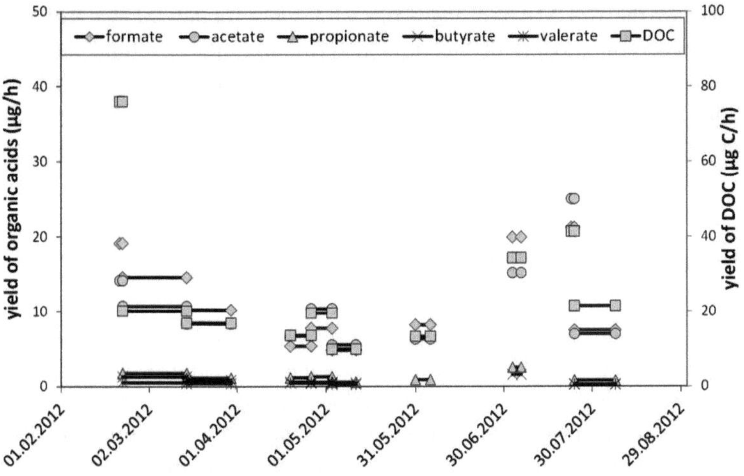

Fig. 11 Yields of LMW organic acids (in µg/h) and DOC (in µg C/h) washed out from produced gas into distilled water (well Ktzi 202; February–October 2012)

the washing experiment or to the observed decrease in distilled water volume in the gas washing bottle. Using size exclusion chromatography DOC was separated into different fractions. Dominant DOC-fractions were low molecular weight (LMW) acids (30–50 %) and LMW neutrals (40–60 %).

DCM extracts from the washing bottle contained a huge amount of phthalates. These esters are commonly used as plasticizers and are derived from the tube connection between riser tube and gas washing bottle. Composition of hydrocarbons (*n*-alkanes from nC_{14} to nC_{26}, hopanes and steranes) in all DCM extracts is comparable. In DCM extracts the distribution of hopanes is comparable to some technical additives that have been used in drilling and injection activities at Ketzin. Thus, it cannot be easily distinguished if the organic compounds detected in gas washing bottle extracts are derived from natural organic matter of reservoir, cap rock or formation fluids or from anthropogenic activities at the pilot site.

5 Modelling, Simulations and Data Management for the Ketzin Project

Sound geological models are a prerequisite to the sustainable and safe use of the subsurface. They consist of the structural geometry of geological bodies and their properties. Like for all models, appropriate simplifications of the in situ conditions are necessary to allow e.g. the simulation of complex dynamic processes. Depending on the purpose of the applied simulation, different models need to be considered, for example for the evaluation of the long- or short-term behavior of the CO_2 distribution in the subsurface or for the evaluation of rock-fluid interactions. In the course of the CO_2MAN project, the geological models of the Ketzin project including reservoir models of the storage formation and structural models of the wider area were updated. The static model was then used for further dynamic reservoir simulations taking into account hydrodynamics as well as geochemical and geomechanical processes.

5.1 Integrative Geological Modelling

For the Ketzin pilot site, like for many other sites worldwide, the challenge is to integrate a diversity of information of both local and regional origin as well as of different data types and formats and combine it for consistent models. Data available for the Ketzin project include reconnaissance data (e.g. regional geological maps, borehole reports, logging data, core data, hydraulic test data, seismic surveys from the 1960s and 1970s) and data collected during development and operation of the storage site. Important details are given for the first evaluation of open-hole logging data by Norden et al. (2010), for the first interpretation of the seismic data by Juhlin et al. (2007), Kazemeini et al. (2009) and Yordkayhun et al. (2009), and for the near-surface groundwater system by Norden (2011). The characterization of the storage formation is presented by Förster et al. (2009, 2010) and hydraulic and monitoring data is presented by Würdemann et al. (2010), Martens et al. (2012, 2013), Wiese et al. (2010) and Ivanova et al. (2012).

Data was integrated using the commercial software package Petrel. As neither the Ketzin seismic 3D data nor the borehole data is supplying sufficient information to resolve the internal structure of the complex fluviatile Stuttgart Formation in any detail and thus do not allow a deterministic modelling of the facies architecture and related properties of the reservoir, geostatistical approaches were used (Norden and Frykman 2013). The resulting geological models were step-wise adapted if necessary, guided by observations based on new drillings at the Ketzin site and based on the results of the monitoring. In return, the geological models supported the interpretation of the monitoring data (e.g. Bergmann et al. 2012, 2014; Chen et al. 2014).

The geological models of the Ketzin pilot site consist of reservoir models of the storage formation and structural models of the wider Ketzin area, including the overburden with its multi-barrier cap rock system. At the top of the anticlinal Ketzin structure, a Graben fault zone is present which is related to the updoming of the Roskow-Ketzin salt pillow (Fig. 12a). The faults of this graben extend from the Base Tertiary (see e.g. Juhlin et al. 2007) to the Buntsandstein and do most likely reach down to the Zechstein salt (although not resolved by the Ketzin 3D seismic data). These faults were not present at the time of deposition of the Stuttgart Formation. The shown petrophysical characterization of the Stuttgart Formation is one possible scenario taken into account the general geological setting and the existing borehole data (facies, porosity, permeability etc.; Fig. 12b). Based on such a scenario, dynamic simulations were performed to model the distribution of CO_2 in the reservoir (e.g. Kempka et al. 2010; Kempka and Kühn 2013).

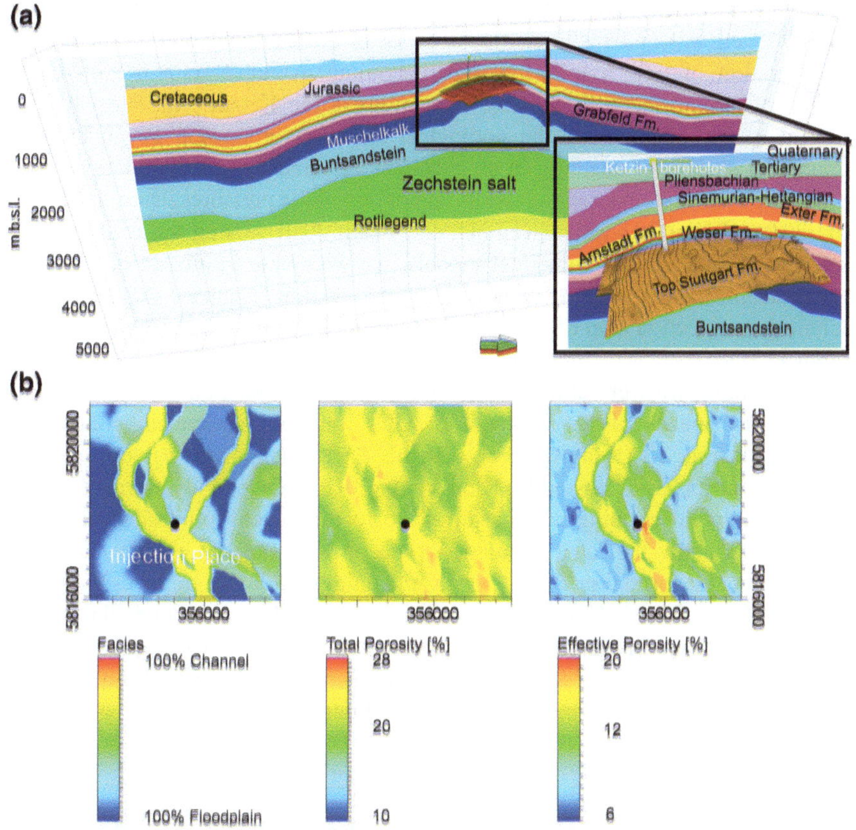

Fig. 12 **a** Regional S–N section of the Ketzin pilot site. Length of profile approx. 40 km. The inlet map shows the Top Stuttgart surface. Drilling paths and offsets along faults could be recognized. Exaggeration is 3-fold. **b** Average facies and porosity maps of the uppermost 20 m of one realization of the Stuttgart facies model. Plotted are mean facies distribution (*left*), average total porosity (*middle*), and effective porosity (*right*). The location of the Ketzin pilot site is indicated by a *black dot*. The size of the reservoir model is 5 km × 5 km (Figure adapted from Norden and Frykman 2013)

5.2 Reservoir Simulations

Numerical multi-phase flow simulations were already carried out in the planning phase of the Ketzin project by different modelling groups to account for sustainable injection rates in terms of reservoir pressure elevation and expected CO_2 arrival times at the two observation wells Ktzi 200 and Ktzi 202. A prediction of CO_2 arrival times at both observation wells was undertaken in advance of the start of injection in June 2008. While the numerical simulators applied showed relatively small differences in simulation results, all numerical modelling groups were only able to match the CO_2 arrival time at the Ktzi 201 well, while that predicted for the Ktzi 202 well was underestimated by a factor of three (Kempka et al. 2010).

Integration of new monitoring data becoming available with the increasing amount of monitoring activities and campaigns at the Ketzin pilot site with time allowed us for revisions of the geological model (Norden and Frykman 2013; Kempka et al. 2013a) resulting in an excellent agreement of the simulation results with the CO_2 arrival times at both wells and of downhole pressures determined at the Ktzi 201 and Ktzi 202 wells (Kempka and Kühn 2013). Using the resulting calibrated (history-matched) numerical models, we were able to predict the Ktzi 201 downhole pressure for more than 1 year of operation (January 2012 onwards) including an injection stop of about 6 months with low deviations only (Class pers. comm.).

Furthermore, we were able to carry out long-term predictions on reservoir stabilization by means of development of CO_2 trapping mechanism contribution illustrated in Fig. 13 (Kempka et al. 2013b). Thereto, innovative hydro-chemical coupling concepts as discussed by Klein et al. (2013) were employed to account for long-term mineral trapping at reservoir scale for the Ketzin pilot site. Our simulation results show that at a simulation time of 16,000 years almost all CO_2 is dissolved in the formation fluid (0.2 % remaining residually trapped in gaseous state), while 11.7–30.9 % precipitate as siderite and dolomite depending on the assumptions underlying the hydro-chemical simulation model (compare Sect. 5.3).

The numerical reservoir simulation results were coupled to hydro-mechanical simulations to account for mechanical system integrity (compare Sect. 5.4).

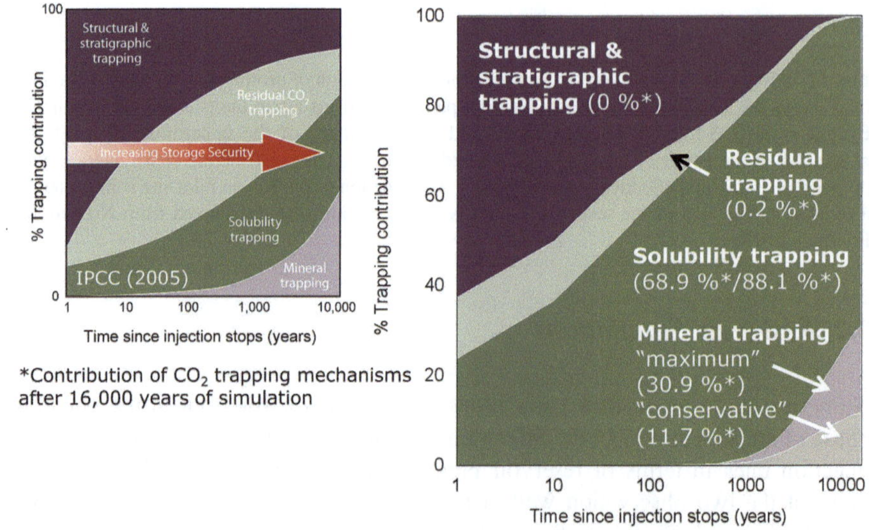

Fig. 13 *Left* General CO_2 trapping mechanisms in geological formations (IPCC 2005). *Right* simulated trapping mechanisms of CO_2 for the Ketzin pilot site

5.3 Geochemical Simulations

The quantification of water-rock interactions in reservoir and cap rock triggered by the injection of CO_2 at the Ketzin pilot site was carried out by means of batch geochemical models and coupled reactive transport simulations based on the available characterization of the Stuttgart formation and the pristine formation fluids (Förster et al. 2006, 2010; Norden et al. 2010 for the mineralogical analyses on core samples and Würdemann et al. 2010 for fluid analyses).

3D reactive transport models on computationally affordable coarse grid predict that mineral trapping of CO_2 at the Ketzin pilot site will be dominated by siderite, dolomite and magnesite, with only transient appearance of calcite (Fig. 14). Significant precipitation in the reservoir will start only about 500 years after injection stop and is expected to continue for at least 10,000 years (the considered simulation time). Clay minerals such as illite and chlorite and anhydritic rock cement will be dissolved by the CO_2-charged fluids. The net predicted result is a negligible loss of porosity, since the largest predicted change in mineral volume amounts to only about 3,000 cm^3/m^3 of rock (=0.3 vol%). However due to the spatially quite large extent of the reservoir affected by chemical reactions due to the CO_2 migration, up to 40 % of the total injected carbon is predicted to be minerally trapped after 10,000 years. However this result could be overestimated due to the coarse grids used for the simulations and consequent poor matching of observation data and migration pattern of CO_2 in the reservoir. Thereto, a novel one-way coupling strategy for multiphase hydrodynamics and chemistry has been developed (Klein et al. 2013) and validated for the Ketzin pilot site. Through this strategy it is possible to consider much larger, heterogeneous grids for multiphase flow and more

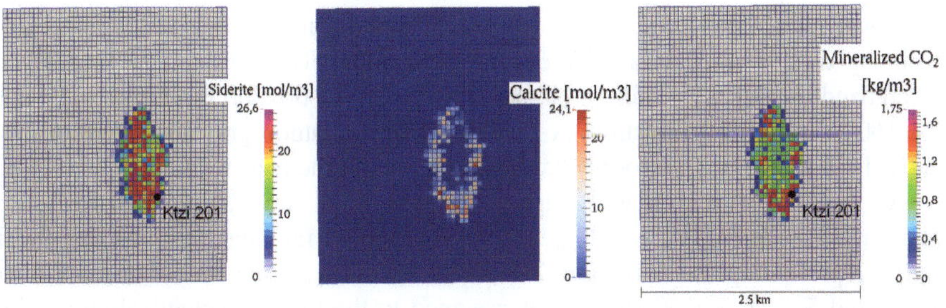

Fig. 14 Top view of 3D reactive transport models 2,200 years after injection stop at Ketzin. The size of the model is 2.5 km × 3.5 km. The injection well Ktzi 201 is marked by a *black dot*. Depicted are the main CO_2 sink minerals, the iron carbonate siderite (*left*), the transient appearance of calcium carbonate calcite (*middle*) and the overall amount of carbon-bearing precipitates in reservoir (*right*). Coarseness of the grid, boundary conditions and spatial heterogeneity of porosity and permeability are in these reactive simulations simplified due to constraint on the number of grid elements allowable for the computationally expensive coupled simulations. This is reflected in differences in the migration pattern of the CO_2 in comparison with seismic monitoring and non-reactive modeling and therefore in the distribution of mineralization

complex chemistry without the oversimplifications necessary for the fully-coupled simulations. The resulting error is acceptable, mainly arising from the assumptions about the minimum CO_2 gas saturation and dissolved CO_2 concentration that can be considered geochemically active. It was confirmed by extensive validation that choosing reasonable cut-offs for these quantities achieves good matching with fully-coupled simulations. This method contributed to the evaluation of the mineral trapping potential using the history matched simulations of Kempka and Kühn (2013) and predicted that between 11.7 and 30.9 % (following a conservative or optimistic cut-off choice, respectively) of the injected CO_2 will be mineralized after 16,000 years (Kempka et al. 2013b). 1D reaction-diffusion models of dissolved CO_2 into the first layers of mudstone cap rock predict that CO_2 would penetrate for about 5 m in the cap rock after 500 years. The models show an overall tendency of loss of porosity through newly precipitated carbonates and thus highlight the self-healing potential of the covering horizon.

The results of 3D coupled reactive transport models point towards a long-term stabilization of the reservoir at the Ketzin pilot site, with no significant mineral alteration during the operational time.

5.4 Geomechanical Modelling and Simulation

Mechanical system integrity taking into account reservoir rock, cap rock and fault integrity was assessed by hydro-mechanical simulations for the Ketzin pilot site (Kempka et al. 2014). For that purpose, history matched reservoir simulations (Kempka and Kühn 2013) were considered as a basis for our one-way hydro-mechanical model simulations. Thereto, a 40 km × 40 km × 5 km hydro-mechanical model with twelve lithological units and 24 discrete major faults was implemented based on a previously elaborated 3D structural geological model considering 2D/3D seismic and well log data as well as available geological maps. The mechanical model was discretized by about 1.4 million elements with sizes of 50–800 m in horizontal and 20–160 m in vertical direction. A refinement of the lateral grid discretization was carried out to capture hydro-mechanical processes in the near-well area.

An initial stress regime with a maximum horizontal stress azimuth of 150° ± 5° and a magnitude of $S_{Hmax} = S_{hmin} = 0.85\ S_v$ (normal faulting stress regime), with S_{Hmax} representing the maximum horizontal stress, S_{hmin} the minimum horizontal stress and S_v the vertical stress, was assigned to the hydro-mechanical simulation model in agreement with the interpretation of sonic logs from the Ktzi 201 well (Sinha et al. 2010). Hereby, S_v was determined from the gravitational load of the overburden using numerical simulations.

Mechanical model parameterization followed the data on the Northeast German Basin given by Ouellet et al. (2010), Nagelhout and Roest (1997) and Kopf (1965). The Mohr-Coulomb plasticity model was applied for the rock matrix and the FLAC3D ubiquitous joint model (Itasca 2012) for fault representation. Thereto, an element-wise representation of fault geometries by means of dip direction angle and

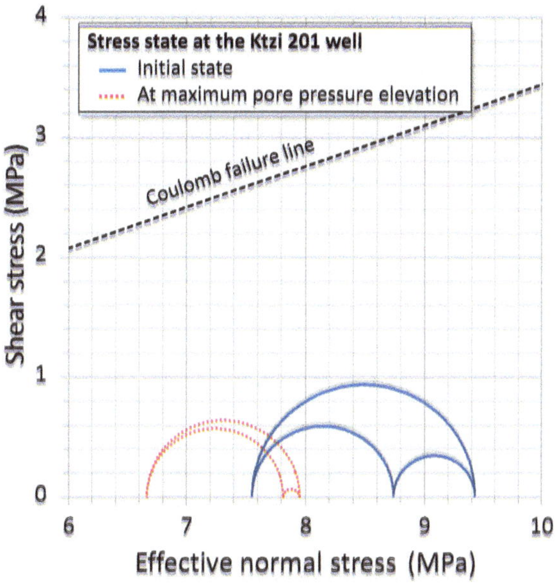

Fig. 15 Initial stress state (*blue*) and state at maximum pore pressure elevation (*red*) at most upper element of storage formation at injection well Ktzi 201 based on numerical hydro-mechanical simulations

dip angle was taken into account to allow for a detailed fault analysis. After running the hydro-mechanical model to a mechanical equilibrium, spatial pore pressure distributions calculated for 14 different time steps by the numerical reservoir simulations (compare Sect. 5.2) were integrated as coupling parameter into the hydro-mechanical simulations.

Simulation results show maximum vertical displacements of about 6 mm at the reservoir top and about 4 mm at the ground surface in March 2010 with a maximum vertical displacement radius of up to 3 km. Both are in good agreement with the hydro-mechanical simulations carried out by Ouellet et al. (2010) considering the first 500 days of CO_2 injection only. Neither shear nor tensile failure is observed at any time step in the rock matrix and ubiquitous joints elements (representing the faults) in the hydro-mechanical model due to the limited increase in pore pressure resulting from the CO_2 injection (Fig. 15). The reservoir and coupled numerical simulation results, therefore, suggest safe and reliable CO_2 storage operation at the Ketzin pilot site.

5.5 Data Management

Comprehensive understanding of subsurface processes of CO_2 storage at the Ketzin pilot site requires reliable geological interpretation and modelling at multiple scales which only become possible by a careful integration of results derived from drilling and different monitoring approaches (Behrends and Conze 2012). Thus, the integrative compilation of geological, operational and scientific monitoring data which had started with the CO_2SINK project was continued and extended in the CO_2MAN project.

Since the first wells for research purposes were drilled at Ketzin in 2007, a data entry system was provided during the acquisition phase of geological field data and continuously extended throughout the subsequent projects. Hence, this system was also used when the wells P300 and Ktzi 203 were drilled in 2011 and 2012, respectively.

A web-based data management system (DMS) contributes to a close cooperation among scientists from different disciplines. The main functions of the DMS implemented for the CO_2MAN project were:

- *Information*: Dissemination of project management material (planning documents, internal reports, conference abstracts, papers etc.) and new science data sets.
- *Archiving*: Provision of metadata-, long-term storage-, reporting- and downloading-services; integration of published data with new unpublished working data sets and legacy data sets from predecessor projects and external partners.
- *Collaboration*: Exchange platform to internal data sets, benchmarking, quality and version control of data sets and models.

The data management system was implemented using a secured web-based platform. Target audiences were all members of the project consortium. Science data sets stored in the system comprise e.g. basic geological data, technical drilling data, down-hole measurements, monitoring time series along the injection history up to geological models of the storage formation. The software development project had to deal with heterogeneous data sets of diverse origin, different scales and data file formats. It is required to extend the workflow of curating science data to later project phases. All these services can be continuously used and will be further developed during the follow up project COMPLETE.

6 Public Outreach

For the Ketzin project, public outreach was a key element from the very beginning and already started in 2005 to ensure an open and transparent dialogue with all stakeholders (Schilling et al. 2009b). During the course of the project, it turned out to be important to provide up-to-date information and keep the dialogue with the public (Martens et al. 2013). Thus, public relations were also a central component of the CO_2MAN project. In order to inform the general public about the background of CO_2 storage and the research results from Ketzin manifold information material was developed and disseminated using various communication tools (Szizybalski et al. 2014).

The main contact point is the information centre at the Ketzin pilot site where a visitor service is offered on a weekly basis. With the expansion of the centre in

2011 larger visitor groups could be welcomed and more permanent exhibits presented. About 2,140 people from Germany and abroad visited the pilot site throughout the CO_2MAN project phase whereof most were students and scientists ($\sim$50 %) and the general public ($\sim$25 %).

Since 2011, an annual open house day was held at the pilot site (Fig. 16). At these events, visitors can inform themselves about the research activities during guided tours and presentations and talk to scientists (Martens et al. 2013). The open day is carried out in close cooperation with the city of Ketzin as we also include the local fire brigade and the local retail, e.g. musicians, bakery, and the Ketzin tourist information. At the annual "Long Night for the Sciences" in Berlin and Potsdam the Ketzin project is also presented already since 2007. For educational activities it is important to provide comprehensible information materials and experiments, for example, demonstrating the presence of CO_2 in everyday life or the concept of CO_2 trapping mechanism. Therefore, five mobile experiments were added as information tools and used, for example, during school visits.

The initial Ketzin project website, which was developed during the CO_2SINK project, was superseded by the public website http://www.co2ketzin.de in 2011. Besides background information on the Ketzin project, the history of the pilot site, the objectives and current status of CO_2MAN were presented in both German and English. In addition, a bilingual Ketzin brochure was regularly updated according to the current status of the project and also made available on the website. The information material was complemented by a film entitled "The geological storage of CO_2" made up of seven short segments (Hübner et al. 2013) funded by the CLEAN project also including an overview of the Ketzin pilot site. A further short film which was produced within the CO_2MAN project documents the drilling activities at Ketzin in 2011 and 2012. These short films are available on the public website, on DVD and are also shown during visits at schools.

Our public outreach activities are reflected by a mainly positive media resonance and a wide public acceptance for the research activities at Ketzin, allowing the research on CO_2 storage, e.g. large-scale seismic measurements, to be conducted without severe restrictions.

Fig. 16 Annual open house day at the Ketzin pilot site (*left*: June 2012; *right*: June 2013)

7 Conclusion

The Ketzin project is one of the world's most significant sites for investigating CO_2 storage on a pilot scale. In the course of the CO_2MAN project, two new boreholes could be drilled at the site and the CO_2 injection was continued and finally ceased after the injection of a total amount of 67 kt of CO_2 on 29 August 2013. The accompanying scientific activities in the work packages "Research Infrastructure", "Geophysical Monitoring", "Reservoir Processes" and "Modelling and Simulations" as well as in the fields of data management, public outreach and project management were completed as planned and afforded a rich learning experience. The results obtained in the joint research project show that:

- the geological storage of CO_2 at the Ketzin pilot site runs safely and reliably,
- a meaningful, site-specific combination of geochemical and geophysical monitoring techniques is able to detect even small amounts of CO_2 and to image its spatial distribution,
- fluid-rock interactions induced by the injected CO_2 have no significant effects at the Ketzin pilot site and do not affect the integrity of the reservoir and cap rocks,
- numerical simulations are able to predict timely and spatial behavior of the injected CO_2 and to provide prognosis on the long-term behavior of the storage formation
- a targeted communication and dissemination programme is able to establish a wide public acceptance for research activities like the Ketzin project and to overcome critical public perception even for highly debated technologies.

Although the CO_2 injection at Ketzin ceased in August 2013 and the CO_2MAN project ended in December 2013, R&D activities on CO_2 storage are still underway in order to address and close the entire life cycle of a storage site. Hence post-injection monitoring and well abandonment are the main focus of the on-going post-injection phase and part of the COMPLETE project which has started in January 2014.

Acknowledgments The authors thank all participants of the CO_2MAN project who made the Ketzin project a success. We would also like to acknowledge the German Federal Ministry of Education and Research (BMBF—Grant numbers 03G0760A to F), Dillinger Hüttenwerke, OMV, RWE, Saarstahl, Statoil and Vattenfall for funding as well as the site owner VGS and the Projektträger Jülich for their continuous support.

References

Baumann G (2013) Determination of displacement and evaporation/precipitation processes via pulsed neutron-gamma (PNG) monitoring for CO_2 storage operations. Dissertation, Technical University Berlin

Baumann G, Henninges J, De Lucia M (2014) Monitoring of saturation changes and salt precipitation during CO_2 injection using pulsed neutron-gamma logging at the Ketzin site. Int J Greenhouse Gas Control 28:134–146

Behrends K, Conze R (2012) Science data management for a co_2-storage project. geophysical research abstracts, vol 14, EGU2012-8501-1, EGU General Assembly 2012

Bergmann P, Schmidt-Hattenberger C, Kiessling D, Rücker C, Labitzke T, Henninges J, Baumann G, Schütt H (2012) Surface-downhole electrical resistivity tomography applied to monitoring of CO_2 storage at Ketzin, Germany. Geophysics 77:B253–B267. doi:10.1190/geo2011-0515.1

Bergmann P, Ivandic M, Norden B, Rücker C, Kiessling D, Lüth S, Schmidt-Hattenberger C, Juhlin C (2014) Combination of seismic reflection and constrained resistivity inversion with an application to 4D imaging of the storage site, Ketzin, Germany. Geophysics 79(2):B37–B50. doi:10.1190/geo2013-0131.1

Bock S, Förster H, Meier A, Pudlo D, Förster A, Gaupp, R (2013) Impact of 4-year CO_2 injection on reservoir-rock integrity at the CO_2 pilot site Ketzin (Germany). In: American geophysical union, fall meeting 2013, abstract #H21L-02

Chen F, Wiese B, Zhou Q, Kowalsky MB, Norden B, Kempka T, Birkholzer JT (2014) Numerical modeling of the pumping tests at the Ketzin pilot site for CO_2 injection: model calibration and heterogeneity effects. Int J of Greenhouse Gas Control 22:200–212. doi:10.1016/j.ijggc.2014.01.003

Daley TM, Freifeld BM, Ajo-Franklin J, Dou S, Pevzner R, Shulakova V, Kashikar S, Miller DE, Goetz J, Henninges J, Lueth S (2013) Field testing of fiber-optic distributed acoustic sensing (DAS) for subsurface seismic monitoring. Lead Edge 32:699–706. doi:10.1190/tle32060699.1

Fischer S (2013) Mineralogical-geochemical effects during geological storage of CO_2—experimental investigations and geochemical modeling. Scientific Technical Report 13/13, GFZ German Research Centre for Geosciences

Fischer S, Liebscher A, De Lucia M, Hecht L, Ketzin Team (2013) Reactivity of sandstone and siltstone samples from the Ketzin pilot CO_2 storage site—laboratory experiments and reactive geochemical modeling. Environmental Earth Sciences, Special issue 1–22. doi:10.1007/s12665-013-2669-4

Förster A, Norden B, Zinck-Jorgensen K, Frykman P, Kulenkampff J, Spangenberg E, Erzinger J, Zimmer M, Kopp J, Borm G, Juhlin C, Cosma CG, Hurter S (2006) Baseline characterization of the CO_2SINK geological storage site at Ketzin, Germany. Environ Geosci 13:145–161

Förster A, Giese R, Juhlin C, Norden B, Springer N, CO_2SINK Group (2009) The geology of the CO_2SINK site: from regional scale to laboratory scale. Energy Procedia 1(1):2911–2918. doi:10.1016/j.egypro.2009.02.066

Förster A, Schöner R, Förster HJ, Norden B, Blaschke AW, Luckert J, Beutler G, Gaupp R, Rhede D (2010) Reservoir characterization of a CO_2 storage aquifer: the upper Triassic Stuttgart Formation in the northeast German basin. Mar Pet Geol 27(10):2156–2172

Freifeld BM, Daley T, Hovorka S, Henninges J, Underschultz J, Sharma S (2009) Recent advances in well-based monitoring of geologic carbon sequestration. Energy Procedia 1:2277–2284

Fry NK, Fredrickson JK, Fishbain S, Wagner M, Stahl DA (1997) Population structure of microbial communities associated with two deep, anaerobic, alkaline aquifers. Appl Environ Microbiol 63(4):1498–1504

Hawthorne SB (1990) Analytical-scale supercritical fluid extraction. Anal Chem 62:633A–642A

Hazen TC, Jimenez L, de Victoria GL, Fliermans CB (1991) Comparison of bacteria from deep subsurface sediment and adjacent groundwater. Microb Ecol 22:293–304

Henninges J, Liebscher A, Bannach A, Brandt W, Hurter S, Köhler S, Möller F, CO_2SINK Group (2011) P-T-rho and two-phase fluid conditions with inverted density profile in observation wells at the CO_2 storage site at Ketzin (Germany). Energy Procedia 4:6085–6090. doi:10.1016/j.egypro.2011.02.614

Hübner A, Kollersberger T, Pilz P, Tesmer M, Kühn M (2013) Public Outreach. In: Kühn M., Münch U (eds) CO_2 large-scale enhanced gas recovery in the altmark natural gas field. GEOTECHNOLOGIEN science report, no. 19, Advanced Technologies in Earth Sciences, Springer, p 199

IPCC (2005) Special report on carbon dioxide capture and storage. Prepared by working group III of the Intergovernmental panel on climate change. In: Metz B, Davidson O, de Coninck HC, Loos M, Meyer LA (eds) Cambridge University Press, Cambridge

Itasca (2012) FLAC3D software version 5.0. User's manual. Advanced three-dimensional continuum modelling for geotechnical analysis of rock, soil and structural support

Ivandic M, Yang C, Lüth S, Cosma C, Juhlin C (2012) Time-lapse analysis of sparse 3D seismic data from the CO_2 storage pilot site at. J Appl Geophys 84:14–28. doi:10.1016/j.jappgeo.2012.05.010

Ivandic M, Juhlin C, Lüth S, Bergmann P, Kashubin A (2013) Geophysical monitoring of CO_2 at the Ketzin storage site: the results of the second 3D repeat seismic survey. In: 75th EAGE conference and exhibition incorporating SPE EUROPEC 2013, London, UK, 10–13 June 2013

Ivanova A, Kashubin A, Juhojuntti N, Kummerow J, Henninges J, Juhlin C, Lüth S, Ivandic M (2012) Monitoring and volumetric estimation of injected CO_2 using 4D seismic, petrophysical data, core measurements and well logging: a case study at Ketzin, Germany. Geophys Prospect 60:957–973. doi:10.1111/j.1365-2478.2012.01045.x

Juhlin C, Giese R, Zinck-Jorgensen K, Cosma C, Kazemeini H, Juhojuntti N, Lüth S, Norden B, Förster A (2007) 3D baseline seismics at Ketzin, Germany: the CO_2SINK project. Geophysics 72(5):8121–8132. doi:10.1190/1.2754667

Kazemeini H, Juhlin C, Zinck-Jorgensen K, Norden B (2009) Application of the continuous wavelet transform on seismic data for mapping of channel deposits and gas detection at the CO_2SINK site, Ketzin, Germany. Geophys Prospect 57(1):111–123. doi:10.1111/j.1365-2478.2008.00723.x

Kempka T, Kühn M (2013) Numerical simulations of CO_2 arrival times and reservoir pressure coincide with observations from the Ketzin pilot site, Germany. Environ Earth Sci 70(8):3675–3685. doi:10.1007/s12665-013-2614-6

Kempka T, Kühn M, Class H, Frykman P, Kopp A, Nielsen CM, Probst P (2010) Modeling of CO_2 arrival time at Ketzin—part I. Int J Greenhouse Gas Control 4(6):1007–1015

Kempka T, Class H, Görke UJ, Norden B, Kolditz O, Kühn M, Walter L, Wang W, Zehner B (2013a) A dynamic flow simulation code intercomparison based on the revised static model of the Ketzin pilot site. Energy Procedia 40:418–427

Kempka T, Klein E, De Lucia M, Tillner E, Kühn M (2013b) Assessment of long-term CO_2 trapping mechanisms at the Ketzin pilot site (Germany) by coupled numerical modelling. Energy Procedia 37:5419–5426

Kempka T, Klapperer S, Norden B (2014) Coupled hydro-mechanical simulations demonstrate system integrity at the Ketzin pilot site for CO_2 storage, Germany. In: 2014 ISRM European rock mechanics symposium (EUROCK 2014), 27–29 May 2014, Vigo/Spain, 6 p

Kharaka YK, Thordsen JJ, Hovorka SD, Seay Nance H, Cole DR, Phelps TJ, Knauss KG (2009) Potential environmental issues of CO_2 storage in deep saline aquifers: geochemical results from the Frio-I Brine Pilot test, Texas, USA. Appl Geochem 24:1106–1112

Kiessling D, Schmidt-Hattenberger C, Schuett H, Schilling F, Krueger K, Schoebel B, Danckwardt E, Kummerow J, CO_2SINK Group (2010) Geoelectrical methods for monitoring geological CO_2 storage: first results from cross-hole and surface-downhole measurements from the CO_2SINK test site at Ketzin (Germany). Int J Greenhouse Gas Control 4(5):816–826 doi:10.1016/j.ijggc.2010.05.001

Klein E, De Lucia M, Kempka T, Kühn M (2013) Evaluation of longterm mineral trapping at the Ketzin pilot site for CO_2 storage: an integrative approach using geochemical modelling and reservoir simulation. Int J Greenhouse Gas Control 19:720–730

Kolak JJ, Burruss RC (2006) Geochemical investigation of the potential for mobilizing non-methane hydrocarbons during carbon dioxide storage in deep coal beds. Energy Fuels 20:566–574

Kopf M (1965) Feldgeologie-Dichtebestimmung. Nord-deutsch-Polnisches Becken, Ergebnisbericht. VEB Geophysik Leipzig. Unpublished Report, 47 p

Liebscher A, Möller F, Bannach A, Köhler S, Wiebach J, Schmidt-Hattenberger C, Weiner M, Pretschner C, Ebert K, Zemke J (2013) Injection operation and operational pressure-temperature monitoring at the CO_2 storage pilot site Ketzin, Germany-Design, results, recommendations. Int J of Greenhouse Gas Control 15:163–173. doi:10.1016/j.ijggc.2013.02.019

Loizzo M, Henninges J, Zimmer M, Liebscher A (2013) Multi-phase equilibrium in a CO_2-filled observation well at the Ketzin pilot Site. Energy Procedia 37:3621–3629. doi:10.1016/j.egypro.2013.06.255

Martens S, Kempka T, Liebscher A, Lüth S, Möller F, Myrttinen A, Norden B, Schmidt-Hattenberger C, Zimmer M, Kühn M (2012) Europe's longest-operating on-shore CO_2 storage site at Ketzin, Germany: a progress report after three years of injection. Environ Earth Sci 67 (2):323–334. doi:10.1007/s12665-012-1672-5

Martens S, Liebscher A, Möller F, Henninges J, Kempka T, Lüth S, Norden B, Prevedel B, Szizybalski A, Zimmer M, Kühn M, Ketzin Group (2013) CO_2 storage at the Ketzin pilot site, Germany: fourth year of injection, monitoring, modelling and verification. Energy Procedia 37:6434–6443. doi:10.1016/j.egypro.2013.06.573

Morozova D, Wandrey M, Alawi M, Zimmer M, Vieth A, Zettlitzer M, Wuerdemann H (2010) Monitoring of the microbial community composition in saline aquifers during CO_2 storage by fluorescence in situ hybridisation. Int J Greenhouse Gas Control 4(6):981–989. doi:10.1016/j.ijggc.2009.11.014

Myrttinen A, Becker V, van Geldern R, Würdemann H, Morozova D, Zimmer M, Taubald H, Blum P, Barth JAC (2010) Carbon and oxygen isotope indications for CO_2 behaviour after injection: first results from the Ketzin site (Germany). Int J Greenhouse Gas Control 4 (6):1000–1006. doi:10.1016/j.ijggc.2010.02.005

Nagelhout ACG, Roest JPA (1997) Investigating fault slip in a model of an underground gas storage facility. Int J Rock Mech Min 34(3–4) Paper No. 212

Norden B (2011) Modelling of the near-surface groundwater flow system at the CO_2SINK site Ketzin, Germany. Zeitschrift der Deutschen Gesellschaft für Geowissenschaften 162(1):63–77. doi:10.1127/1860-1804/2011/0162-0063

Norden B, Frykman P (2013) Geological modelling of the triassic Stuttgart Formation at the Ketzin CO_2 storage site, Germany. Int J Greenhouse Gas Control 19:756–774

Norden B, Förster A, Vu-Hoang D, Marcelis F, Springer N, Le Nir I (2010) Lithological and petrophysical core-log interpretation in CO_2SINK, the European CO_2 onshore research storage and verification project. SPE Reserv Eval Eng 13:179–192

Nowak M, Myrttinen A, Van Geldern R, Becker V, Mayer B, Barth JAC (2013) A brief overview of isotope measurements carried out at various CCS pilot sites worldwide. In: Hou MZ, Xie H, Were P (eds) Clean energy systems in the subsurface: production, storage and conversion. Springer, Heidelberg, pp 75–87

Ouellet A, Bérard T, Frykman P, Welsh P, Minton J, Pamucku Y, Hurter S, Schmidt-Hattenberger C (2010) Reservoir geomechanics case study of seal integrity under CO_2 storage conditions at Ketzin, Germany. In: Ninth annual conference on carbon capture and sequestration. 10–13 May 2010, 16 p

Parker T, Shatalin S, Farhadiroushan M (2014) Distributed acoustic sensing—a new tool for seismic applications. First Break 32(2):61–69

Prevedel B, Wohlgemuth L, Legarth B, Henninges J, Schütt H, Schmidt-Hattenberger C, Norden B, Förster A, Hurter S (2009) The CO_2SINK boreholes for geological CO_2 storage testing. Energy Procedia 1(1):2087–2094. doi:10.1016/j.egypro.2009.01.272

Rempel KU, Liebscher A, Heinrich W, Schettler G (2011) An experimental investigation of trace element dissolution in carbon dioxide: applications to the geological storage of CO_2. Chem Geol 289(3–4):224–234. doi:10.1016/j.chemgeo.2011.08.003

Scherf AK, Zetzl C, Smirnova I, Zettlitzer M, Vieth-Hillebrand A (2011) Mobilisation of organic compounds from reservoir rocks through the injection of CO_2—comparison of baseline characterization and laboratory experiments. Energy Procedia 4:4524–4531

Schilling F, Borm G, Würdemann H, Möller F, Kühn M, CO_2SINK Group (2009a) Status report on the first European on-shore CO_2 storage site at Ketzin (Germany). Energy Procedia, 1(1) 2029–2035

Schilling F, Ossing F, Würdemann H, CO_2SINK Team (2009b) Public acceptance for geological CO_2-storage. Geophysical Research Abstracts 11, EGU2009-13801-2, EGU General Assembly 2009

Schmidt-Hattenberger C, Bergmann P, Labitzke T, Schröder S, Krüger K, Rücker C, Schütt H (2012) A modular geoelectrical monitoring system as part of the surveillance concept in CO_2 storage projects. Energy Procedia 23:400–407. doi:10.1016/j.egypro.2012.06.062

Sinha BK, Ouellet A, Bérard T (2010) Estimation of principal horizontal stresses using radial profiles of shear slownesses utilizing sonic data from a CO_2 storage site in saline aquifer in Germany. In: SPWLA 51st annual logging symposium, 19–23 June 2010, 16 p

Stroink L, Gerling JP, Kühn M, Schilling FR (2009) Die geologische Speicherung von CO_2—Aktuelle Forschungsergebnisse und Perspektiven. GEOTECHNOLOGIEN Science Report No. 14, Potsdam, 138 p

Szizybalski A, Kollersberger T, Möller F, Martens S, Liebscher A, Kühn M (2014) The Ketzin pilot site, Germany—a communication concept that supports the research on CO_2 storage. Energy Procedia 51:274–280. doi:10.1016/j.egypro.2014.07.032

Wandrey M, Morozova D, Zettlitzer M, Wuerdemann H (2010) Assessing drilling mud and technical fluid contamination in rock core and brine samples intended for microbiological monitoring at the CO_2 storage site in Ketzin using fluorescent dye tracers. Int J Greenhouse Gas 4(6):972–980. doi:10.1016/j.ijggc.2010.05.012

Wiese B, Böhner J, Enachescu C, Wuerdemann H, Zimmermann G (2010) Hydraulic characterisation of the Stuttgart Formation at the pilot test site for CO_2 storage, Ketzin, Germany. Int J Greenhouse Gas Control 4(6):960–971. doi:10.1016/j.ijggc.2010.06.013

Würdemann H, Möller F, Kühn M, Heidug W, Christensen NP, Borm G, Schilling F (2010) CO_2SINK—from site characterisation and risk assessment to monitoring and verification: one year of operational experience with the field laboratory for CO_2 storage at Ketzin, Germany. Int J Greenhouse Gas Control 4(6):938–951. doi:10.1016/j.ijggc.2010.08.010

Yordkayhun S, Tryggvason A, Norden B, Juhlin C, Bergmann B (2009) 3D seismic traveltime tomography imaging of the shallow subsurface at the CO_2SINK project site, Ketzin, Germany. Geophysics 74(1):G1–G15. doi:10.1190/1.3026553

Zimmer M, Pilz P, Erzinger J (2011a) Long-term surface carbon dioxide flux monitoring at the Ketzin carbon dioxide storage test site. Environ Geosci 18:119–130

Zimmer M, Erzinger J, Kujawa C, CO_2SINK Group (2011b) The gas membrane sensor (GMS): a new method for gas measurements in deep boreholes applied at the CO_2SINK site. Int J Greenhouse Gas Control 5:995–1001

MONACO—Monitoring Approach for Geological CO_2 Storage Sites Using a Hierarchical Observation Concept

Claudia Schütze, Karin Bräuer, Peter Dietrich, Viktoria Engnath, Michael Gisi, Gunnar Horak, Carsten Leven, Alexander Lübben, Ingo Möller, Michael Nierychlo, Stefan Schlömer, Andreas Schuck, Ulrich Serfling, Arno Simon, Thomas Streil and Uta Sauer

Abstract The reliable detection and assessment of potential CO_2 leakages from storage formations require the application of assurance monitoring tools at different spatial scales. Such tools also play an important role in helping to establish a risk assessment strategy at carbon dioxide capture and storage (CCS) facilities. Within the framework of the MONACO project ("Monitoring approach for geological CO_2 storage sites using a hierarchical observation concept"), an integrative

C. Schütze (✉) · P. Dietrich · U. Sauer
Department Monitoring and Exploration Technologies, UFZ – Helmholtz Centre for Environmental Research, Permoserstraße 15, 04318 Leipzig, Germany
e-mail: claudia.schuetze@ufz.de

K. Bräuer
Department Catchment Hydrology, UFZ – Helmholtz Centre for Environmental Research, Permoserstraße 15, 04318 Leipzig, Germany

I. Möller · S. Schlömer
Department 1.5 Resource Geochemistry, Federal Institute for Geosciences and Natural Resources (BGR), Stilleweg 2, 30655 Hannover, Germany

C. Leven · A. Lübben
Center for Applied Geoscience, University of Tübingen, Hölderlinstr. 12, 72076 Tübingen, Germany

A. Schuck · U. Serfling
GGL Geophysik und Geotechnik Leipzig GmbH, Bautzner Straße 67, 04347 Leipzig, Germany

M. Nierychlo
AXIO-NET GmbH, Osterstraße 24, 30159 Hannover, Germany

G. Horak · T. Streil
SARAD GmbH, Wiesbadener Str. 20, 01159 Dresden, Germany

M. Gisi · A. Simon
Bruker Optik GmbH, Rudolf-Plank-Str. 27, 76275 Ettlingen, Germany

V. Engnath
MapConcept Ltd., Gohliser Straße 13, 04105 Leipzig, Germany

A. Liebscher and U. Münch (eds.), *Geological Storage of CO_2 – Long Term Security Aspects*, Advanced Technologies in Earth Sciences,
DOI 10.1007/978-3-319-13930-2_2

hierarchical assurance monitoring concept was developed and validated with the aim of establishing a modular observation strategy including investigations in the shallow subsurface, at ground surface level, and in the atmosphere. Numerous methods and technologies from different disciplines (such as chemistry, hydrogeology, meteorology, and geophysics) were either combined or used complementarily to one another, with results subsequently being jointly interpreted. Patterns of atmospheric CO_2 distributions in terms of leakage detection can be observed on large scales with the help of infrared spectroscopy or micrometeorological methods, which aim to identify zones with unexpected or anomalous atmospheric CO_2 concentrations. On the meso-scale, exchange processes between ground surface level and subsurface structures need to be localized using geophysical methods and soil gas surveys. Subsequently, the resulting images and maps can be used for selecting profiles for detailed in situ soil gas and geophysical monitoring, which helps to constrain the extent of leakages and allows us to understand controlling features of the observable fluid flow patterns. The tools utilized were tested at several natural and industrial analogues with various CO_2 sources. A comprehensive validation of the opportunities and limitations of all applied method combinations is given and it shows that large spatial areas need to be consistently covered in sufficient spatial and temporal resolutions.

1 Introduction

In recent years, global concerns about greenhouse gas emissions have stimulated considerable interest in carbon capture and storage (CCS) as a climate change mitigation option which can be used to reduce man-made CO_2 emissions. This is achieved by separating and capturing CO_2 from emission sources, then injecting and storing it in the subsurface. While the public perception of CCS nowadays is rather negative, the IPCC states that the majority of CCS deployment will occur in the second half of this century (IPCC 2005). Therefore, techniques are needed to measure the amount of CO_2 stored at a specific sequestration site, to monitor the site for leakages and storage integrity over time, and to verify that the CO_2 is safely stored and not harmful to the host ecosystem (Hovorka 2008).

The IPCC also states that CO_2 storage risks are comparable to those associated with similar industrial operations, such as underground natural-gas storage (UNEP 2006). However, the greatest environmental risk associated with CCS technology is gradual leakage through undetected faults, fractures or wells, or the potential problems caused by leakages due to injection well failure or leakages up through an abandoned well. These potential leakages could negate the initial environmental benefits of capturing and storing CO_2 emissions and may have harmful effects on human health (Georgiou et al. 2007). Successful monitoring plans need to cover different areas at different scales to enable detection of any significant irregularities, or CO_2 migration paths and any leakages at the surface. The detection of atmospheric releases is especially necessary to establish an early warning system and to plan mitigating actions. Benson (2006) stated that an effective monitoring program should first of all focus on detecting whether or not emissions are occurring. Once any actual or possible emissions are detected, more detailed investigations are necessary for precise localization and quantification. Therefore, a monitoring concept combining appropriate methods is needed to gain timely information about the location of migration paths, seepages, and the CO_2 distribution in the shallow subsurface.

2 Application of a Hierarchical Monitoring Approach

There are two distinct purposes for undertaking monitoring at CO_2 storage sites: (1) to ensure conformance by tracking the pressure buildup and CO_2 inside the storage complex, thereby helping to indicate the long term security of the site ('integrity monitoring') and (2) to ensure containment by triggering timely control measures to mitigate any unexpected leakage, helping to demonstrate the current security situation, especially in the area surrounding the storage complex ('assurance monitoring') (Bourne et al. 2014). Several geochemical and geophysical (such as time lapse seismics) techniques allow for monitoring of the regional distribution of CO_2 in the storage complexes, seal integrity and the pressure evolution in response to injection. They can therefore be used to verify storage conformance and are valuable tools for integrity monitoring (IEA 2012). Assurance monitoring is used to compare pre- and post-injection properties to verify containment and the absence of any environmental effects outside the storage complex. These assurance monitoring tools must consider various monitoring zones (atmosphere, biosphere, ground surface, aquifer/vadose zone, and storage formation), their lateral variabilities and transport-relevant flow paths.

The aim of the MONACO project was to apply and validate a near-surface monitoring concept covering different scales to enable reliable detection of CO_2 migration and seepage. Our approach focuses on the development of assurance monitoring techniques—especially in the atmosphere, at surface level and in the vadose zone or the saturated zone. Applied groundwater, soil, soil gas, and atmospheric monitoring tools provide data about environmental integrity at increasing distances from the reservoir. Large-scale atmospheric monitoring methods are applied to investigate air composition to help determine unexpected CO_2 levels. Subsequently applied meso- and point-scale surface and near-surface monitoring techniques focus on structural settings in the subsurface and CO_2 interaction processes with the aim of identifying areas of risk for human beings and ecosystems.

According to Bourne et al. (2014), a successful monitoring plan complies with regulatory requirements (e.g., requirement to perform adequate pre-injection characterization and baseline monitoring), clearly defines monitoring objectives for risk assessment (risk based monitoring); selects appropriate monitoring tools for the site (site-specific monitoring); and continuously evaluates the monitoring systems (adaptive monitoring) (Bourne et al. 2014). Appropriate site monitoring requires a suitable and modular design to select the right tool, to meet the right need, at the right phase of the implementation (Fig. 1).

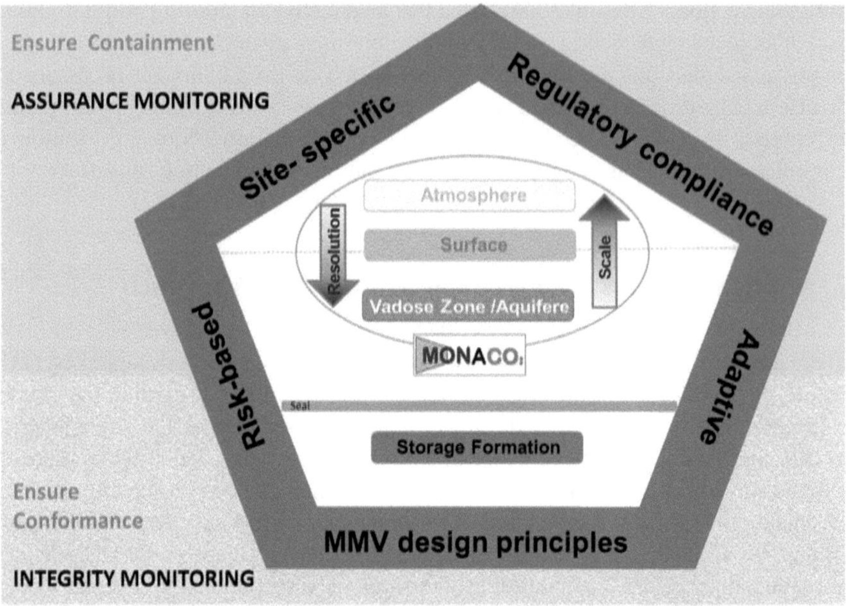

Fig. 1 Illustration of measurement, monitoring and verification (*MMV*) principles (Bourne et al. 2014) and different monitoring zones to monitor CO_2 accumulation, possible migration paths and CO_2 leakages. The project MONACO considers near-surface monitoring zones and applies monitoring methods with different resolution and applicable at different scales

Results and lessons learned from the MONACO approach were primarily obtained by applying the integrative monitoring concept including the practical field work and the necessary data processing. Field work was carried out on several test sites with normal ambient CO_2 conditions in the Altmark region (Northern Germany) and on two natural CO_2 degassing sites in the Cheb Basin (Czech Republic) and Starzach (Baden-Württemberg, Germany).

The Cheb Basin (NW Bohemia) is a CO_2 leaking natural analogue and is a promising location for directly investigating processes along preferential migration paths and verifying monitoring tools. Here, mantle-derived CO_2 is emitted from both isolated gas vents (mofettes) and from extensive diffuse degassing zones. This is caused by a structural fault as preferential pathway (Weinlich et al. 1999). The degassing vents are in some cases characterized by vegetation anomalies. Similar conditions concerning enhanced natural CO_2 exhalations were found at the Starzach site which was used for CO_2 mining in previous times.

By using a web based information system established within the MONACO project, different web map services (WMS), digital elevation models, aerial photographs, borehole information and processed monitoring data enable a comprehensive database of these sites for data interpretation.

2.1 *Tools for Large-Scale Monitoring—Atmospheric Monitoring*

Methods applied at large scales can provide key information about CO_2 leakage occurrences and therefore help identifying potential areas for further meso-scale investigation. The impact on the land surface and near-surface atmosphere caused by elevated CO_2 concentrations may even alter spectral reflectance or emissivity characteristics and can be detected using remote sensing techniques. Examples of such techniques include multi- and hyperspectral airborne remote sensing, as well as ground-based remote sensing infrared or laser spectroscopy (Shuler and Tang 2005).

Within a CCS site, atmospheric monitoring in the vicinity of the storage project is designed to detect and quantify emissions from potential leakage sources (e.g., permeable faults, abandoned wells). An effective atmospheric monitoring tool should satisfy the following requirements: (1) be capable of large-scale observation with sufficient spatial and temporal resolution, (2) fast application and rapid data interpretation, and (3) have sufficient sensitivity to increased atmospheric CO_2 concentrations and fluxes, triggering control mechanisms for subsequent steps. Sensors that can measure atmospheric CO_2 anomalies over open paths which are hundreds of meters long are especially useful in helping us to obtain an initial overview and first assessment of leakages, and provide the required information so that further efficient observations can be made.

2.1.1 Open-Path Fourier-Transform Infrared (OP FTIR) Spectroscopy

A promising approach for detecting elevated CO_2 concentrations along an open optical path is the measurement of absorption loss using OP FTIR spectroscopy and open-path tunable diode laser absorption spectroscopy (TDLAS) (Seto and McRae 2011; Etheridge et al. 2011; Shuler and Tang 2005; Reiche et al. 2014). These ground-based remote sensing methods are proven to be flexible long-path techniques for the characterization of larger areas, and are able to simultaneously detect various volatile atmospheric compounds relevant for environmental assessment with a single rapid measurement.

OP FTIR spectroscopy is based on the analysis of ambient (passive mode) or artificial infrared radiation (active mode) in the 700–4,000 cm^{-1} wave-number range along optical pathways (in km-range). Many greenhouse gas molecules (e.g., CO_2, H_2O, CH_4) have unique signatures (absorption or emission bands) in the spectral range under consideration. IR spectroscopy allows spatial characterization of emissions and can be applied non-invasively as an automated surveillance method in large and potentially inaccessible areas. It is proven to be a powerful technique, enabling online monitoring of fugitive emissions for industrial, environmental and health applications (Griffith et al. 2002; Harig et al. 2006; Harig and Matz 2001; EN_15483 2008).

The application of both active and passive ground-based OP FTIR spectroscopy was validated within this project as one possible method for achieving large-scale scanning of atmospheric composition, in terms of identifying areas of higher leakage vulnerability where detailed subsurface investigations (on meso- or point-scale levels) are subsequently required. Based on investigation of natural CO_2 degassing sites and analysis of industrial emissions, OP FTIR spectroscopy proved itself to be a robust and suitable monitoring method. To ensure reliable results, certain 'best practice' recommendations have to be taken into account:

- OP FTIR spectroscopy is an optical technique. Hence, unobstructed optical pathways to target zones are required. In denser industrial or urban areas, this restriction might pose a significant challenge.
- The measurements result in integral concentration values along the optical pathway. The integrative character of the measurement needs to be considered when designing monitoring schemes and when interpreting measurements with respect to the localization and quantification of emissions.
- The detection of small scale sources (e.g., point emissions) might be challenging. A dense grid of optical pathways (resulting in a large data amount) is required. However, when measuring along large distances, the ability to identify emission sources improves with increasing concentration variations. The sensitivity of the method can also be increased when considering relative temporal variability instead of absolute values.
- Site-specific influences including parameters such as principal wind direction, meteorological conditions, topographic influences, infrastructure, other artificial emission sources, and biological background need to be monitored prior to and during atmospheric monitoring.
- Atmospheric dispersion effects can have a strong impact on the detectability of CO_2 anomalies. Mixture and dilution processes in the near-surface atmosphere have to be considered and can be simulated using atmospheric dispersion models, in order to assess observed data (Flesch et al. 2005; Gal et al. 2012; Leuning et al. 2008).
- Passives open-path measurements offer the chance of achieving robust surveys in various arbitrary measurement directions, which are useful for gaining an overview. Furthermore, the large optical path lengths which can be achieved represent a key advantage when surveying large areas (several km^2). However, a passive system is not best suited for the retrieval of high-precision quantitative gas concentration data. Reasons for this include: an undefined path length and width for long pathways, complex signal behavior due to the combined emission and absorption behavior of the target gas, and problems caused by weak signals due to low temperature differences between the target gas and the background environment.
- To improve quantitative analysis in the case where weak sources are present, the application of a robust active open-path spectrometer is recommended.

In contrast to weak passive IR-radiation emitted in the background, an active source of radiation is used, which emits a constantly high signal level leading to outstanding detection capabilities. Since the radiation path length is known, the spectrometer, by design, is sensitive to its own artificially emitted radiation only; high-precision gas concentration measurements are possible.

In our study, OP FTIR spectroscopy was evaluated and is considered to be a suitable tool for use as part of an early warning monitoring concept. A fully automated high resolution active OP FTIR spectrometer system was designed within the frame of the MONACO project, fulfilling the requirements for reliable large-scale atmospheric monitoring at CCS sites (Bruker 2014). Monitoring operation and spectral analysis can be carried out permanently and automatically with high temporal resolution. While concentration retrieval is performed in real-time, reliable interpretation of concentration values with respect to stored CO_2 leakages may require expert knowledge for each specific site. A combination with other atmospheric monitoring techniques is recommended (Fig. 2).

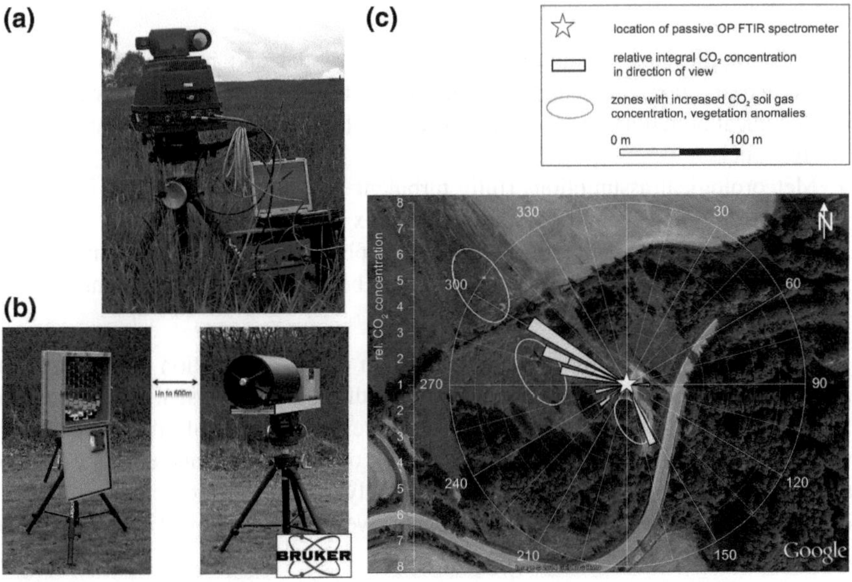

Fig. 2 Validation of OP FTIR spectroscopy to identify atmospheric CO_2 anomalies based on MONACO project results. **a** Passive monitoring equipment including passive OP FTIR spectrometer with stand-alone power supply and controlling notebook. **b** New active OP FTIR spectrometer system consisting of retroreflector (*left*) and open-path spectrometer with collimation optics (*right*) with a maximum investigation distance of 600 m. **c** Results of passive monitoring scan at a natural carbon dioxide degassing site (Czech Republic). Zones with distinctly increased atmospheric CO_2 concentration can be observed in the direction of known soil gas anomalies (modified after Schütze et al. 2013; aerial photo: Google Earth 2014)

2.1.2 Eddy Covariance Method

Looking back on more than 30 years of experience in micrometeorological and ecological studies, the eddy covariance method (EC) has often and consistently been proposed times as a potential suitable method for the monitoring of geologic CO_2 storage sites (e.g., Leuning et al. 2008). The main reason for this is the technique's capability to derive accurate gas fluxes as spatially-integrated expressions of the related exchange between the ground surface and the atmospheric boundary layer (spatial coverage range: from several hundred m^2 to a few km^2, temporal resolution: from several minutes to hours). However, previous studies have also shown that a relatively high leakage rate would be required for leakage detection via EC (Lewicki et al. 2009; Etheridge et al. 2011).

For technical and methodological comparisons, complementary near-surface CO_2 monitoring methods were deployed, along with EC equipment—namely CO_2 accumulation chambers, permanent soil CO_2 monitoring stations and air CO_2 monitoring sensors.

All aspects considered, deduction of gas exchange rates using the EC method is a complex statistical approach that is based on several restricting model assumptions that form boundary conditions for the deployment of this method (Burba 2013). Key constraints include:

- Topographical pre-conditions (necessity for a flat and homogeneous ground surface, the measurement point (location of tower) represents an upwind area)
- Technical prerequisites (instruments are able to detect minimal variations at high frequencies)
- Meteorological assumptions (fully turbulent flux, total vertical flow is negligible, steady state conditions during the flux averaging interval, air flow convergences or divergences as well as atmospheric gravity waves are negligible, air density fluctuations do not exist or can be corrected, measurements capture the boundary layer of interest).

The EC technique requires careful selection of the observation site and special precautions when undertaking technical handling of the essential instruments. Problems related to equipment setup include several sources of errors, e.g., selection of the measuring height, a possible tilting of the instruments, effects of sensor separation (namely distance between gas analyzer and anemometer), as well as distortions caused by the installations themselves. Other technical and operational demands to be considered include:

- The installed technical equipment should be constructed and set up in a way that does not disturb the existing nature of turbulence.
- The equipment must be environmentally robust and suitable for performing remote operations (e.g., low power consumption, assured power supply), while the setup and maintenance of system components (cleaning, calibration, replacements etc.) should be as easy as possible, in order to facilitate and ensure an accurate configuration of the instruments and their performance during operation.

The demands the EC method imposes regarding instrumentation, survey design and implementation, as well as data processing, are still great. However, the instrumental and data processing aspects of this approach have reached a high level of maturity. This method can be classified as being robust and reliable, even if it still benefits from technical processes concerning the instruments involved and software applications. If the boundary conditions are met, the EC technique produces accurate flux information on a medium scale with good temporal and spatial resolution. During the field experiments in our project, the instruments showed a very high availability and any system downtime was almost exclusively caused by common maintenance work or due to external factors, such as power cuts or data transfer problems. The quality of the calculated flux data depended notably on the prevailing wind conditions and, in some cases, on the influence of air moisture (relative humidity >90 %, fog, drizzle, rain). Thus, the meteorological constraint "no wind = no flux data" can unfortunately be extended to mean "weak wind = poor data quality". Nonetheless, typical data coverage in literature is in the range 65–75 % over the course of a year (Falge et al. 2001), although our experiments had slightly better temporal coverage. As a rule of thumb, the general total EC measurement error seems to be between 5 and 20 %, while Baldocchi (2003) indicates that an error of less than 7 % occurs during the day and less than 12 % at night.

Currently, there is no overall agreement on a single, standardized methodology for the EC technique, although much work towards harmonization has already been carried out by the international EC research community and its networks. Since each observation site has its own characteristics, almost every EC experiment needs to consider different parameters. Thus, the applicability of EC technique remains, to a certain extent, a site-specific monitoring approach. Use of the EC method demands expert knowledge and it is presently not a "simple, transparent technique for day-to-day monitoring practice".

If the monitoring concept focuses on CO_2 fluxes on a medium scale, the EC technique has a unique position, despite the limitations mentioned above. It has the capability to act as a methodological link between integral large-scale monitoring efforts on one hand, and small-scale approaches distributed over large areas on the other—even if some methodological developments still must be made in order to ensure a logical combination of data from different scales.

However, with regard to CO_2 storage practice, one important question arises: At what stage of storage operations is the *quantification* of CO_2 fluxes required? For all intents and purposes, extended baseline flux quantification, i.e., the identification of the natural background and its variability, is mandatory—otherwise, additional CO_2 flux caused by potential leaks cannot be quantified. Furthermore, once leakage is detected by any monitoring approach, flux quantifications are also needed in order to obtain information on how much CO_2 escapes from the storage site into the atmosphere and to identify potentially hazardous areas, initiate project remediation strategies, and to verify the success of the corrective measures. In all other phases of normal storage operations, near-surface monitoring might routinely rely on other indicators; indicators that can be more easily determined and in a more transparent way than the complex computation required for CO_2 fluxes using the EC approach.

Monitoring of CO_2 concentrations instead of fluxes would, for example, cover such an easy-to-use indicator. Strategically well-placed and fitted with elementary wind sensors, simple air CO_2 sensors can measure CO_2 concentrations straightforwardly. They provide sufficient information to ensure continuous near-surface monitoring by identifying recurring data patterns including their normal variations on one hand, but also detect potential anomalies on the other.

Both OP FTIR spectroscopy and the EC technique have been validated in our study as suitable monitoring techniques. They are near-surface atmosphere monitoring methods that work on larger scales. However, it needs to be noticed that geologic CO_2 storage is realized deep underground. It is therefore evident that it cannot cover all monitoring aspects of industrial CCS operations and must be an integral part of a comprehensive monitoring concept consisting of methods that focus on other environmental compartments or on different temporal and spatial scales.

2.2 *Tools for Meso-scale Monitoring—Surface-Based Monitoring*

Within our study, patterns of atmospheric CO_2 variability were observed with the help of FTIR and EC methods. In zones of increased CO_2 concentration, the source processes (man-made, natural) need to be clarified and the surface or near subsurface areas should be monitored in detail. Ground-based deformation studies, geophysical methods such as electromagnetics (EMI), electrical resistivity tomography (ERT) and self-potential (SP), soil gas concentration, flux measurements and soil moisture and temperature mapping are efficient methods for identifying near-surface structures which favor gas accumulation and migration applied in the MONACO approach. Combinations of various geophysical methods and soil-gas investigations (CO_2 concentration and flux rate) provide insights into the physical properties of sediments (e.g. resistivity variations), structural features (e.g., secondary traps) and transport processes (e.g., migration of fluids, CO_2 solution). Such an integrative approach derives information about preferential degassing pathways (e.g., Buselli and Lu 2001; Byrdina et al. 2009; Lamert et al. 2012; Pettinelli et al. 2010; Schütze et al. 2012).

2.2.1 Geophysical Methods

The application of geophysical methods is motivated by two main processes. Firstly, CO_2 (dissolved, volatile) in the pore space has an impact on physical sediment properties (e.g., electrical resistivity). Secondly, fluid movement may induce dynamic, time-dependent processes (e.g., temporal variations in geophysical parameters, generation of electro-kinetic effects). Within the MONACO project the supposed variation in geophysical parameters due to the presence of CO_2 was investigated using a combination of several geophysical methods—such as SP monitoring and mapping, EMI mapping, ERT survey and refraction seismic measurements.

The self-potential (SP) method measures a natural electrical potential field distribution at the ground surface. It is used to map distinct anomalies or to monitor temporal changes caused by dynamic processes. For field applications, it is often difficult to separate the variously superimposed sources of SP signals in the measured data induced by a combination of electrokinetic effects, electrochemical potential differences and thermoelectric coupling effects. However, the determination of streaming potentials (electrokinetic effects) could be a possible parameter to feature CO_2 migrations or could at least be an indicator for fluid transport in the subsurface (Byrdina et al. 2009; Revil et al. 1999a; Smaczny et al. 2010; Sprunt et al. 1994). It must be considered that SP anomalies are influenced by soil structure, rock variations, meteorological conditions and/or groundwater flow. Streaming potentials are sensitive to variations in hydrological parameters, which are expressed in considerable time dependence (Ernstson and Scherer 1986). Furthermore, the effect of more intense chemical reactions due to higher CO_2 concentrations in the subsurface can encourage evolution of an increased electrochemical effect on the SP values (Zlotnicki and Nishida 2003). In our measurements, the observed anomalies are potentially driven by gas flow associated with transport of a water phase within the permeable zones (Fig. 3). These effects are also observed by Byrdina et al. (2009), Revil et al. (1999b), Sauer et al. (2014), Sandig et al. (2014). Following 'best practice' recommendations, the reliable identification of influencing subsurface properties (e.g., porosity, gas or water saturations, conductivity) on the SP signal requires additional geophysical methods (e.g., ERT) and environmental data (e.g., soil gas concentration measurements) (Flechsig et al. 2008; Jardani et al. 2007).

Resistivity and electrical conductivity can act as geophysical indicators for the presence of CO_2 in pore space. Variations in resistivity depend on physical sediment properties such as conductive mineral components, porosity, clay content, water saturation, and electrolyte concentration (Flechsig et al. 2008; Knödel et al. 2007; Reynolds 2011). Recent field tests, laboratory experiments, and numerical simulation studies show that electrical resistivity is highly sensitive to the presence of CO_2 (Bergmann et al. 2012; Börner et al. 2013; Kharaka et al. 2009; Lamert et al. 2012). Gaseous CO_2 intrusion into shallow groundwater systems generally causes increased gas phase content in the soil pore space, which accordingly leads to increased bulk resistivity. However, subsequent dissolution of CO_2 in partly-groundwater saturated sediments leads to the occurrence of carbonic acid followed by generally decreased pH values and increased alkalinity. These circumstances lead to decreased bulk resistivity.

Electromagnetic induction mapping (EMI) is a non-invasive method for measuring the apparent electrical conductivity. EMI methods are considered as being a promising approach for monitoring CO_2 storage (Börner et al. 2010). Due to the small expected differences in electrical conductivity in the shallow subsurface, careful device calibration, operation, and interpretation is necessary. Our results indicate that EMI can be used as an appropriate tool for a fast and rough outline survey of the recent main geological structures. For more detailed insights, the subsequent application of geoelectrical investigations is recommended.

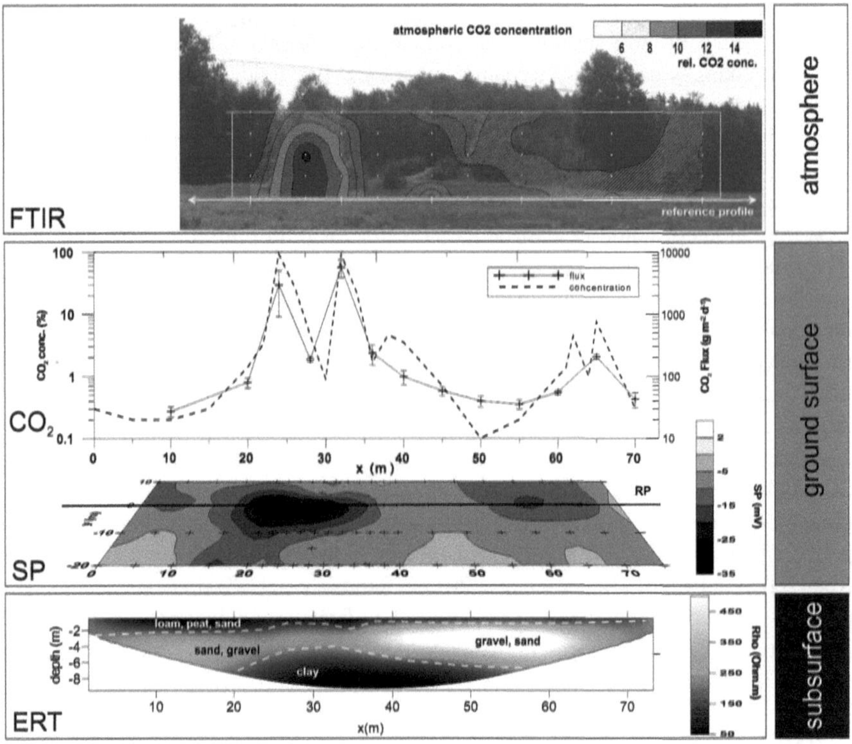

Fig. 3 Example for validation of the hierarchical approach at a natural analogue site in the Cheb basin (CZ). **Atmospheric monitoring**: Scanning passive OP FTIR spectroscopy was applied. The image shows atmospheric CO_2 concentration in relation to normal conditions. Atmospheric CO_2 concentration maximum value occurs above the main soil CO_2 degassing anomaly. **Ground surface monitoring**: Soil gas concentration distribution and soil flux measurement displays an anomaly indicating the main degassing zone. The CO_2 flux measurements using the accumulation chamber method show two distinct flux maxima above the anomaly threshold of 50 g m^{-2} d^{-1}. Soil CO_2 concentration values measured along the same soil gas profile show also two distinct concentration maxima with nearly 100 % CO_2 concentration. SP distribution at ground level: negative SP anomaly correlates with CO_2 concentration anomaly in the atmosphere and near surface. **Subsurface monitoring**: In the ERT results a distinct disturbance in the resistivity layers is obvious beneath the main degassing zone. Lithological units shown were derived from a drilling log

Electrical resistivity tomography (ERT) is a non-invasive geophysical method providing information on the subsurface resistivity pattern. Knowledge about the resistivity distribution can be applied to map shallow subsurface structures depending on the investigation depth—and is subject to electrode spacing, electrode configuration, and the resistivity distribution of the subsurface. The determination of resistivity anomalies is considered to be useful when investigating disturbances caused by variations in lithological parameters and fluid content (Flechsig et al. 2010; Schütze et al. 2012). Within the frame of the MONACO project, ERT surveys were used to reveal internal structures responsible for fluid migration or trapping in the shallow sedimentary layers. It turned out that the ERT method is a valuable tool for monitoring temporal and spatial changes in resistivity patterns due to variations in fluid transport processes. Furthermore, the analysis of resistivity anomalies is crucial when it comes to understanding the observed SP patterns (Fig. 3).

Additional shallow refraction seismic investigations were validated in the project as suitable method which helped us obtain supplementary structural information from the seismic velocity models. This method yields high resolution images concerning the structural geological setup and provides valuable data to minimize the ambiguity of geophysical models.

2.2.2 Soil Gas Surveys

Surface-based measurements of CO_2 concentration and CO_2 flux provide reliable insights in leakage processes, which can allow us to constrain the extent of potential leakages and to understand the controlling features of the observable fluid flow patterns (Schütze et al. 2012).

Reliable soil gas sampling requires a thorough sampling technique. The soil gas CO_2 concentration is measured in shallow depths (minimum sampling depth 0.5 m below ground level) and the mixture of gas samples with fresh air have to be avoided. Flux measurements are typically based on the accumulation chamber method (Chiodini et al. 1998). Both investigation techniques were validated at the test sites in the Cheb Basin and Starzach. These techniques are considered to be valuable tools for the mapping and quantifying of CO_2 seepage to the surface via preferential pathways. However, detailed soil flux and concentration investigations of larger areas are time-consuming. Changes in meteorological conditions during measurement need to be carefully considered during subsequent interpretation.

In different measuring campaigns during the project, we observed a high spatial variance of the soil gas concentration and flux on small scales with respect to location, spatial extent and amplitude. Land use, meteorological and soil moisture conditions especially influence gas migration and seepage. Therefore, the joint interpretation of soil gas measurements with geophysical data and soil moisture/temperature data is important, to gain a realistic site-specific overview for risk analysis (Fig. 3).

2.2.3 Soil Parameter Measurements

Soil temperature and soil moisture are influencing parameters on geophysical and geochemical parameters (SP, ERT, EMI and CO_2 concentration) and, as an immediate response to meteorological conditions, must be considered when carrying out data interpretation. These parameters can be mapped and monitored using standard devices such as temperature probes and soil moisture sensors based on time domain reflectometry (TDR).

Results achieved from integrative measurement taken at our test site in the Cheb basin indicate that the hierarchical monitoring approach represents a successful multidisciplinary modular concept (Sauer et al. 2013; Schütze et al. 2013). The application of OP FTIR spectroscopy in combination with soil gas surveys and ERT investigations has proven to be a valuable tool for comprehensive characterization of the atmospheric and near-surface CO_2 distribution, as well as subsurface structural features (Fig. 3).

2.2.4 Ground-Based Deformation Studies

Precise measurements of ground deformation (uplift or subsidence) can be acquired remotely using geodetic techniques e.g., radar satellites positioned above CO_2 storage sites can be used to determine the surface-level impacts of injection. It can be seen that the amount of geomechanical deformation caused as a result of CO_2 injection is likely to be a function of the volume of CO_2 injected (Verdon et al. 2013). The satellite-based Interferometric Synthetic Aperture Radar (InSAR) has proven to be successful for monitoring (to centimeter level degree of accuracy) volcanic and earthquake deformation (Zhao et al. 2012; Biggs et al. 2009) and was especially valuable when used at the In Salah CCS site (Ringrose et al. 2013; White et al. 2014).

Differential Global Navigation Satellite System (DGNSS) observations were evaluated within the project for high precision monitoring of surface deformation and movement. GNSS observations obtained from a four station reference network were post-processed using GNSS networking software. One of the reference stations was replaced by a monitoring unit consisting of a GNSS receiver and a crank unit with a mounted GNSS antenna. By using the crank unit, the height of the monitoring station were adjusted to the required level. Monthly height adjustments that were undertaken were recorded and compared with the height results obtained by GNSS post-processing analyses. Applying a special filter for the determination of coordinates for the monitoring station, the standard deviation of the time series was reduced down to 1 mm (height). Figure 4 shows the post-processing results, indicating the detectability level of the manual height changes of the crank unit.

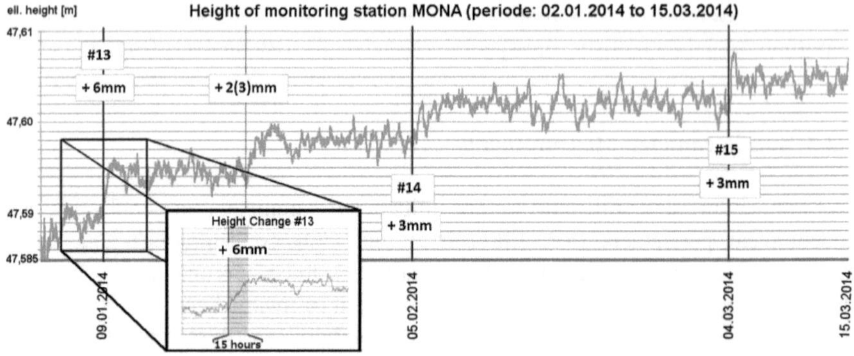

Fig. 4 Post-processing results—monitoring station MONA, The manual height changes of the crank unit are seen in the coordinate time series very clearly. Event #13 shows that the manual height change was detected in its whole magnitude after around 15 h

The results of the analyses show that when using DGNSS measurements, surface deformations of even a few millimetres can be reliably detected. Compared to other satellite based observation methods, permanent coordinate monitoring of a CCS site is possible using the GNSS approach. Deformations can be detected within a few hours.

2.3 Tools for Small-Scale Monitoring—Subsurface in Situ Monitoring

The distribution of geophysical indicators in conjunction with observed characteristic CO_2 concentration and flux patterns is useful for identifying site locations for detailed in situ monitoring. These can then be used for a further detailed determination of the lithological setting and spatial distribution of the site's permeability, while simultaneously assessing the spatial and temporal migration behavior of CO_2. Geophysical methods (e.g., ERT and seismics), in situ installations (e.g., for CO_2 concentration and flux measurements and isotope analysis of sampling probes) are examples of monitoring tools which can be used to identify deeper geological structures responsible for gas migration and trapping, and to characterize the CO_2 source.

2.3.1 In Situ Measurements and Sampling

Direct Push technology (DP) is a minimally invasive and highly efficient tool used for in situ measurement of different physical and chemical parameters (e.g., electrical conductivity logging, hydraulic conductivity logging) and the installation of monitoring sensors into depths of up to 30 m in unconsolidated to weakly-consolidated sediments (Leven et al. 2011; Zschornack and Leven 2012; Dietrich and Leven 2006). Direct Push can also be used for retrieving soil, gas and water samples needed for chemical analysis. Soil samples provide especially valuable lithological information, which can be used to validate geophysical data.

2.3.2 In Situ Installations

At a field site in Starzach (Baden-Württemberg, Germany), a location with enhanced natural CO_2 exhalation and which was used for CO_2 mining in previous times, soil sensors designed to measure soil temperature (T), volumetric water content, and electrical conductivity (σ) were installed at two different depths (0.3 and 0.6 m below ground level). The reason for doing so was to investigate the influence of mofettes (focused CO_2 degassing) on soil parameters such as electrical conductivity and soil CO_2 concentration at different locations with characteristic low, intermediate, and high soil CO_2 concentrations.

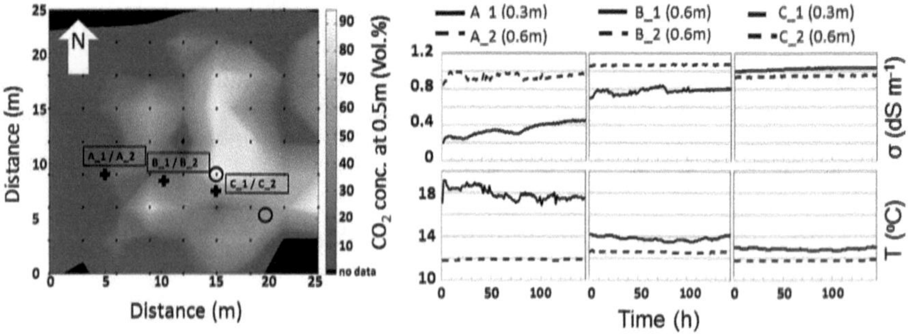

Fig. 5 *Left* Measured CO_2 concentration in 0.5 m depth on a 3 × 3 m grid at the field site in Starzach. Location of two mofettes (*circles*), location of six soil sensors (A_1, A_2, B_1,B_2,C_1, C_2; _1 indicate depth 0.3 m; _2 indicate 0.6 m) (*black crosses*), sample locations (*black dots*). *Right* Measured soil temperature (T) and electrical conductivity (σ) with in situ sensors placed in areas with low soil CO_2 concentration (10–30 vol%: soil sensors A_1 and A_2), intermediate concentration (31–60 vol%: soil sensors B_1 and B_2) and high concentration (61–80 vol%: soil sensors C_1 and C_2)

Figure 5 shows the measured CO_2 concentration at a depth of 0.5 m and some of the sensor data (T, σ) recorded at the site. The shallow sensors (0.3 m) exhibit a general trend of decreasing temperature and increasing electrical conductivity near the mofettes. In contrast, the deeper sensors show almost no variations. The following processes could be responsible for the observed trends: (1) Movement of groundwater along the degassing channels, which reduces the soil temperature and which in general has a higher electrical conductivity than rain water infiltrating the soil (Flechsig et al. 2010; Hölting and Coldeway 2013). (2) Higher dissolution effects in the soil near the mofettes. The higher amount of CO_2 leads to more dissolution of CO_2 in the soil water and the resulting increase in ion load occurs due to reactions of the CO_2 with the aquifer and soil matrix, causing an increase in electrical conductivity.

Based on the project experiences, a network of soil sensors placed above a CO_2 storage site can serve as a suitable monitoring tool. In the event of a leakage, the in situ sensors are able to identify subsurface areas of increased soil CO_2 concentrations. Although temperature and electrical conductivity could be detected in areas of increased CO_2 concentration at our investigated field site, the hydrogeological situation influencing the soil temperature is site-specific. Therefore, only electrical conductivity can be recommended as a more general parameter for indicating areas with potential CO_2 leakages.

To investigate the effect of variations in atmospheric parameters on intensity of mass fluxes from the two mofettes, a hood-shaped metal sheet with an opening for free gas outflow was placed over a mofette with an airtight seal. Gas fluxes out of mofettes are mainly driven by advection and therefore are much stronger than diffusive fluxes (Kämpf et al. 2013). The total amount of free gas outflow from a single mofette was estimated by simultaneously measuring flow velocity, static

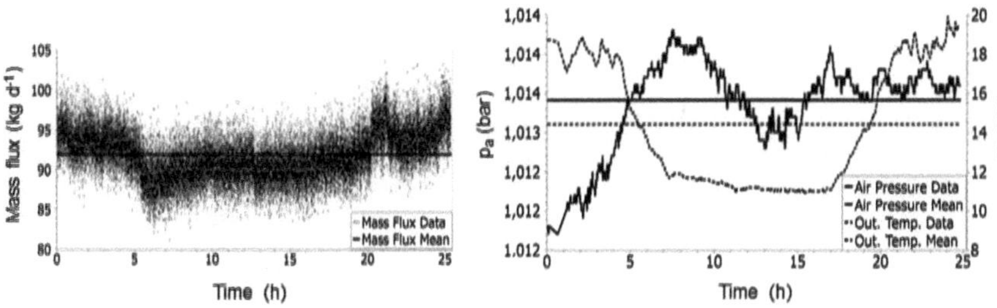

Fig. 6 *Left* Calculated mass flux from a mofette at the field site. The mean mass flux is 92 kg/d. *Right* Outside temperature (T_a) and air pressure (p_a) recorded at a nearby meteorological station over 24 h. Mean T_a for this period is 14.4 °C and mean p_a 1,013.4 mbar

pressure, and gas temperature. Measurements of the volumetric CO_2 content revealed a concentration of almost 100 vol.%. The resulting mass flux of CO_2 gas is shown in Fig. 6 for a period of 24 h compared with air temperature (T_a) and pressure (p_a), recorded at a nearby meteorological station. Variations in these meteorological parameters have no obvious influence on mass flux. However, the trend in air temperature correlates with the trend in mass flux. For smaller or more diffuse mass fluxes, changes in outside temperature and possibly also in air pressure could have a larger impact on mass flux intensity (Vodnik et al. 2009; Chiodini et al. 1998).

As a consequence, monitoring technologies should aim to measure small and diffusive-dominated mass fluxes, where natural variations that occur due to atmospheric parameters are known, so that anomalies can be clearly detected.

The project partner SARAD GmbH developed a new device for in situ flux measurements so that CO_2 flow can be directly measured in the ground and over large measurement distances.

2.3.3 Analysis of Isotopic Composition

Studies of the isotope signature of the soil CO_2 are essential to help distinguish between different CO_2 sources and to characterize the origin of soil CO_2.

The usefulness of analyzing the isotopic composition of soil CO_2 was tested at two sites with natural channel-like CO_2 degassing with CO_2 concentrations up to 90 vol%. In the Cheb Basin, the deep source of CO_2 (mantle derived) is confirmed by the isotope signature (Bräuer et al. 2008, 2011), whereas the origin of the CO_2 degassing in Starzach is still under discussion.

The measurement of the isotopic composition was validated to be an appropriate method to characterize the origin of CO_2 and was successfully used to confine clearly the boundaries of CO_2 soil degassing from areas with degassing of deep originated CO_2 in shallow subsurface (Fig. 7). However, it is a time-intensive, complex analysis method which requires profound expert knowledge.

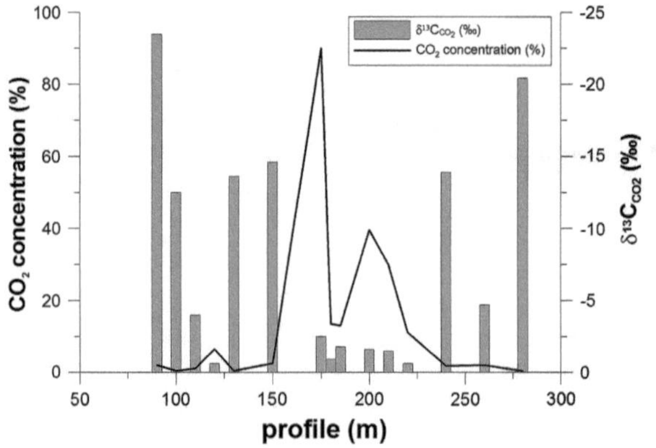

Fig. 7 Project results of investigations at the Cheb Basin site. A natural degassing zone was determined by increased CO_2 soil gas concentration and decreased $\delta\ ^{13}C_{co_2}$ values. The isotopic signature with a $\delta\ ^{13}C_{co_2}$ value of −2 ‰ measured in the degassing area indicates the deep magmatic source of CO_2

2.4 Validation of Methods for Atmospheric, Surface and Subsurface Monitoring

The methods applied for monitoring were evaluated by means of several criteria: robustness, availability, reliability, data accuracy, spatial resolution, spatial integration, temporal resolution, and effort based on practical requirements (e.g., equipment handling, data processing). For an assessment of the different monitoring techniques, the applicability of methods was classified into three groups: (−) limited compliance, (o) acceptable compliance, (+) appropriate compliance (Table 1).

2.4.1 Evaluation and Conclusion

The effectiveness of monitoring methods for CO_2 leakages depends on several factors including the contrast between the physical properties of CO_2 and the pore fluid displaced by CO_2, the lithology and structure of the reservoir, pore fluid pressure and temperature, field setups and surveys, well spacing, and injection patterns (Hagrey et al. 2013; Hoversten and Myer 2000). All methods applied within the MONACO project were evaluated and classified into three criteria:

- (o) **Useful supplementing method**: The application of methods is affiliated with various limitations with respect to the criteria and stand-alone application cannot guarantee suitable results. However, the method can provide additional information for joint interpretation.

Table 1 Validation of integrative methods applied in the MONACO project

	Atmospheric Monitoring				Surface Monitoring							Subsurface Monitoring						
Monitoring method	OP-FTIR Spectroscopy		Eddy Covariance	Atmospheric CO_2 conc. recording	Geophysical methods		Land Deforma-tion Monitoring	Mapping of soil properties		Measurement of soil gas		Geophysical methods		Measurement of soil gas			In- situ measurements	
	active	passive			Electro-magnetics	Self Poten-tial	Differential Global Satellite System (DGNSS)	Tempera-ture	Soil moisture	Soil gas conc. map.	Near-surface flux	Geoelectrics	Seismics	Isotope analyses	Radon conc.	Soil CO_2 conc. monitoring	Tempera-ture	Electical conductivity
Robustness	o	+	+	+	o	+	+	+	+	+	o	+	+	o	o	+	+	+
Availability	o unobstruc-ted sight	o unobstruc-ted sight	+	+	o	o/+	o	+	+	+	+	o/+	o/+	o	o	+	+	+
Reliability	+	+	+	+	+	+	o	o	o	+	o	+	+	+	o	+	+	o
Data accuracy and precision	+ calculated, integral value	o calculated, integral value	o calculated, integral value	+ measured	o integral apparent value	o various sources	+	o high spatial variance on small scales	o high spatial variance on small scales	o high spatial variance on small scales	+ high spatial variance, depends on meteorological conditions	+	+	+	o	+	o +/- 1°C	o +/- 10%
Spatial resolution	o large / medium scale	o large / medium scale	o medium scale	- medium scale	o medium scale	o medium scale	+ large scale	o medium/ small scale	o medium/ small scale	o medium/s mall scale	o medium/ small scale	+ large/ medium scale	+ large/ medium scale	- small scale	- small scale	+ large scale	+/- small scale	+/- small scale
Spatial integration	+ +	+ +	+ depends on quality of model	+	+	o	+	o	o	o	-	+	+	-	-	-/o	-	-
Temporal resolution	+	+	+	+	o	o	+	o	o	o	o/+	o	-	-	-	+	+	+
Effort																		
Equipment handling	+	+	+ needs exact setup and more maintenance	+	o Calibration important	+	o	+	+	+	+	+	+	o	+	-/+ demands subsurface installation, but easy to maintain	+/- demands subsurface installation	+/- demands soil specific calibration and subsurface installation
Data processing	+ complex data interpreta-tion including meteoro-logical modeling	o spectral analysis, complex data interpreta-tion including meteoro-logical modeling expert knowledge	o complex processing, lots of data expert knowledge	+ direct registration basics	o correction, filtering expert knowledge	+ fast and direct	o filtering procedures important, filtering of meteoro-logical influences	+ fast and direct	+ fast and direct	+ fast and direct	o flux calculation expert knowledge	o inversion, time lapse expert knowledge	-/o complex processing, time lapse expert knowledge	- time consu-ming, high laboratory effort expert know-ledge	+	+ direct, less data	+ fast and direct	+ fast and direct

- (+) **Appropriate method**: Methods provide valuable information. However, the application has some restrictions/limitations, which have to be taken into account for interpretation.
- (++) **Favored method**: Applications of methods are suitable and provide significant results concerning assurance monitoring.

This classification for the common assurance monitoring tasks is shown in Table 2.

OP FTIR spectroscopy and EC method are considered to be suitable tools as part of an early warning monitoring concept, enabling detection of diffuse or focused CO_2 degassing over larger scales. OP FTIR spectroscopy can supply information on the distribution of CO_2 concentration under natural air conditions as an input parameter for the subsequent quantification of emission rates. The requirements for EC regarding wind conditions, instrumentation, survey design, implementation, and data processing are still very high. If the measuring requirements are met, the EC technique can provide flux information on a medium scale with sufficient temporal and spatial resolution. In addition, continuous wind speed measurements in combination with air CO_2 concentration registration at different heights, as well as CO_2 concentration monitoring with a handheld infrared gas analyzer, have some potential as an appropriate basic monitoring technique. However, due to the scale of

Table 2 Classification of the methods applied in MONACO project for nine pre-defined assurance monitoring tasks

Assurance monitoring tasks (near-surface applications)	Atmosphere				Ground surface								Subsurface			
	Optical remote sensing	Eddy Covariance Method	Handheld infrared gas analyzer	Distributed CO2 sensors	Ground based deformation studies	Soil gas concentrations	Soil gas composition	Soil flux measurement	Electromagnetics	Self-Potential	Soil moisture	Temperature mapping	Electrical resistivity tomography	Seismics	In-situ CO2 installations	Fluid sampling
Diffuse Degassing into the atmosphere	++	+	+	++				o								
Focussed degassing into the atmosphere	+	+	+	++				o								
Near surface soil contamination			o	o		++	++	++			o	o			++	++
Secondary seal integrity					o	+	++	++					+	+	++	++
Migration in subsurface					+	+	++	++		o		o	+	+	++	++
Preferential pathways					+	+	++	++	o	+	o	o	+	+	o	+
Structural Trapping					+			o	o	o			+	++		
Geological structures favoring degassing and accumulation									+	+			++	++		
Groundwater contamination with CO_2 or brine							+		+	o			++	o	++	++

++ favored method
\+ appropriate method
o useful supplementing method

CO_2 storage, these atmospheric methods could only be an integral part of a comprehensive monitoring concept consisting of methods that focus on other environmental domains or on different temporal and spatial scales.

Within the MONACO hierarchical approach the GNSS approach for detecting ground deformation was tested. It was shown that GNSS is suitable for continuous monitoring and investigation of contemporary ground surface deformation in mm range. However, information about height variation is based on distinct measurements at monitoring stations and cannot provide extensive information such as that obtained when using InSAR data.

Interpretation of various geophysical data can yield important information regarding fluid flow and transport processes in permeable near-surface layers, as well as structural and hydraulic properties. The selection of an appropriate method for the fulfillment of assurance monitoring tasks (such as detection of migration and preferential pathways, investigation of structural trapping) depends on site characteristics. The combination of geophysical methods, soil gas investigations and in situ installations concerning e.g., chemical properties of the groundwater, soil gas composition, soil gas ratios, and isotope ratios are necessary to improve the understanding of both gas migration processes and the resultant geophysical response functions, and to minimize ambiguities concerning the investigated geophysical parameter distribution. Furthermore, the influence of environmental conditions on geophysical and soil gas parameters needs to be considered when attempting to establish 'best practice' recommendation for CO_2 storage monitoring concepts.

The results of the MONACO project demonstrated that a successful near-surface monitoring plan should base on a hierarchical approach to cover different areas at different scales enabling a reliable detection of CO_2 migration paths and any leakages at the surface.

Acknowledgments This study was carried out within the framework of the "MONACO—Monitoring approach for geological CO_2 storage sites using a hierarchical observation concept" (grant ID: 03G0785) research project. Financial support provided by the German Federal Ministry of Education and Research (BMBF) in the frame of the Priority Program "Geotechnologien" is gratefully acknowledged. We cordially thank our native speaker C. Higgins for proofreading the manuscript. The authors would also like to thank C. Seeger, J. Poggenburg, A. Schossland, H. Kotas, G. Schmidt, K. Faiß and C. Sandig for technical support and T. Foken (Univ. Bayreuth), G. Burba, G. Fratini, F. Griessbaum and P. Martin (LI-COR Biosciences, Lincoln, Bad Homburg etc.) for theoretical input, fruitful exchange and technical advice.

References

Baldocchi D (2003) Assessing the eddy covariance technique for evaluating carbon dioxide exchange rates of ecosystems: past, present and future. Glob Change Biol 9(4):479–492. doi:10.1046/j.1365-2486.2003.00629.x

Benson SM (2006) Monitoring carbon dioxide sequestration in deep geological formations for inventory verification and carbon credits. https://pangea.stanford.edu/research/bensonlab/presentations/Monitoring%20Carbon%20Dioxide%20Sequestration%20in%20Deep%20Geological%20Formations%20for%20Inventory%20Verification%20and%20Carbon%20Credits.pdf

Bergmann P, Schmidt-Hattenberger C, Kiessling D, Rücker C, Labitzke T, Henninges J, Baumann G, Schütt H (2012) Surface-downhole electrical resistivity tomography applied to monitoring of CO_2 storage at Ketzin, Germany. Geophysics 77:B253–B267. doi:10.1190/geo2011-0515.1

Biggs J, Anthony E, Ebinger C (2009) Multiple inflation and deflation events at Kenyan volcanoes. East Afr Rift Geol 37(11):979–982. doi:10.1130/G30133A

Börner JH, Herdegren V, Börner R-U, Spitzer K (2010) Electromagnetic monitoring of CO_2 storage in deep saline aquifers—laboratory experiments and numerical simulations. In: Workshop on electromagnetic induction in the earth, Giza, Egypt, 18–24 Sept 2010

Börner JH, Herdegen V, Repke J-U, Spitzer K (2013) The impact of CO_2 on the electrical properties of water bearing porous media—laboratory experiments with respect to carbon capture and storage. Geophys Prospect 61(s1):446–460. doi:10.1111/j.1365-2478.2012.01129.x

Bourne S, Crouch S, Smith M (2014) A risk-based framework for measurement, monitoring and verification of the Quest CCS project, Alberta, Canada. Int J Greenhouse Gas Control 26:109–126. doi:http://dx.doi.org/10.1016/j.ijggc.2014.04.026

Bräuer K, Kämpf H, Niedermann S, Strauch G, Tesař J (2008) Natural laboratory NW Bohemia: comprehensive fluid studies between 1992 and 2005 used to trace geodynamic processes. Geochem Geophys Geosyst 9(4, Q04018):1–30. doi:10.1029/2007gc001921

Bräuer K, Kämpf H, Koch U, Strauch G (2011) Monthly monitoring of gas and isotope compositions in the free gas phase at degassing locations close to the Nový Kostel focal zone in the western Eger Rift, Czech Republic. Chem Geol 290(3–4):163–176. doi:10.1016/j.chemgeo.2011.09.012

Bruker (2014) Open path air monitoring system. Bruker optics GmbH. http://www.bruker.com/products/infrared-near-infrared-and-raman-spectroscopy/remote-sensing/ops/overview.html. Accessed 24 July 2014

Burba G (2013) Eddy covariance method for scientific, industrial, agricultural and regulatory applications (Personal communication at EGU 2014, Apr 2014). Lincoln, NE

Buselli G, Lu K (2001) Groundwater contamination monitoring with multichannel electrical and electromagnetic methods. J Appl Geophys 48(1):11–23. doi:10.1016/S0926-9851(01)00055-6

Byrdina S, Revil A, Pant SR, Koirala BP, Shrestha PL, Tiwari DR, Gautam UP, Shrestha K, Sapkota SN, Contraires S, Perrier F (2009) Dipolar self-potential anomaly associated with carbon dioxide and radon flux at Syabru-Bensi hot springs in central Nepal. J Geophys Res 114 (B10):1–14. doi:10.1029/2008jb006154

Chiodini G, Cioni R, Guidi M, Raco B, Marini L (1998) Soil CO_2 flux measurements in volcanic and geothermal areas. Appl Geochem 13(5):543–552. doi:http://dx.doi.org/10.1016/S0883-2927(97)00076-0

Dietrich P, Leven C (2006) Direct push technologies. In: Kirsch R (ed) Groundwater geophysics, vol 548. Springer, Berlin, pp 347–366. doi:10.1007/3-540-29387-6_11

EN_15483 (2008) Ambient air quality—atmospheric measurements near ground with FTIR spectroscopy. European Standard, CEN, Brussels

Ernstson K, Scherer U (1986) Self-potential variations with time and their relation to hydrogeologic and meteorological parameters. Geophysics 51(10):1967–1977. doi:10.1190/1.1442052

Etheridge D, Luhar A, Loh Z, Leuning R, Spencer D, Steele P, Zegelin S, Allison C, Krummel P, Leist M, van der Schoot M (2011) Atmospheric monitoring of the CO_2 CRC Otway project and lessons for large scale CO_2 storage projects. Energy Procedia 4:3666–3675. doi:http://dx.doi.org/10.1016/j.egypro.2011.02.298

Falge E, Baldocchi D, Olson R, Anthoni P, Aubinet M, Bernhofer C, Burba G, Ceulemans R, Clement R, Dolman H, Granier A, Gross P, Grünwald T, Hollinger D, Jensen N-O, Katul G, Keronen P, Kowlaski A, Lai C, Law B, Meyers T, Moncrieff J, Moors E, Munger J, Pilegaard K, Rannik Ü, Rebmann C, Suyker A, Tenhunen J, Tu K, Verma S, Vesala T, Wilson K, Wofsy S (2001) Gap filling strategies for defensible annual sums of net ecosystem exchange. Agric For Meteorol 107:43–69

Flechsig C, Bussert R, Rechner J, Schütze C, Kämpf H (2008) The Hartoušov Mofette Field in the Cheb Basin, Western Eger Rift (Czech Republic): a comperative geoelectric, sedimentologic and soil gas study of a magmatic diffuse degassing structure. Z Geol Wiss 36(3):177–193

Flechsig C, Fabig T, Rücker C, Schütze C (2010) Geoelectrical investigations in the Cheb Basin/ W-Bohemia: an approach to evaluate the near-surface conductivity structure. Stud Geophys Geod 54(3):443–463. doi:10.1007/s11200-010-0026-6

Flesch TK, Wilson JD, Harper LA (2005) Deducing ground-to-air emissions from observed trace gas concentrations: a field trial with wind disturbance. J Appl Meteorol 44:475–484. doi:10.1175/ JAM2214.1

Gal F, Brach M, Braibant G, Bény C, Michel K (2012) What can be learned from natural analogue studies in view of CO_2 leakage issues in carbon capture and storage applications? Geochemical case study of Sainte-Marguerite area (French Massif Central). Int J Greenhouse Gas Control 10:470–485. doi:10.1016/j.ijggc.2012.07.015

Georgiou P, Quick H, Jenkins H, Hayes C, Jensen D, Kelly J, Tollner D, Vale D, Price R, Thomson K (2007) Between a rock and a hard place: the science of geosequestration, Australian Government response report of the House of Representatives Standing Committee on Science and Innovation. Parliament of the Commonwealth of Australia, Canberra

Google Earth (2014) Hartousov 50.133494N 12.462030E, Date of image: 01 Jan 2004. Accessed 13 Aug 2014

Griffith DWT, Leuning R, Denmead OT, Jamie IM (2002) Air–land exchanges of CO_2, CH_4 and N_2O measured by FTIR spectrometry and micrometeorological techniques. Atmos Environ 36:1833–1842. doi:10.1016/S1352-2310(02)00139-5

Hagrey SAa, Strahser M, Rabbel W (2013) Seismic and geoelectric modeling studies of parameters controlling CO_2 geostorage in saline formations. Int J Greenhouse Gas Control 19:796–806. doi:10.1016/j.ijggc.2013.01.041

Harig R, Matz G (2001) Toxic cloud imaging by infrared spectrometry: a scanning FTIR system for identification and visualization. Field Anal Chem Technol 5(1–2):75–90. doi:10.1002/fact.1008

Harig R, Matz G, Rusch P (2006) Infrarot-Fernerkundung für die chemische Gefahrenabwehr. Zivilschutz-Forschung. Zivilschutz-Forschung, vol 58. Bundesamt für Bevölkerungsschutz und Katastrophenhilfe, Bonn

Hölting B, Coldeway WG (2013) Hydrogeologie: Einführung in die Allgemeine und Angewandte Hydrogeologie. Springer Spektrum, Berlin

Hoversten GM, Myer LR (2000) Monitoring of CO_2 sequestration using integrated geophysical and reservoir data. In: 5th international conference on greenhouse gas control technologies, Collingwood, Victoria, Australia, pp 305–310

Hovorka SD (2008) Surveillance of a geologic sequestration project: monitoring, validation, accounting. NETL webinar for the American Water Works Association

IEA (2012) Best practices for validating CO_2 geological storage: observations and guidance from the IEAGHG Weyburn-Midale CO_2 monitoring and storage project. Geoscience Publishing

IPCC (2005) Carbon capture and storage: Summary for policy makers and technical summary. Cambridge University Press

Jardani A, Revil A, Santos F, Fauchard C, Dupont JP (2007) Detection of preferential infiltration pathways in sinkholes using joint inversion of self-potential and EM34 conductivity data. Geophys Prospect 55(5):749–760. doi:10.1111/j.1365-2478.2007.00638.x

Kämpf H, Bräuer K, Schumann J, Hahne K, Strauch G (2013) CO_2 discharge in an active, non-volcanic continental rift area (Czech Republic): characterisation ($\delta^{13}C$, $^3He/^4He$) and quantification of diffuse and vent CO_2 emissions. Chemical Geology 339:71–83. doi:10.1016/j.chemgeo.2012.08.005

Kharaka YK, Thordsen JJ, Kakouros E, Ambats G, Herkelrath WN, Beers SR, Birkholzer JT, Apps JA, Spycher NF, Zheng L, Trautz RC, Rauch HW, Gullickson KS (2009) Changes in the chemistry of shallow groundwater related to the 2008 injection of CO_2 at the ZERT field site, Bozeman. Montana. Environ Earth Sci 60(2):273–284. doi:10.1007/s12665-009-0401-1

Knödel K, Lange G, Voigt H-J (2007) Environmental geology-handbook for field methods and case studies. Springer, Berlin

Lamert H, Geistlinger H, Werban U, Schütze C, Peter A, Hornbruch G, Schulz A, Pohlert M, Kalia S, Beyer M, Großmann J, Dahmke A, Dietrich P (2012) Feasibility of geoelectrical monitoring and multiphase modeling for process understanding of gaseous CO_2 injection into a shallow aquifer. Environ Earth Sci 67(2):447–462. doi:10.1007/s12665-012-1669-0

Leuning R, Etheridge D, Luhar A, Dunse B (2008) Atmospheric monitoring and verification technologies for CO_2 geosequestration. Int J Greenhouse Gas Control 2(3):401–414. doi:http://dx.doi.org/10.1016/j.ijggc.2008.01.002

Leven C, Weiß H, Vienken T, Dietrich P (2011) Direct-Push-Technologien – Effiziente Untersuchungsmethoden für die Untergrunderkundung. Grundwasser 16(4):221–234. doi:10.1007/s00767-011-0175-8

Lewicki JL, Hilley GE, Fischer ML, Pana L, Oldenburg CM, Dobeck L, Spangler L (2009) Detection of CO_2 leakage by eddy covariance during the ZERT project's CO_2 release experiments. Energy Procedia 1(1):2301–2306. doi:10.1016/j.egypro.2009.01.299

Pettinelli E, Beaubien S, Zaja A, Menghini A, Praticelli N, Mattei E, Di Matteo A, Annunziatellis A, Ciotoli G, Lombardi S (2010) Characterization of a CO_2 gas vent using various geophysical and geochemical methods. Geophysics 75(3):B137–B146. doi:10.1190/1.3420735

Reiche N, Westerkamp T, Lau S, Borsdorf H, Dietrich P, Schütze C (2014) Comparative study to evaluate three ground-based optical remote sensing techniques under field conditions by a gas tracer experiment. Environ Earth Sci 72(5):1435–1441. doi:10.1007/s12665-014-3312-8

Revil A, Pezard PA, Glover PWJ (1999a) Steaming potential in porous media: 1. Theory of the zeta potential. J Geophys Res 104(B6):2156–2202. doi:10.1029/1999JB900089

Revil A, Schwaeger H, Cathles LM, Manhardt PD (1999b) Streaming potential in porous media: 2. Theory and application to geothermal systems. J Geophys Res 104(B9):2156–2202. doi:10.1029/1999JB900090

Reynolds JM (2011) An introduction to applied and environmental geophysics. Wiley, Baffins Lane, Chichester, West Sussex PO19 IUD, England

Ringrose PS, Mathieson AS, Wright IW, Selama F, Hansen O, Bissell R, Saoula N, Midgley J (2013) The In Salah CO_2 storage project: lessons learned and knowledge transfer. Energy Procedia 37:6226-6236. doi:http://dx.doi.org/10.1016/j.egypro.2013.06.551

Sandig C, Sauer U, Bräuer K, Serfling U, Schütze C (2014) Comparative study of geophysical and soil–gas investigations at the Hartoušov (Czech Republic) natural CO_2 degassing site. Environ Earth Sci 72(5):1421–1434. doi:10.1007/s12665-014-3242-5

Sauer U, Schütze C, Leven C, Schlömer S, Dietrich P (2013) An integrative hierarchical monitoring approach applied at a natural analogue site to monitor CO_2 degassing areas. Acta Geotech 9(1):127–133. doi:10.1007/s11440-013-0224-9

Sauer U, Watanabe N, Singh A, Dietrich P, Kolditz O, Schütze C (2014) Joint interpretation of geoelectrical and soil-gas measurements for monitoring CO_2 releases at a natural analogue. Near Surface Geophysics 12(2007):165–178. doi:10.3997/1873-0604.2013052

Schütze C, Sauer U, Beyer K, Lamert H, Bräuer K, Strauch G, Flechsig C, Kämpf H, Dietrich P (2012) Natural analogues: a potential approach for developing reliable monitoring methods to understand subsurface CO_2 migration processes. Environ Earth Sci 67(2):411–423. doi:10.1007/s12665-012-1701-4

Schütze C, Dietrich P, Sauer U (2013) Diagnostic monitoring to identify preferential near-surface structures for CO_2 degassing into the atmosphere: Tools for investigations at different spatial scales validated at a natural analogue site. Int J Greenhouse Gas Control 18:285–295. doi:10.1016/j.ijggc.2013.07.006

Seto CJ, McRae GJ (2011) Reducing risk in basin scale CO_2 sequestration: a framework for integrated monitoring design. Environ Sci Technol 45(3):845–859. doi:10.1021/es102240w

Shuler P, Tang Y (2005) Atmospheric CO_2 monitoring systems. In: Thomas DC, Benson SM (eds) Carbon dioxide capture for storage in deep geologic formations—results from the CO_2 capture project geologic storage of carbon dioxide with monitoring and verification, vol 2. Elsevier, pp 1015–1030

Smaczny J, Clauser C, Klitzsch N, Blaschek R (2010) Numerical interpretation of self-potential (SP) data in the context of CO_2 storage. Paper presented at the EGU 2010 (May 2-7, 2010), Vienna, Austria

Sprunt ES, Mercer TB, Djabbarah NF (1994) Streaming potential from multiphase flow. Geophysics 59(5):707–711. doi:10.1190/1.1443628

UNEP (2006) Can carbon dioxide storage help cut greenhouse emissions? A Simplified guide to the IPCC's Special report on carbon dioxide capture and storage
Verdon JP, Kendall J-M, Stork AL, Chadwick RA, White DJ, Bissell RC (2013) Comparison of geomechanical deformation induced by megatonne-scale CO_2 storage at Sleipner, Weyburn, and in Salah. Proc Natl Acad Sci 110(30):E2762–E2771. doi:10.1073/pnas.1302156110
Vodnik D, Videmšek U, Pintar M, Maček I, Pfanz H (2009) The characteristics of soil CO_2 fluxes at a site with natural CO_2 enrichment. Geoderma 150(1–2):32–37. doi:10.1016/j.geoderma.2009.01.005
Weinlich FH, Bräuer K, Kämpf H, Strauch G, Tesař J, Weise SM (1999) An active subcontinental mantle volatile system in the western Eger rift, Central Europe: gas flux, isotopic (He, C, and N) and compositional fingerprints. Geochimica et Cosmochimica Acta 63(21):3653–3671. doi: http://dx.doi.org/10.1016/S0016-7037(99)00187-8
White JA, Chiaramonte L, Ezzedine S, Foxall W, Hao Y, Ramirez A, McNab W (2014) Geomechanical behavior of the reservoir and caprock system at the in Salah CO_2 storage project. Proc Natl Acad Sci 111(24):8747–8752. doi:10.1073/pnas.1316465111
Zhao W, Amelung F, Dixon T (2012) Monitoring ground deformation on carbon sequestration reservoirs in North America. In: Fringe 2011 workshop, Frascati, Italy, 2012. vol SP-697. ESA
Zlotnicki J, Nishida Y (2003) Review on morphological insights of self-potential anomalies on volcanoes. Surv Geophys 24(4):291–338. doi:10.1023/B:GEOP.0000004188.67923.ac
Zschornack L, Leven C (2012) Minimal invasive methods. In: Kästner M, Braeckevelt M, Döberl G et al. (eds) Model-driven soil probing, site assessment and evaluation—guidance on technologies, vol 307. Sapienza Università Editrice Rome, Italy, pp 337–388

Advances in Stable Isotope Monitoring of CO_2 Under Elevated Pressures, Temperatures and Salinities: Selected Results from the Project CO_2ISO-LABEL

Johannes A.C. Barth, Michael Mader, Anssi Myrttinen, Veith Becker, Robert van Geldern and Bernhard Mayer

Abstract The BMBF project CO_2ISO-LABEL (Carbon and Oxygen ISOtopes under extreme conditions LABoratory EvaLuations for CO_2-storage monitoring) investigated stable isotope methods in laboratory studies for transferral to carbon capture and storage (CCS) field sites including enhanced gas and oil recovery (EGR and EOR). The isotope composition of injected CO_2 and water are useful tracers for migration and water-rock-gas interactions during such operations. However, quantification of carbon and oxygen equilibrium isotope effects at elevated pressures and temperatures are so far scarce. They thus need more investigations under p/T conditions that are characteristic for reservoirs and overlying aquifers. With this, the main objective of the project was to improve stable carbon and oxygen isotope methods for monitoring CO_2 storage sites and their impact of injected CO_2 on reservoir geochemistry under controlled laboratory settings. An important finding was that isotope fractionations of carbon between CO_2 and dissolved inorganic carbon (DIC) were not significantly different from each other in experiments with pure CO_2 and pressures between 59 and 190 bar. Furthermore, influences of rock types (limestone, dolomite and sandstone) and fluid salinities were found to be negligible for carbon isotope fractionation between CO_2 and DIC. Another finding was that water oxygen isotope ratios changed systematically in response to different CO_2/H_2O molar ratios in closed system equilibration experiments. This helps to reconstruct the amounts of CO_2 that equilibrated with formation waters. Results of the project will enable better assessment of geochemical conditions in underground carbon storage sites or other subsurface systems where large amounts of CO_2 interact with water and rocks.

J.A.C. Barth (✉) · M. Mader · A. Myrttinen · R. van Geldern
Department of Geography and Geosciences, GeoZentrum Nordbayern,
University of Erlangen-Nuremberg (FAU), Schlossgarten 5, 91054 Erlangen, Germany
e-mail: johannes.barth@fau.de

V. Becker · B. Mayer
Applied Geochemistry Group, Department of Geoscience, University of Calgary,
2500 University Drive NW, Calgary, AB T2N 1N4, Canada

A. Liebscher and U. Münch (eds.), *Geological Storage of CO_2 – Long Term Security Aspects*, Advanced Technologies in Earth Sciences,
DOI 10.1007/978-3-319-13930-2_3

1 Project Description and Its Most Important Aims

The main objective of this work was to improve stable isotope methods for monitoring CO_2 and dissolved inorganic carbon (DIC) migration within CO_2 storage sites. In order to test application of isotope fractionation effects to reservoir conditions under elevated temperatures, pressures and salinities that are encountered in field conditions, investigations were carried out in the laboratory under controlled conditions. Such new experimental results are important for tracing the movement and fate of CO_2 during and after geological storage of CO_2.

Specifically, the aim of the project was to improve carbon and oxygen stable isotope methods as monitoring tools in carbon storage projects at p/T ranges of close to 200 bar and over 100 °C with increased fluid salinities. The target of these investigations were changes in carbon stable isotope compositions of CO_2 and dissolved inorganic carbon (DIC) as well as shifts in $^{18}O/^{16}O$ rations in water. One important pre-requisite of the project was to obtain original fluids from well-known field pilot projects such as the Canadian Weyburn and Pembina EOR sites. This enabled comparison of results from experiments conducted with original reservoir fluids from field sites with artificial analogue fluids. In this chapter we cover two important aspects of stable isotope applications for CCS with artificial fluids:

(i) effects of p/T conditions of up to 105 °C and 190 bar on carbon isotope ratios of CO_2 under variable reservoir fluid salinities and rock types (limestone, dolomite and sandstone) and
(ii) changes of oxygen isotope ratios of water in response to large amounts of CO_2.

Temperature-dependence of isotope fractionation between the species of dissolved inorganic carbon ($H_2CO_3^*$, HCO_3^- and CO_3^{2-}), CO_2 and precipitated carbonate minerals have previously been investigated in various studies (Chacko and Deines 2008; Deines 2004; Halas et al. 1997; Inoue and Sugimura 1985; Mook et al. 1974; Zhang et al. 1995). However, these investigations were usually conducted at lower salinities that are typical for fresh waters and under ambient pressure conditions. Higher pressures and temperatures conditions for CO_2 of over 100 °C and up to 190 bar represent a knowledge gap for the application of stable isotope methods in carbon storage monitoring. With this, one important question for stable isotope investigations at CO_2 storage sites (including EGR and EOR) is whether CO_2 is subject to different stable isotope fractionation effects under higher p/T conditions. Moreover, it was unknown if high salinities influence carbonate chemistry and associated carbon isotope fractionations. Such knowledge is important, if we systematically want to apply stable isotope measurements as monitoring tools during and after CO_2 injection into geological formations.

Another useful aspect of stable isotope applications in CCS operations is that CO_2 and water are known to equilibrate their $^{18}O/^{16}O$ ratios. For CCS operations this has been investigated in a few selected studies (Johnson and Mayer 2011; Johnson et al. 2011a, b; Lécuyer et al. 2009). These studies are promising to yield new tools that indicate how formation waters become influenced by large amounts of CO_2. For instance, if oxygen isotope changes of water depend on the amount of present CO_2, they may also help to reconstruct the molar amounts of gas and water that have equilibrated. Johnson and Mayer (2011) conducted experiments with CO_2 and water at high pressures of up to 190 bar at 50 °C. They worked with ^{18}O-enriched water in order to avoid that isotope effects are masked by similar isotope ratios of waters and CO_2.

Here, we present first results that evaluate oxygen isotope exchanges between CO_2 and H_2O at natural abundance (i.e. without addition of ^{18}O-labelled water or CO_2) and at ambient pressures. Even though CCS operations have higher p/T-boundary conditions, our results afford comparison of this isotope technique with changes of water isotopes at room temperature and normal atmospheric pressures. This comparison can reveal if the observed systematic alterations of oxygen isotopes in water also apply to lower temperature- and pressure environments. If so, p/T conditions that lie between CCS storage sites and ambient ones can also apply this isotope technique. These may occur for instance in potential leakage scenarios, in which large amounts of CO_2 can enter near-surface aquifers. In addition, the data presented here enable comparisons to these previous techniques by Johnson et al. (2011a, b) at higher p/T conditions and introduce a method that relies on water oxygen isotope changes alone.

2 Methods

2.1 *High p/T CO_2 Experiments*

A total of 32 experiments were carried out to test carbon isotope changes at various salinities and p/T conditions (Table 1). They were performed in metal vessels that were filled with water and CO_2 directly from a gas cylinder with starting pressures of 59 bar (Fig. 1a). Tests were carried out with distilled water and also with saline solutions of 50 g L^{-1} NaCl and 500 mg L^{-1} $NaHCO_3$ and with 30 g of ground rock (limestones, dolomites, and sandstone). Grain sizes of the ground rock samples were 0.5–2 mm and the fluid to gas ratio was 400–100 mL. Once the vessels were pressurised, they were placed into an oven at temperatures of 25, 50, 75 and 105 °C in separate experiments. Resulting pressures increased to maximum values of 190 bar after the metal reaction cells reached temperatures of up to 105 °C. In a similar study, Myrttinen et al. (2014) showed that equilibration between CO_2 and DIC was reached after 5 hours. Once this equilibrium was established, the fluid was first sampled into an attached 50 mL steel vessel and subsequently transferred into a gas sampling bag that contained 6 mL of 85 % phosphoric acid (H_3PO_4). This gas bag was then agitated for at least one hour on an orbital shaker with 60 rotations per minute. This was necessary to convert all DIC into CO_2 gas for further isotope analyses. Once the fluid sampling was complete, the remaining gas in the 500 mL

Table 1 Carbon isotope results of pressure experiments using vessel type A (Fig. 1) and solutions with 50 g of NaCl and 500 mg of $NaHCO_3$

Fluid	Rock type	Temp (°C)	pCO_2 (bar)	$\delta^{13}C_{DIC}$ (‰) VPDB	$\delta^{13}C_{CO2}$ (‰) VPDB	ln α (‰)
aq. dest.	Only water	25	59	−35.2	−34.6	−0.68
aq. dest.	Only water	50	88	−35.7	−35.3	−0.40
aq. dest.	Only water	75	105	−35.2	−34.8	−0.39
aq. dest.	Only water	105	158	−35.0	−34.3	−0.71
aq. dest.	Sandstone	25	60	−34.9	−34.4	−0.52
aq. dest.	Sandstone	50	95	−35.7	−34.4	−1.36
aq. dest.	Sandstone	75	123	−35.3	−34.5	−0.80
aq. dest.	Sandstone	105	152	−34.6	−34.2	−0.41
aq. dest.	Limestone	25	65	−34.2	−34.2	0.00
aq. dest.	Limestone	50	98	−35.3	−34.9	−0.41
aq. dest.	Limestone	75	130	−34.1	−34.3	0.24
aq. dest.	Limestone	105	158	−34.3	−33.9	−0.35
aq. dest.	Dolomite	25	60	−34.1	−33.5	−0.56
aq. dest.	Dolomite	50	88	−35.3	−34.8	−0.51
aq. dest.	Dolomite	75	110	−34.8	−34.5	−0.29
aq. dest.	Dolomite	105	190	−34.6	−34.2	−0.36
brine	Only water	25	60	−34.7	−34.2	−0.56
brine	Only water	50	80	−35.4	−35.2	−0.19
brine	Only water	75	120	−34.8	−34.7	−0.05
brine	Only water	105	150	−35.0	−34.1	−0.97
brine	Sandstone	25	60	−34.8	−34.2	−0.68
brine	Sandstone	50	100	−34.6	−34.6	0.00
brine	Sandstone	75	110	−34.6	−34.4	−0.25
brine	Sandstone	105	130	−32.9	−34.1	1.18
brine	Dolomite	25	60	−34.5	−34.0	−0.50
brine	Dolomite	50	78	−35.1	−35.0	−0.10
brine	Dolomite	75	115	−34.7	−34.4	−0.31
brine	Dolomite	105	150	−34.5	−34.0	−0.47
brine	Limestone	25	60	−34.6	−33.9	−0.63
brine	Limestone	50	98	−34.9	−34.8	−0.11
brine	Limestone	75	105	−34.2	−34.0	−0.26
brine	Limestone	105	160	−33.7	−33.6	−0.13

vessel was also transferred into a separate pre-evacuated gas-sampling bag for carbon isotope analyses of the CO_2. For these experiments, carbon isotope ratios of DIC and the CO_2 headspace were analysed on a Thermo Fisher MAT 253 with measurement uncertainties of less than ±0.2 ‰.

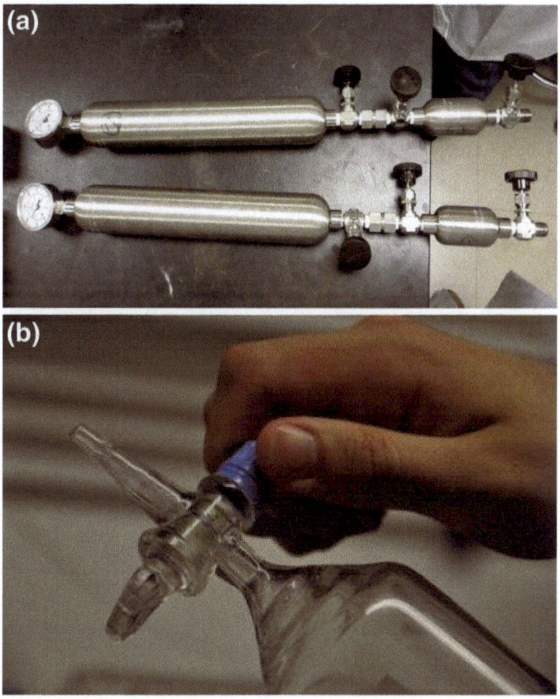

Fig. 1 Reaction vessels for experiments with CO_2-water-rock interaction: **a** 500 mL stainless steel vessel for water-rock-CO_2 interaction experiments with starting pressures of 59 bar attached to a 50 mL steel sampling vessel, **b** glass flask for experiments at ambient pressure and temperature

2.2 CO_2-Water Experiments with Oxygen Stable Isotopes

A total of 27 experiments were carried out for investigations of oxygen isotope exchanges between CO_2 and water. Each of these experiments had 3–5 sub-samples and the averages are displayed in Table 2. For the experiments 1-L glass-flasks with septa and two inlet faucets were used (Fig. 1b). These are known as gas-mouse-vessels. For these equilibration tests various amounts of water (i.e. 5, 10, 30, 50 and 100 mL) were exposed to pure CO_2. The vessels were shaken for up to 222 h on an orbital shaker with 60 rotations per minute at room temperature (22 °C). They were then sampled for CO_2 directly from the septum with a gas-tight syringe and transferred into helium-flushed 12-mL-Exetainer® vials. Water from these experiments was also subsampled in 5-mL open wide mouth vessels. From these the water was transferred immediately by pipette into 1.5-mL vials and then closed with butyl rubber caps for further analyses.

For stable isotope measurements of the CO_2 in these experiments, a Delta V Advantage mass spectrometer (Thermo Fisher Scientific, Bremen, Germany) was modified at its Gasbench® inlet system with additional helium dilution to accommodate CO_2 concentrations of 100 %. Water samples were analysed on a Picarro L1102-*i* cavity ring-down stable isotope analyser (Picarro Inc., Santa Clara, USA). All data were corrected for instrumental drift, memory, and linearity effects (i.e. changes of isotope ratios in dependence of peak size).

Table 2 Water isotope values after various equilibration times with 100 % CO_2 in 1 L flasks on a shaker with average oxygen isotope values of the water

Time (h)	H_2O in 1 L flask (mL)	$\delta^{18}O_{H2O}$ (‰) SMOW
2.0	5	−10.7
9.2	5	−13.6
22.1	5	−14.8
46.8	5	−14.9
66.5	5	−15.0
67.6	5	−14.9
93.9	**5**	**−14.6**
138.8	**5**	**−14.4**
171.5	**5**	**−15.1**
192.0	**5**	**−14.5**
2.0	10	−10.2
8.6	10	−11.9
22.0	10	−12.3
56.3	10	−12.2
94.5	**10**	**−12.1**
120.0	**10**	**−12.1**
138.9	**10**	**−11.9**
194.1	**10**	**−12.1**
2.1	30	-9.5
12.8	30	-9.8
19.9	30	-9.8
66.5	30	-9.8
94.5	**30**	**-9.9**
138.8	**30**	**-9.9**
196.1	**30**	**-9.8**
222.0	**50**	**-9.6**
71.3	**100**	**-9.0**

Starting values were δ^{18} O_{H2O} = −8.7 ‰ and δ^{18} O_{CO2} = + 9.3 ‰ VSMOW. Cells marked in bold were used to calculate the averages for Fig. 4, because their equilibration times were longer than 94 hours.

2.3 Isotope Measurements

All isotope measurements were normalized to international scales by analyses of multiple laboratory standards that were directly calibrated against international reference materials. The isotope ratios of water, dissolved carbon and CO_2 are expressed in the delta notation as a per mille (‰) deviation from a standard. These are the Vienna Standard Mean Ocean Water (VSMOW) for water samples and Vienna Pee Dee Belemnite (VPDB) for DIC and CO_2 samples. The delta notation is defined as

$$\delta = \left(\left(R_{\text{sample}}/R_{\text{reference}}\right) - 1\right) \tag{1}$$

where R is the ratio of the numbers (n) of the heavy and light isotopes of an element (e.g. $n(^{18}O)/n(^{16}O)$) in the sample and the reference (Coplen 2011). This delta notation systematically expresses the stable isotope ratios of O ($^{18}O/^{16}O$) or C ($^{13}C/^{12}C$) in water, CO_2 and DIC, respectively. According to this definition samples with more negative values are relatively depleted with respect to heavy isotopes and those with more positive values are enriched in heavy isotopes. When two phases are involved (i.e. $CO_{2(g)}$ and DIC) it is useful to express the isotope fractionation between them as a difference. This difference between delta-values (Δδ) of two phases A and B is defined by

$$\Delta\delta = \delta_A - \delta_B \tag{2}$$

The difference in isotope distribution between these two phases is also known as the isotope enrichment factor, ε, and can be expressed as

$$\varepsilon_{(A-B)} = \alpha_{(A-B)} - 1 \tag{3}$$

where A and B denote the phases and α is the isotope fractionation factor (Coplen 2011). Thus, the isotope fractionation and isotope enrichment factors are related to the isotope difference by the following approximations

$$\Delta\delta \approx \varepsilon_{(A-B)} \tag{4}$$

and

$$\Delta\delta \approx \ln\alpha \tag{5}$$

All values are usually multiplied by 1,000 to enable better comparison of the per mille values.

3 Selected Results and Discussion

3.1 *High p/T CO_2 Experiments*

Results of the investigations for carbon isotope changes between DIC and CO_2 are listed in Table 1. Unlike known temperature dependencies of stable isotope fractionation, influences of elevated pressures of up to 190 bar CO_2 on isotope fractionations have hardly been investigated to date. This is however a necessity for using stable isotope approaches as tracers for injected CO_2 in geological formations. In addition, reservoir fluids in geological CO_2 storage formations are often saline. This in turn reduces CO_2 solubilities and with this mineral dissolution

(Portier and Rochelle 2005). However, the effects of higher salinity have hardly been tested for stable isotope systematics of carbon when CO_2 is present in overabundance such as during CCS operations.

Figure 2 shows differences in carbon isotope values between pressure experiments using reaction vessel type 1A with CO_2 and DIC at pressures between 59 and 190 bar, salinities of 50 g L^{-1}, and with limestone, dolomite and sandstone, respectively. The ε values between CO_2 and DIC (i.e. $\delta^{13}C_{DIC}$-$\delta^{13}C_{CO2} \approx \ln\alpha$DIC-$CO_2$) ranged around −0.5 ‰ and no significant differences between the experiments with distilled water and brines or varying rock types were detected (Fig. 2a, b). Moreover, no noticeable differences in ε between CO_2 and DIC were found at the various tested pressures between 59 and 190 bar. This implies that for carbon isotopes the exchange between CO_2 and DIC did not depend on the tested p/T conditions of the CO_2 when CO_2 was the dominant phase. These results confirm that if the injected CO_2 is present in excess, it dominates isotope equilibrations between CO_2 and DIC.

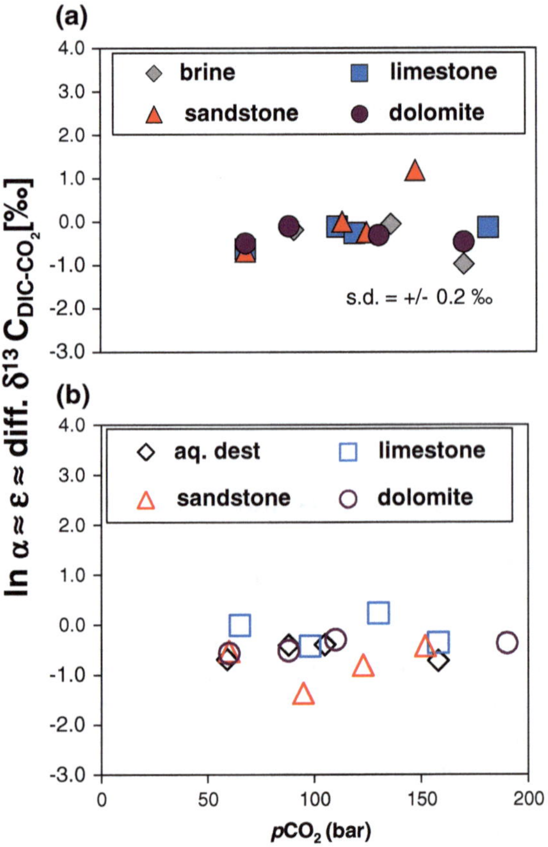

Fig. 2 Carbon isotope differences between DIC and CO_2 from rock-water-CO_2 experiments carried out at temperatures between 25 and 105 °C and pressures between 59 and 190 bar with vessel type A (cf Fig. 1). **a** experiments carried out with brine (50 g L^{-1} NaCl and 500 mg of $NaHCO_3$) **b** experiments carried out with distilled water differences between experiments using brine and distilled water are shown to be negligible under the p/T tested. They are also not influenced by the different rock types tested (i.e. limestone, dolomite and sandstone)

3.2 CO_2-Water Experiments

Unlike the above homogeneous carbon isotope distributions, the CO_2-H_2O equilibration experiments with the gas mouse vessels revealed systematic changes in the oxygen isotope composition of the water due to the presence of various amounts of CO_2 (Fig. 3). The initial $\delta^{18}O$ value of the water was −8.7 ‰ and the one of the CO_2 was +9.3 ‰. Generally, changes in $\delta^{18}O$ values of the water were larger for smaller volumes of water. These experiments had relatively larger amounts of CO_2 that could influence the lower number of water molecules. With this, the $\delta^{18}O$ of the water changed from its original value by 6.0, 3.4, 1.1, 0.9 and 0.3 ‰ for corresponding oxygen contributions from CO_2 of 22.8, 12.8, 4.6, 2.7, and 1.3 % (Table 3). The latter was determined as the percentage of total oxygen (i.e. from H_2O and CO_2) present within the closed experimental systems. They derive from molar ratios of CO_2 and H_2O with the 5, 10, 30, 50 and 100 mL of water in the 1-L flasks.

The isotope composition of the water started to stabilise after about 50 h of shaking, however more constant values were reached after about 94 h. Note that CO_2 injection operations may cause similar turbulent interactions between CO_2 and formation waters in the vicinity of the injection well. These may compare to shaking the glass vessels in the laboratory. However, during most times at field injection sites the interaction between CO_2 and H_2O may be less dynamic. Therefore, further experiments are required to determine equilibration times under static or almost-static conditions.

When using the difference between equilibrated and initial $\delta^{18}O_{H2O}$ values, we can use the relationship in Fig. 4 to determine how much CO_2 equilibrated with the

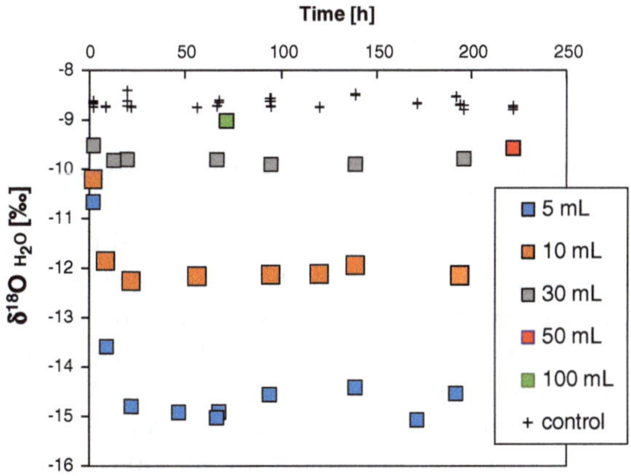

Fig. 3 Changes of the oxygen isotope composition of water ($\delta^{18}O_{H2O}$) after exposure to pure CO_2 in a 1-L glass flask (gas mouse vessel, i.e. type B in Fig. 1) at 22 °C and ambient pressures. Control samples were exposed to air during the experiment and showed no change of initial $\delta^{18}O_{H2O}$ values. The *different colours* correspond to the different volumes of water added to the 1-L vessels

Table 3 Average changes of $\delta^{18}O_{H2O}$ from water baseline $\delta^{18}O_{H2O}$ = −8.7 ‰ VSMOW for equilibration times longer than 94 h and corresponding percent-contributions of oxygen from CO_2

H_2O in 1-L flask (mL)	Change in $\delta^{18}O_{H2O}$ from baseline value (‰) SMOW	Oxygen contribution of CO_2 in 1-L flask (%)	Calculated oxygen contribution of CO_2 after Johnson et al. (2011a) (%)
5	6.02	22.8	26.4
10	3.40	12.8	14.9
30	1.18	4.6	5.2
50	0.89	2.7	3.9
100	0.34	1.3	1.5

water. In similar experiments under higher pressures and with ^{18}O-labelled water, Johnson et al. (2011a) introduced a formula with the initial and final isotope compositions of CO_2 and H_2O and the temperature-dependent isotope enrichment, ε, between both phases with

$$X^{O}_{CO_2} = \frac{\delta^{18}O^{i}_{H_2O} - \delta^{18}O^{f}_{H_2O}}{\delta^{18}O^{i}_{H_2O} + \varepsilon - \delta^{18}O_{CO_2}} \tag{6}$$

whereas

$X^{O}_{CO_2}$ proportion of oxygen in the system that stems from CO_2 (in % when multiplied by 100)

$\delta^{18}O^{i}_{H_2O}$ measure for $^{18}O/^{16}O$ ratio of the water before contact with CO_2

$\delta^{18}O^{f}_{H_2O}$ measure for $^{18}O/^{16}O$ ratio of the water after equilibration with CO_2

ε temperature-dependent isotope fractionation between oxygen isotope ratios of H_2O and CO_2

$\delta^{18}O_{CO_2}$ measure for $^{18}O/^{16}O$ ratio of the injected CO_2

When applying formula 6 with the used $\delta^{18}O_{CO2}$ values of +9.3 ‰ and $\delta^{18}O_{H2O}$ of −8.7 ‰ (both in VSMOW), we obtained values of 26.4, 14.9, 5.2, 3.9 and 1.5 % for $X^{O}_{CO_2}$ for the experiments with 5, 10, 30, 50 and 100 mL water, respectively. These results show good agreement with the values obtained by the method that used differences between initial and final water isotope values (Table 3). For instance, the maximum offset between both methods of determination was 3.6 % for the experiment that used the lowest volume of water (5 mL). The experiments with such small volumes of water were more difficult to control and analytical results showed more scatter. This may also have caused the larger offset between the two methods for this experiment.

The above comparison shows that natural abundance experiments at ambient pressures in the laboratory can transfer to reactions of CO_2 in formation waters to determine if interactions occur between the gas- and fluid phases. When differences in the isotope composition of water can be detected after equilibration, this

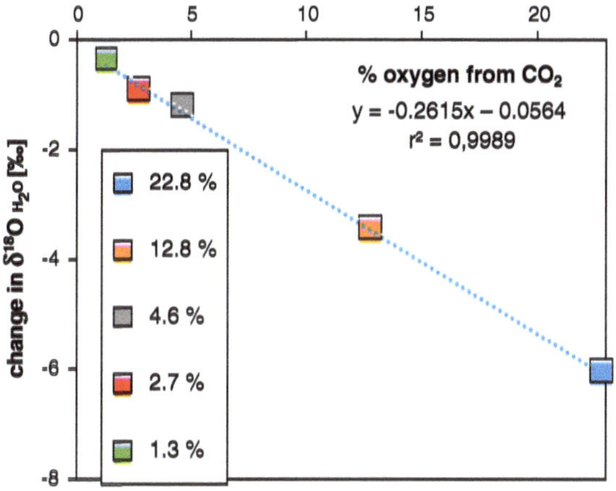

Fig. 4 Average changes (for $t > 94$ h) of the oxygen isotope composition of water from their initial values (i.e. before the start of the experiment) versus the percentage of oxygen that is bound in CO_2 with respect to all present oxygen in the system (i.e. number of O atoms in H_2O and CO_2 molecules in the 1-L flask experiments). The linear relationship helps to reconstruct molar ratios between water and CO_2 that have equilibrated their oxygen isotopes under closed system conditions

technique can help to evaluate how much CO_2 has been present relative to formation fluids in the pore space. Note that the experiments were carried out at room temperature and under closed system conditions. The good match with the data calculated using Eq. 6 shows that this method is independent of pressure and temperature conditions. Our results also show that similar results can be achieved with water that has not been labelled with heavy isotopes. Such natural abundance waters usually occur in CCS scenarios. The method can thus transfer from the laboratory to CCS field sites. Further experiments should test for other boundary conditions such as variable amounts of salt in formation waters and possible longer equilibration times when samples are not agitated.

4 Conclusions

This chapter outlined the project CO_2-ISOLABEL that helped to expand knowledge on isotope systematics and stable isotope fractionation effects for oxygen and carbon. These are important monitoring tools during CCS, EOR, and EGR operations.

One key finding is that pressure, salinity and various rock types tested had negligible effects on carbon isotope fractionation magnitudes at p/T conditions between 25 and 105 °C and 59 and 190 bar. This holds true when CO_2 is the dominant phase. This dominance of CO_2 caused the magnitude of carbon isotope fractionation to become independent of salinity and rock type. These findings are important because stable isotope tools can apply in CCS operations without concerns about their validity under different reservoir conditions. In addition, transition of CO_2 from the reservoir to more shallow aquifers as a result of potential leakage will not compromise the application of the same isotope principles.

Changes of water isotope compositions as a result of contact with CO_2 have so far been investigated in few studies and also offer an emerging monitoring tool to determine molar ratios of CO_2 and water that have equilibrated isotopically. When using water isotopes alone, this offers an easy to use mapping tool to identify parts of an underground reservoir that have been affected by large amounts of injected CO_2. This method requires that the oxygen isotope ratios of the water and the CO_2 are sufficiently different to identify isotope changes. Alternatively, injected CO_2 or the formation waters could be labelled in their oxygen isotope composition. However, this is difficult to achieve with constant inputs during large-scale injection operations. In case changes in the water isotope composition are evident when compared to baseline data (i.e. before injection of CO_2), they enable reconstruction of the amounts of CO_2 that have equilibrated with the water.

Overall, the project expanded possibilities of stable carbon and oxygen isotope techniques for assessing water and carbon cycles in the subsurface. Its results demonstrate that isotope techniques are valuable tools that can help in monitoring programmes during and after CCS.

Acknowledgments This study was conducted as part of CO_2ISO-LABEL project that was funded by the German Federal Ministry of Education and Research (BMBF) in the Geotechnologien Program (Grant No: 03G0801A). This research was also carried out in international collaboration with the University of Calgary with the Carbon Management Canada Inc. (CMC-NCE) under the project Storage Geochemistry (C01). We are indebted to S. Hintze, C. Hanke, I. Wein, S. Meyer, and M. Hertel from the Friedrich-Alexander-University Erlangen-Nuremberg as well as to M. Nightingale, S. Taylor and M. Shevalier form the University of Calgary for help with laboratory set-ups and analyses.

References

Chacko T, Deines P (2008) Theoretical calculation of oxygen isotope fractionation factors in carbonate systems. Geochim Cosmochim Acta 72:3642–3660. doi:10.1016/j.gca.2008.06.001

Coplen TB (2011) Guidelines and recommended terms for expression of stable-isotope-ratio and gas-ratio measurement results. Rapid Commun Mass Spectrom 25:2538–2560. doi:10.1002/rcm.5129

Deines P (2004) Carbon isotope effects in carbonate systems. Geochim Cosmochim Acta 68:2659–2679

Halas S, Szaran J, Niezgoda H (1997) Experimental determination of carbon isotope equilibrium fractionation between dissolved carbonate and carbon dioxide. Geochim Cosmochim Acta 61:2691–2695

Inoue H, Sugimura Y (1985) Carbon isotopic fractionation during the CO_2 exchange process between air and sea water under equilibrium and kinetic conditions. Geochim Cosmochim Acta 49:2453–2460

Johnson G, Mayer B (2011) Oxygen isotope exchange between H_2O and CO_2 at elevated CO_2 pressures: implications for monitoring of geological CO_2 storage. Appl Geochem 26:1184–1191

Johnson G, Mayer B, Nightingale M, Shevalier M, Hutcheon I (2011a) Using oxygen isotope ratios to quantitatively assess trapping mechanisms during CO_2 injection into geological reservoirs: the Pembina case study. Chem Geol 283:185–193

Johnson G, Mayer B, Shevalier M, Nightingale M, Hutcheon I (2011b) Quantifying CO_2 pore-space saturation at the Pembina Cardium CO_2 Monitoring Pilot (Alberta, Canada) using oxygen isotopes of reservoir fluids and gases. Energy Procedia 4:3942–3948. doi:10.1016/j.egypro.2011.02.333

Lécuyer C, Gardien V, Rigaudier T, Fourel F, Martineau F, Cros A (2009) Oxygen isotope fractionation and equilibration kinetics between CO_2 and H_2O as a function of salinity of aqueous solutions. Chem Geol 264:122–126

Mook WG, Bommerson JC, Staverman WH (1974) Carbon isotope fractionation between dissolved bicarbonate and gaseous carbon dioxide. Earth Planet Sci Lett 22:169–176

Myrttinen A, Becker V, Mayer B, Barth JAC (2014) Stable carbon isotope fractionation data between H_2CO_3* and CO_2(g) extended to 120 °C. Rapid Commun Mass Spectrom 28:1691–1696. doi:10.1002/rcm.6950

Portier S, Rochelle C (2005) Modelling CO_2 solubility in pure water and NaCl-type waters from 0 to 300 °C and from 1 to 300 bar: application to the Utsira formation at Sleipner. Chem Geol 217:187–199. doi:10.1016/j.chemgeo.2004.12.007

Zhang J, Quay PD, Wilbur DO (1995) Carbon isotope fractionation during gas-water exchange and dissolution of CO_2. Geochim Cosmochim Acta 59:107–114

CO_2BioPerm—Influence of Bio-geochemical CO_2-Transformation Processes on the Long-Term Permeability

Nils Hoth, Claudia Gniese, Jana Rakoczy, Anne Weber, Steffen Kümmel, Susan Reichel, Carsten Freese, Michaela Hache, Andrea Kassahun, Alexandra Schulz, Heike Fischer, Martin Mühling, Robert Starke, Rene Kahnt, Carsten Vogt, Hans-Hermann Richnow, Martin Krüger, Axel Schippers and Michael Schlömann

Abstract The RECOBIO projects (Hoth et al. in Recycling of sequestrated CO_2 by microbial—biogeochemical transformation in the deep subsurface—RECOBIO 2009a; Geotechnol Sci Rep 14:58–65, 2009b; Untersuchung der biogeochemischen transformation von im tiefen Untergrund gespeichertem CO_2—RECOBIO 2 2011) have shown the relevance of biogeochemical processes, related to CO_2 injection. These processes represent an additional pathway for biogeochemical CO_2 storage. The main result was the microbial transformation (binding) of injected CO_2 (formation of organic compounds). This can also influence the pressure behaviour of the system. Furthermore the organic layers can act as nucleation sites and so catalyse the carbonate solid formation. So the main focus of the CO_2BIOPERM project was now to investigate the influence of these processes on the permeability behaviour of the system. Furthermore other aquifer structures, not related to natural

N. Hoth (✉)
Department of Mining, Technical University Bergakademie Freiberg (TUBAF), Zeunerstr. 1A, 09599 Freiberg, Germany
e-mail: Nils.Hoth@tu-freiberg.de

C. Gniese · C. Freese · M. Mühling · M. Schlömann
Technical University Bergakademie Freiberg (TUBAF), Freiberg, Germany

J. Rakoczy · M. Krüger · A. Schippers
Federal Institute for Geosciences and Natural Resources Hanover (BGR), Hanover, Germany

A. Weber · M. Hache · A. Kassahun
Dresden Groundwater Research Centre (DGFZ), Dresden, Germany

S. Kümmel · A. Schulz · R. Starke · C. Vogt · H.-H. Richnow
Helmholtz Centre for Environmental Research (UFZ), Leipzig, Germany

S. Reichel · H. Fischer · R. Kahnt
G.E.O.S. Ingenieurgesellschaft mbH, Halsbrücke, Germany

A. Liebscher and U. Münch (eds.), *Geological Storage of CO_2 – Long Term Security Aspects*, Advanced Technologies in Earth Sciences,
DOI 10.1007/978-3-319-13930-2_4

gas fields, were characterised by microbiological, molecular genetic investigations. The biocenosis is also often dominated, like in natural gas fields, by sulphate reducers and fermenting bacteria. The study of CO_2 effects to the cultivation of microorganisms showed for deep aquifer microorganisms a strategy to survive the CO_2 stress by spore forming. The proteomic analysis gave a first view how many and which proteins were down and up regulated under CO_2 stress. A part of the flow experiments, which were operated in discontinuously flowed batch mode, are presented in detail. There is no strong influence of the processes on the permeability behaviour for high permeable reservoir sandstones. Nevertheless the sequential extractions on the solid materials, after the tests, underline the ongoing biogeochemical reactions.

1 Introduction—Previous Work and Overall Goals

Related to CCS the topic of geochemical interactions, induced by large scale injection of CO_2 is clearly on the agenda. But the influence of microbial catalyzed biogeochemical transformation processes is often neglected.

The RECOBIO projects showed for the first time the importance of these processes in a wide range of investigations (Hoth et al. 2009a, b; Ehinger et al. 2009; Gniese et al. 2013).

- There was evidence for the presence of active Bacteria and Archaea in produced formation waters of the natural gas field Schneeren. All Archaea belonged to the group of methane producing microbes. Moreover, 75 % of these Archaea use the CO_2-fixing metabolic pathway (*Methanoculleus* spp.).
- The fast microbial mediated transformation of injected CO_2 was the main output from RECOBIO. Autoclave experiments under ideal conditions (with H_2) with formation waters, the microbial community and milled rock materials resulted in a fast build-up of dissolved organic carbon (DOC) from CO_2. Therefore, this process represents an additional pathway for biogeochemical CO_2 storage. It is important to know about this mechanism, because it has an influence on the pressure behaviour of the system. Without this knowledge decreasing pressures could be otherwise erroneously interpreted as leaky systems. These effects are known also from former town gas operations in the 1970s (Frei et al. 1986; Smigan et al. 1990).
- Autotrophic (means CO_2 transforming) sulphate reduction, which is coupled to sulphide formation, has also be kept in mind. And the importance of biofilm interactions has been demonstrated, in connection to the evidence of long-term H_2-generation on silicate rock materials.

The main results of RECOBIO can be summarised as follows:

(a) A biogeochemical, microbial transformation (binding) of injected CO_2 is relevant for deep storage formations.
(b) These reactions result in the formation of organic compounds, which on the one hand dissolve in the formation waters and on the other hand appear as biofilms on the surfaces of the formation rocks.
(c) These organic layers act as nucleation sites in the catalysis of carbonate solid phase formations.

From these results the question arises of the long-term permeability behaviour related to these processes. The interaction of biogeochemical transformed CO_2 as DOC or biofilms with silicates or cements, can lead not only to a reduction but also an increase in permeability, under pressure conditions typical of reservoirs.

While organic acids accelerate the dissolving processes of silicates, the formation of organosilicate gel coatings, can reduce permeability.

So the idea of the CO_2BIOPERM project was now to transfer this "RECOBIO-knowledge" into the investigation of core flow experiments.

Beside these core flow experiments high-pressure incubations were also carried out, with already existing enrichment cultures from relevant formation sites, pure cultures of ecological relevant microorganisms, as well as biocenosis from new investigated sites. So the idea was to look, if the RECOBIO results of specialised microorganisms (related to formations waters of natural gas fields) are transferable to "normal" deep aquifer systems.

2 Microbial Communities in Deep Aquifers—On Basis of the Geothermal Waters of Molasse Basin, Germany

2.1 Experimental Approach

Understanding microbial communities (and their metabolic capabilities), which live in designated CO_2 storage sites, is an important condition to evaluate the long-term safety of these sites. This is because microorganisms might be able to transform CO_2 into permeability-altering compounds (e.g. carbonates, short fatty acids). In order to gain insight into the microbial life of deep aquifers, the composition of microbial communities at two sites with geothermal usage of formation waters was investigated, along with some geochemical data.

The formation waters were extracted at two geothermal plants located in the Bavarian Molasse Basin near Munich. The water samples were analysed for total cells numbers using SybrGreen for DNA staining with subsequent cell counting. Thereby the microbial community composition was investigated using PCR amplification of the 16S rRNA gene and subsequent cloning and sequencing.

Additionally, sampling depth, water temperature and pH value were recorded. The microbial communities were compared among themselves and with samples taken from deep saline aquifers in the North German Basin.

2.2 Results

The formation waters were extracted from aquifers at depths of 3,300 m (site 1) and 1,700–2,000 m (site 2) with temperatures of 100 and 65 °C. At both sites, a strong sulphidic odour was noticeable, indicating sulphate reduction in these formations. The pH values, measured directly on-site after sampling, were around neutral, 6.8 (site 1) and 7.3 (site 2). As typical for the Molasse Basin, the salt content was low for both sites (<1 % of weight).

The bacterial communities of the water samples of the Molasse Basin strongly differed from each other (see Fig. 1). Site 1 was dominated by *Thermodesulfovibrio* (*Nitrospira*, 80 % relative abundance in clone library), a thermophilic sulphate-reducing genus. At site 2, *Proteobacteria* (mainly α and β) were the most abundant group (65 %) whereas *Thermodesulfovibrio* made up for only 6 % of the total

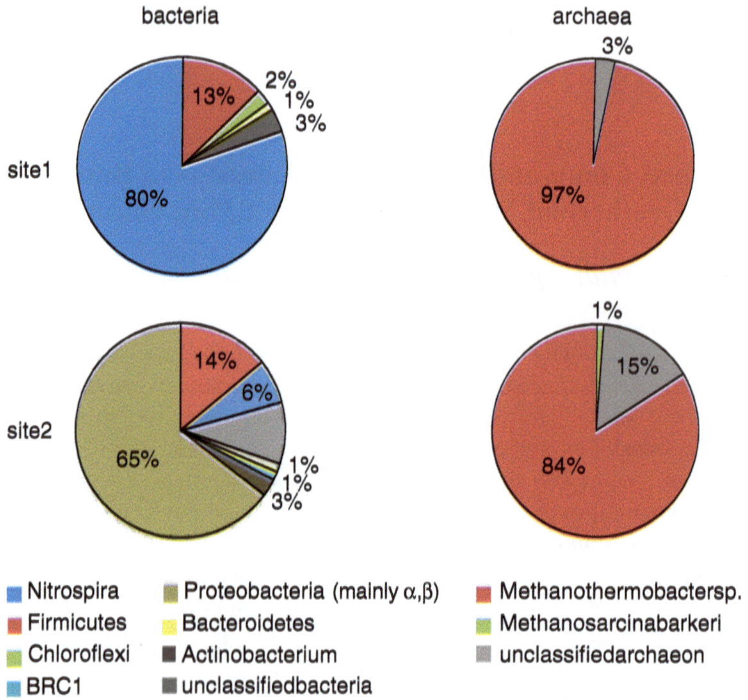

Fig. 1 Phylogenetic composition of microbial communities from deep aquifers located in the Bavarian Molasse Basin

bacterial community. The second most dominant group at both sites was *Firmicutes*, with an equal share of the total bacterial community of 13–14 %. At site 1, the *Firmicutes* group was entirely formed by members of spore-forming *Clostridia* (fermenting or sulphate-reducing bacteria) whereas site 2 mainly comprised members of *Bacillus* (aerobic or facultative bacteria). These two dominant groups were followed by less abundant members of *Bacteroidetes*, *Chloroflexi*, *Actinobacteria* (1–3 % each), and unclassified bacteria (3–9 %).

In contrast, archaeal communities at the sampling sites were almost identical. Both sites were dominated by the same genus, *Methanothermobacter* (>84 %), which comprises thermophilic hydrogenotrophic methane-forming archaea.

Prior to this project, the microbial community from a deep saline aquifer in the North German Basin was investigated (Hoth et al. 2009a, 2011; Ehinger et al. 2009). The most distinctive feature of the water from deep aquifers in the North German Basin is its high salinity. The microbial community clearly differed from the samples taken from the deep ('low-saline') aquifers in the Molasse Basin. The most dominant group was *Firmicutes* (42 %), followed by *Deltaproteobacteria* (21 %, typically sulphate reducers), and *Bacteroidetes* (19 %). Additionally, *Thermotogae* (10 %) were detected, a thermophilic phylum which was not detected in the samples from the Molasse Basin.

Despite the differing microbial communities, all three sites have in common a quiet rather low total cell number (10^3–10^5 cells per ml water).

In general, the comparison of the microbial communities of the different sampling sites reveals that the phylogenetic composition not only differs between distinct geological settings (i.e., North German Basin versus Molasse Basin) but also between sampling sites within the same formation (i.e., site 1 and 2 of the Molasse Basin). The phylogenetic data collected from the sampling sites investigated here imply that theoretically the same processes could happen, for example, sulphate-reducers have been detected at all sites, hence, sulphate reduction *may* take place at all sites. However, such a conclusion about actual metabolic activities should not be drawn from phylogenetic data only but rather from further laboratory growth and/or activity studies which could demonstrate whether, and to what extent, certain metabolic processes may take place.

In conclusion, these data emphasize that each geological setting, which will be designated for CO_2 storage, should be considered individually in terms of its microbial community.

3 Effects of CO_2 on Different Microorganisms

3.1 Cultivation of Petrotoga Sp.

The following set of experiments was performed in order to understand how CO_2 affects microorganisms on a molecular level. The experiments were performed using the thermophilic thiosulphate-reducing *Petrotoga* sp. (*Thermotogae*), which was isolated from a natural gas field rendering it a suitable model organism to study CO_2 effects in potential storage formations. To simulate the high-pressure conditions of the deep underground, *Petrotoga* sp. was incubated in high-pressure vessels. 100 bar of CO_2 or N_2 or as hydrostatic pressure was applied. Furthermore, different pH values were adjusted, in order to mimic the lowered pH-values caused by high CO_2 concentrations. All incubations were conducted at 60 °C and maltose was used as a carbon source.

The highest growth yield (related to the optical density measured at a wavelength of 600 nm—OD600) and maltose consumption was observed under 100 bar hydrostatic pressure and pH 7 (the control experiment). Also, neither high gas pressure (100 bar N_2) nor slightly lowered pH (pH 6, at 100 bar hydrostatic pressure) inhibit growth or substrate consumption (see Fig. 2).

In contrast, the addition of CO_2 (10 or 100 bar) fully inhibited both growth and maltose consumption and neither of these activities could be detected after CO_2 had

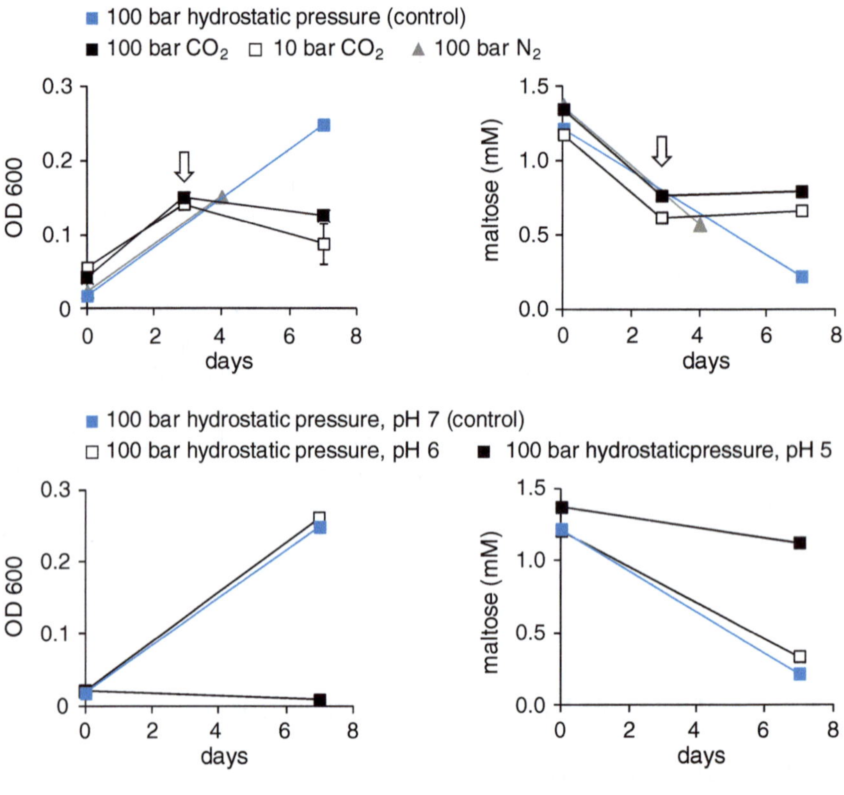

Fig. 2 Cultivation of *Petrotoga* sp. at high gas pressure (CO_2, N_2), hydrostatic pressure, and pH variation. The *arrows* in the *upper panel* indicate addition of CO_2 (before: atmospheric pressure). *Left panel* = growth curves (OD 600—optical density at a wavelength of 600 nm), *right panel* = maltose consumption. Results from two replicates are shown (except for N_2 which is a single on). Small error bars may be hidden by symbols

been released and the cells had been transferred to fresh medium (data not shown). The reason for this irreversibly inhibiting effect will be clarified during proteomic and transcriptomic analyses (see Sect. 3.2). So far it seems that the low pH value plays an important role as the incubation at low pH 5 partly mirrored the inhibition effects of high pressure CO_2 (i.e., no cell growth). Yet, little maltose consumption was observed which indicates that the cells might use the energy gained for housekeeping functions (i.e., pH stability) rather than for cell growth.

3.2 *Microbial Response to Different Levels of* CO_2

3.2.1 Proteomic Investigations

Indigenous microbes can be present in many compartments of the subsurface (Lovley and Chapelle 1995; Chivian et al. 2008), even in hot deep reservoirs due to the remarkable temperature adaptation of thermophiles and hyperthermophiles (Hendry 2006; L'Haridon et al. 1995). Consequently, microorganisms play a major role in subsurface carbon cycling, and thus may severely affect geochemical processes in CO_2 storage formations, if they can cope with extraordinary high CO_2 concentrations. Hence, the impact of different CO_2 concentrations (0–80 % CO_2 in water under atmospheric conditions) on microbial growth and metabolism was investigated with the genome-sequenced model subsurface microorganisms *Pseudomonas putida* F1, *Bacillus subtilis* 168 (both aerobic) and *Desulfovibrio vulgaris* Hildenborough (sulphate reducing). Increasing CO_2 concentrations inhibited the growth of all investigated strains. However, cells were still alive and could divide under CO_2 stress. Verification of microbial stress response, towards increased CO_2 partial pressures, was done by analyzing the microbial proteome and comparing the protein expression patterns by a volcano plot analysis (Fig. 3). The most significant change was observed for *B. subtilis*. Here, the expression of 83 proteins varied when *B. subtilis* was cultivated in the presence of 50 % CO_2; 21 proteins were down-regulated and 62 were up-regulated. This pattern shifted in the presence of 80 % CO_2, leading to a change in the expression of 105 proteins of which 103 were down-regulated. Identified down-regulated proteins were mainly involved in DNA regulation and replication, as well as protein biosynthesis. In comparison, the effect for *P. putida* was less pronounced. At 50 % CO_2 the expression of 46 proteins was changed; 42 proteins were down-regulated and 4 up-regulated. This pattern was nearly the same for *P. putida* in the presence of 80 % CO_2. Then a total of 57 proteins showed a varying expression whereat 54 proteins were down-regulated. Here, identified proteins were involved in the DNA replication and the general metabolism (citric acid cycle). A significant lower influence of CO_2 on protein expression was observed for *D. vulgaris*, the anaerobic model microorganism. In the presence of 50 or 80 % CO_2, the expression of only 16 or 14 proteins was changed.

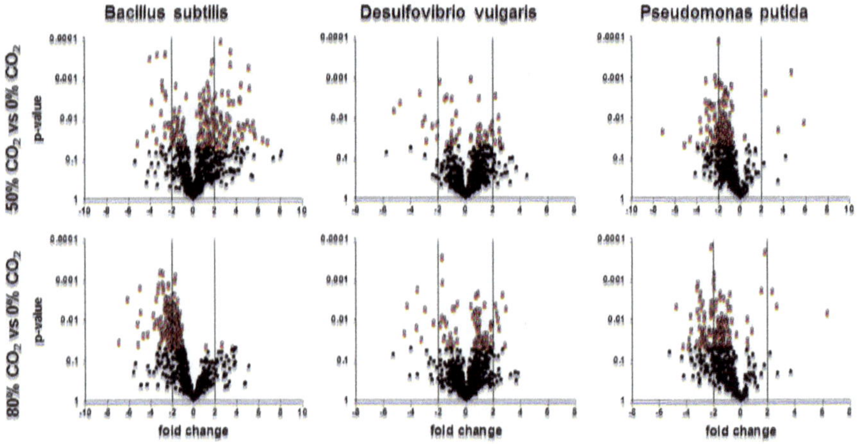

Fig. 3 Volcano plot comparative analysis of the protein expression pattern of *B. subtilis, D. vulgaris and P. putida* cultivated in the presence of different CO_2 partial pressures. The significance determined by Welch's t-test (*p*-value) was plotted versus the fold change (log2 ratio). Changes in protein expression were regarded as being significant when the *p*-value was lower than 5 % (*red dots*) and the log2 ratio either higher than +2 indicating up-regulation or lower than −2 in terms of down-regulation

In summary, the protein expression pattern of the aerobic strains *B. subtilis* and *P. putida* indicated that several metabolic pathways were affected by CO_2 stress. In contrast, the strictly anaerobic *D. vulgaris*, which is more representative for deep anoxic geological formations, seems to be more resistant to high CO_2 concentrations, indicated by a less pronounced proteomic stress response.

3.2.2 Transcriptomic Investigations

Flow through experiments with sediment cores (see Sects. 4.1 and 4.2) were designed to evaluate whether microbial growth potentially impacts on the permeability within the cores by, for instance, clogging pores, microfaults or fissures. A prerequisite for such a scenario is, however, that the microorganisms survive the changing environmental conditions. Studies into the growth behaviour and the survival of various microbial strains under CO_2 stress indicated that microorganisms are indeed able to deal with drastically increasing CO_2 pressures (see Sect. 3.1). Further experiments were designed with the aim to better understand at the molecular (i.e. genetic) level this adaptive potential of the microorganisms to deal with CO_2 stress. Such a level of detail is of relevance when predictions are to be made as to the long-term behaviour of microorganisms in a changing environment.

Petrotoga sp. strain Sch_Z2_3 was chosen as model microorganism since it was isolated from produced formation waters at the natural gas field Schneeren (Lower Saxony, Germany). The experimental approach decided on involves exposure of cultures of *Petrotoga* sp. strain Sch_Z2_3 under various conditions.

This was followed by extraction of the total cellular RNA from which the information on expressed genes (i.e. the messenger RNA—mRNA) can be extracted *via* nucleotide sequencing (i.e. a RNA-Seq approach). Experimental hurdles of this approach, however, require a minimum amount of biomass from which to extract RNA. This demand on biomass, in turn, prevents the use of sediment cores (see Sects. 4.1 and 4.2) for the exposure due to insufficient pore and, hence, culture volume. Therefore incubations need to be conducted in bioreactors specifically designed for conditions of high pressure and high temperature (see Sect. 3.1). Since increasing CO_2 concentration and pressure also result in increasing acidity of the growth medium environmental conditions for the exposure were defined as explained in Sect. 3.1, so that they permit differentiation between pH and CO_2 effect: 100 bar N_2 pressure at pH 7 *versus* 100 bar N_2 pressure at pH 6 versus 100 bar CO_2/N_2-pressure at pH 6 whereby the pH is adjusted purely *via* the fraction of CO_2 added to the gas mix (87 % N_2, 13 % CO_2).

3.3 Potential Mobilization of Organic Carbon from Rocks by CCS—Extraction with Supercritical Carbon Dioxide ($ScCO_2$)

In order to investigate the potential mobilization of organic carbon from geological storage formations, $scCO_2$ experiments with low mature organic matter were performed. So a brown coal sample was extracted with $scCO_2$. A CO_2 mass flow of 10 kg/h was streamed through 250 g grinded brown coal for 2 h at 400 bars and 40 °C. Pre-tests showed that $scCO_2$ by self doesn't produce a mobile organic phase. Therefore methanol with a mass flow of 0.75 kg/h was added as modifier. The extract obtained consisted of a dark yellow solution. Subsequently, the remaining methanol was evaporated and the extract was dried yielding 3 g of an ochre humic acid like solid material. For a qualitative analysis of the chemical composition, dichloromethane (DCM) was added to the extract, as well as to the original brown coal sample and the soluble components were analyzed by a gas chromatograph coupled to a mass spectrometer. The major quantities of the $scCO_2$ extract consisted of humic like materials and were not amenable to gas chromatography. Identification of compounds was done by analyzing the mass spectra of the obtained peaks (Table 1). For the DCM-extract a total of 53 different compounds were detected. Monocyclic aromatic hydrocarbons (toluene, ethylbenzene, xylene) as well as polycyclic aromatic hydrocarbons (naphthalene) were identified in low amounts. In addition, higher alkanes (from hexacosane—$C_{26}H_{54}$ to hentriacontane—$C_{31}H_{64}$) and hopanoids were found in small quantities. However, the number of detected organic compounds increased to 75 for the $scCO_2$/DCM extraction. In comparison to the original brown coal sample, the fraction of low-molecular aromatic hydrocarbons was rather low. In contrast, the portion of higher alkanes ranging from heneicosane ($C_{21}H_{44}$) to tetratriacontane ($C_{34}H_{68}$) and hopanoids was increased. In conclusion, the results indicate that $scCO_2$ can potentially mobilize organic carbon

Table 1 Detected hydrocarbons in brown coal samples extracted with dichloromethane (DCM) or extracted with supercritical CO_2 ($scCO_2$) and subsequently dissolved in DCM

Compounds	Brown coal treated with DCM	Brown coal extracted with $scCO_2$/treated with DCM	Biodegradability under in situ conditions
Saturated hydrocarbons/alkanes	$C_{26}H_{54}$–$C_{31}H_{64}$	$C_{21}H_{44}$–$C_{34}H_{68}$	Yes (Widdel and Rabus 2001; Gieg et al. 2010)
Aromatic hydrocarbons	Derivatives of BTEX and PAH	Derivatives of BTEX and PAH	Yes (Davidova et al. 2007; Foght 2008; Vogt et al. 2011)
Pentacyclic hydrocarbons/hopanoids	C_{31}–C_{32}	C_{31}–C_{33}	No

Based on literature the biodegradability of extracted hydrocarbons under in situ conditions was evaluated

that might be used by microorganisms as carbon and energy source for their metabolism, thus altering the biogeochemical conditions in CO_2 storage formations.

4 Permeability Behaviour in Discontinuous Flow Experiments With and Without Microbial Influence

4.1 Experimental Setup and Performance of Flow Experiments

The idea of these experiments was to investigate the influence of the biogeochemical processes on the permeability behaviour of the flow system.

So each flow experiment consisted of one biogeochemical setup (=BIO) and one geochemical setup (=GEO). Both were performed in parallel, with core plugs of Ø = 10 cm and L = 10 cm. For the first tests Postaer sandstone were used as model sandstone for high permeable reservoirs. Later tests were carried out with reservoir samples from the CO_2 test site Ketzin (courtesy German Research Centre for Geosciences GFZ; see Martens et al., this volume). The previously sterilized cores were put into a rubber sleeve and fixed vertically in a temperature-controlled retainer (60 °C). To avoid edge flow, the cores were subjected to a confining pressure (see Fig. 4).

The sterilized BIO-cores were once inoculated with microorganisms at the beginning of the experiment, while the GEO-cores remained sterile. Hence, variations in permeability behaviour and chemical results can be traced back to either geochemical or biogeochemical processes. The flow medium was aseptically placed in the piston accumulator. The red parts in Fig. 4 correspond to the sterile facility area.

The flow experiments were generally performed as batch tests. So a sterile medium was discontinuously flowed through the cores every 8–14 days and the out

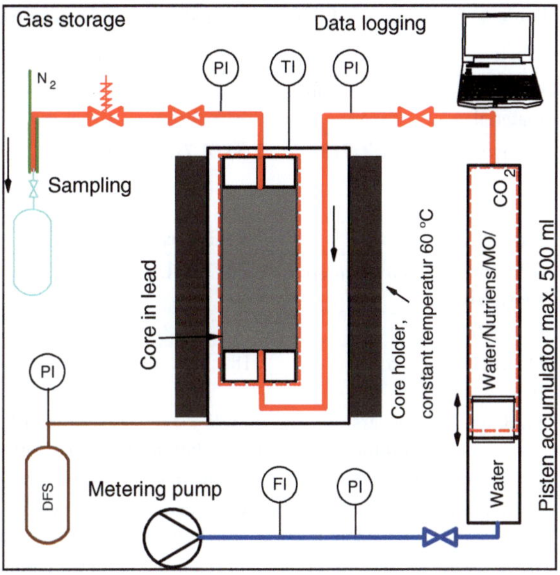

Fig. 4 Experimental setup of the batch flow experiments with core plugs. The *red* coloured parts were in contact with the medium during the experiment and sterilised before the tests

flow was sampled and analysed. The sampling was carried out under inert gas (anaerobic) or pressure-holding to prevent out gassing of the dissolved gas. Inlets and outlets of the cores were closed pressure-tightly during the intermediate phases, when no flow was performed. So within these periods always new semi-equilibrium conditions are developed.

A flow experiment was started after the core plugs were saturated, using at least four pore volumes of sterile inflow medium. Temperature and pressure within the cores were recorded over the whole test duration. The inlet and the outlet pressure were used to calculate the permeability behaviour during the flow cycles.

The volumetric rate of the flow regime was selected, so that required sample volumes were possible. So the flow rate was held at 10 ml/min, which corresponds unfortunately to a high velocity of 1 cm/min.

In the CO_2-step of the batch tests CO_2 was added dissolved in water, at a saturation state of 5 bar and 20 °C.

Table 2 summarizes the specific characteristics and main findings of four flow experiments.

4.2 *Microbial Effect on Permeability Behaviour of High Permeable Rock Materials*

The flow experiments one to three (E1–E3, Table 2) comprised *Postaer* sandstone as core material and *Petrotoga* sp. Sch_Z2_3 as microbial component in the bio-geochemical setups.

Table 2 Summary of characteristics and main findings of four flow experiments (E1–E4)

	E1	E2	E3	E4
Core material	*Postaer* sandstone			Reservoir rock Ketzin (GFZ Potsdam)
Inflow	DSMZ 718	DSMZ 718 (modified)		Produced formation water Schneeren (sterile filtration)
Labelling	–	^{13}C-Glucose, ^{15}N-NH_4Cl		^{13}C-CO_2/^{13}C-$NaHCO_3$
Microbes (only BIO)	*Petrotoga* sp. Sch_Z2_3 (GU938475)			Microbial community Schneeren
CO_2 stress	–	–	+(43d)	+(56d)
Duration [d]	176	118	142	190
Finding	Successful infiltration of *Petrotoga*	Active fermentation of ^{13}C-glucose	Decline of fermentation due to CO_2 stress	Microbial community sustains CO_2 stress
Conclusion	The microbial- or CO_2-depending influence on the permeability behaviour of high permeable rocks is small and more in the range of measurement uncertainties			

DSMZ German Collection of Microorganisms and Cell Cultures; *DSMZ 718 Petrotoga* medium; *DSMZ 718 (modified)* DSMZ 718 without Na_2S, lower concentration of TOC (0.75 g/l tryptone, 0.2 g/l yeast, 2.0 g/l glucose)

Postaer sandstone represents a rock material that serves as analogue to high permeable reservoir sandstones (K of around 10^{-14} m^2). *Petrotoga* sp. had been isolated from produced formation water of the natural gas field Schneeren (Lower Saxony, Germany) and was also detected with molecular-genetic methods in formation water of the almost depleted natural gas field Altmark (Saxony-Anhalt, Germany), which would be suitable as CO_2 storage site in Germany (Hoth et al. 2009a, 2011). That's why, *Petrotoga* sp. was applied as test organism. *Petrotoga* species are anaerobic and thermophilic bacteria with the ability to ferment various sugars or to reduce thiosulphate or elemental sulphur to hydrogen sulphide (Davey et al. 2001).

The first flow experiment demonstrated the successful infiltration of *Petrotoga* sp. Sch_Z2_3 into the biogeochemical core plug for the first time. In addition, using ^{13}C-labelled glucose in the second flow experiment, it was proven that *Petrotoga* sp. actively fermented glucose within the biogeochemical setup (Fig. 5).

The third flow experiment showed that the fermentation of ^{13}C-glucose by *Petrotoga* declined during CO_2 stress. However, it was not proven if the fermenting metabolism of *Petrotoga* sp. could have been recovered after the CO_2 stress.

Figure 6 shows the permeability measurements for test 3. There is a small increase in the permeability for the BIO-setup. But these changes are, compared to the initial permeability, quite small and more in the range of the measuring accuracy.

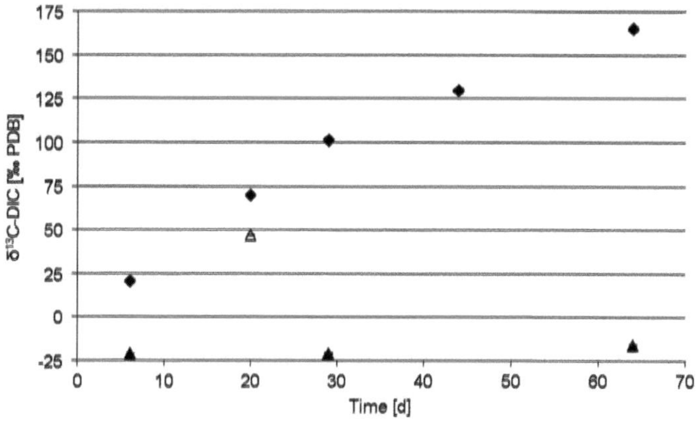

Fig. 5 Fermentation of ^{13}C-glucose to ^{13}C-DIC by *Petrotoga* sp. Sch_Z2_3 in the biogeochemical setup of flow experiment two. *Symbols diamonds*—δ^{13}C-DIC signature of flow-through of biogeochemical core, *closed triangles*—δ^{13}C-DIC of sterile inflowing medium, *open triangle*—δ^{13}C-DIC of the inoculum of the biogeochemical core comprising *Petrotoga* sp. Sch_Z2_3, which was incubated with ^{13}C-glucose for 5 days prior to inoculation

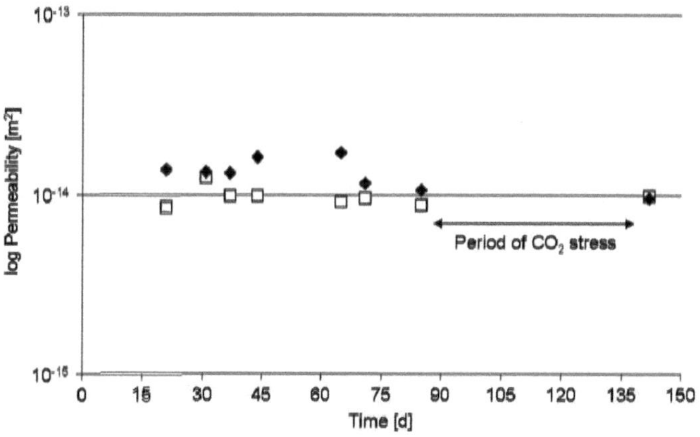

Fig. 6 Permeability behaviour of the biogeochemical (*diamonds*) and the geochemical core (*open squares*) in flow test 3

For flow experiment 4, core plugs of reservoir rock (initial permeability 10^{-14} m^2) from the CO_2 pilot site Ketzin (Brandenburg, Germany) were used (E4 in Table 2). In addition, formation water produced freshly from the borehole Z3 of the natural gas field Schneeren was used as inflowing medium, after sterile filtration and as inoculum for the biogeochemical setup after incubation at 60 °C with a headspace of N_2:H_2:CO_2 (75:20:5) for 7 days.

The incubation of the formation water resulted in an enrichment of specific bacteria. In particular, the relative abundance of *Petrotoga* increased by 11 % compared to the formation water before the incubation. In addition, *Anaerobaculum*,

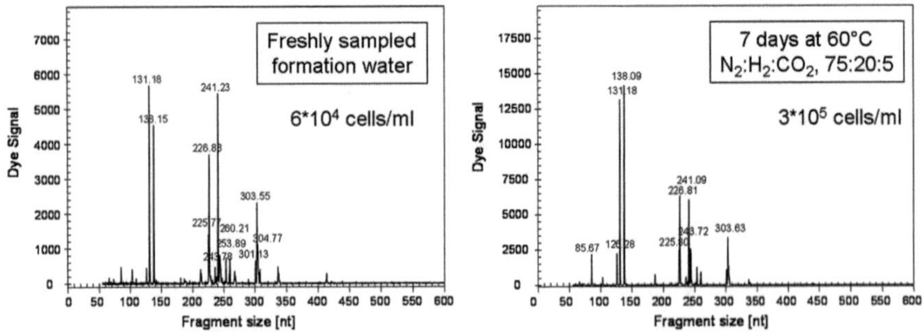

Fig. 7 TRFLP profiles of the fresh formation water from a borehole of gas field Schneeren before and after incubation at 60 °C with a headspace of N_2:H_2:CO_2 (75:20:5) for 7 days. Whole cell numbers were determined by fluorescence microscopy using the DNA-stain DAPI (Diamidinphenylindol)

an anaerobic, thermophilic (up to 65 °C) and halophilic (up to 70 g/l NaCl) genus, was enriched. *Anaerobaculum* reduces thiosulphate, elemental sulphur and L-cystine and is able to produce hydrogen from sugars, amino acids and other organic acids (Maune and Tanner 2012). Beside *Petrotoga* and *Anaerobaculum*, *Thermoanaerobacter* and a sulphate-reducing bacterium whose metabolic lifestyle is not yet known were also assigned to peaks in the TRFLP[1] profile of the formation water (Fig. 7). The genus *Thermoanaerobacter* comprise anaerobic, thermophilic (up to 62 °C) and heterotrophic species, which are able to disproportionate thiosulphate to elemental sulphur and hydrogen sulphide (Wagner et al. 2008).

The permeability of both setups in test 4 started, after saturation with sterile formation water, at a similar level—BIO: 1.4×10^{-14} m^2, GEO: 2.9×10^{-14} m^2 (Fig. 8).

The inoculation of the BIO-core caused initially a drop of its permeability to 1.7×10^{-16} m^2 that recovered after 20 days to the permeability level of the geochemical core. During the period of CO_2 stress (days 41–97), the calculation of permeability was not possible, because of interference by the CO_2 pressure (5–6 bar) at the outlet. Interestingly, after the CO_2 stress, the permeability of both cores was almost equal to the permeability before the CO_2 stress and increased to values of 2.3×10^{-14} m^2 for the BIO-core and 3.0×10^{-14} m^2 for the GEO-core until the end of flow experiment four (day 190). The release of CO_2 and the dissolving of substances that had been precipitated during the CO_2 stress might be of concern when interpreting the permeability behaviour after the CO_2 stress.

Whole cell counts showed an increase from 1.5×10^4 cells/ml to a maximum of 1.5×10^5 cells/ml at the end of the period of CO_2 stress. This indicates that the microbial community was actively living within the BIO-core, but also that the

[1] TRFLP—Terminal Restriction Fragment Length Polymorphism, a molecular-genetic method.

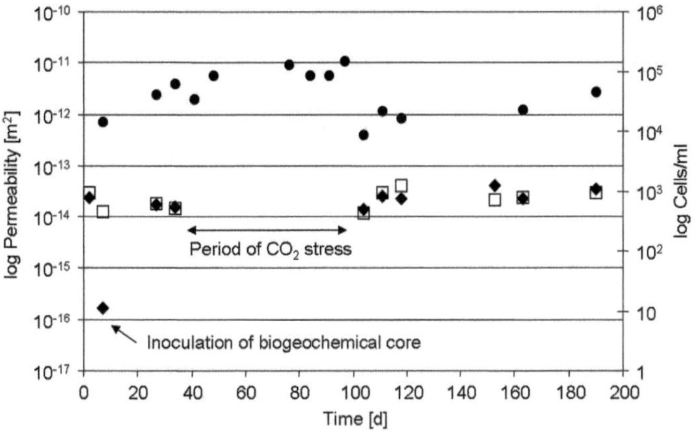

Fig. 8 Permeability behaviour of the biogeochemical (*diamonds*) and the geochemical (*open squares*) core in flow experiment four. *Circles* depict whole cell counts from flow-through of the biogeochemical core using fluorescence microscopy and the DNA-stain DAPI

microbes were constantly discharged from the BIO-core. The lowest cell number (8.6×10^3 cells/ml) was counted after the CO_2 stress when the CO_2 pressure was relieved. But the microbial community recovered to 4.5×10^4 cells/ml in the flow-through until the end of flow experiment four.

TRFLP analyses of the out flowing water revealed that *Thermoanaerobacter* predominated ($\sim$75 % relative abundance) the biogeochemical core, 27 days after inoculation. However, species of *Petrotoga* were most abundant (up to 60 %) during the period of CO_2 stress (days 41–97). After the CO_2 stress, the microbial community became more divers and comprised *Petrotoga*, *Thermoanaerobacter* and a sulphate-reducing bacterium. This recovery of the microbial community demonstrated that it was able to sustain the CO_2 stress.

4.3 Microbial Impact on Mineral Alteration of Reservoir Rock

Sequential extraction (SE) is a common method to get an overview of (reactive) mineral phases and element binding forms in soil or rock samples (e.g. Rao et al. 2008). By applying SE to rock samples prior and after the flow through experiments, the influence of microbial activity on mineral transformations was investigated. 5–10 g of rock material was extracted, according to the procedure cited in (Graupner et al. 2007) (Table 3). Sequential extractions were performed at unaltered rock and at the rock core samples of test 1–3 after the experiments, taken from the inflow, middle and outflow region.

Table 3 Sequential extraction protocol modified from (Zeien 1995; Graupner et al. 2007)

Step	Solvent	Extracted mineral fraction
I	Distilled water	Pore water, water soluble salts
II	1 M NH_4NO_3	Ion exchanger
IIIa	1 M NH_4OAc; pH 6	Carbonates, specifically sorbed complexes
IIIb	1 M $NH_4OAc + NH_4OCl$; pH 6	Easily reducible (Mn-)phases
IV	0.025 M NH4-EDTA; pH 4.7	Organic matter
Va	0.2 M NH_4-oxalate; pH 3.3	Poorly crystalline Fe oxyhydroxides
Vb	Va + 0.1 M ascorbic acid; pH 3	Crystalline Fe oxides, Al oxyhydroxides, alumosilicates
VII	8.8 M H_2O_2; pH 2	Sulphides, alumosilicates
VI	Aqua regia; pH 0	Sulphides enclosed in silicates, alumosilicates

Figure 9 shows amounts of selected elements extracted from unaltered rock and the three samples of the Geo- and Bio-Cores after experiments 1 and 3, respectively. The assessment of extracted elemental amounts per extraction step indicated the following processes as being influenced by activity of *Petrotoga* sp.:

- Sulphur added as Na_2S in test 1 and was partly precipitated as sulphide in Geo-Core 1 and as more readily available possibly organic sulphur in Bio-Core 1.
- High amounts of S were extracted from Bio-Core 3 which were not balanced by equivalent amounts of a cation, supporting the finding of accumulation of organically bound or elemental sulphur as concluded from experiment 1.
- Al and Si (congruent behaviour, not shown) became more readily available through activity of *Petrotoga* in experiment 1. It seems to be that Al and Si were transferred from feldspars in fraction VI to less stable feldspars/phases in fraction Vb and to solution in fraction I.
- The microbial influenced transformation of alumosilicates and feldspars was supported by findings in experiment 3, where microbes enhance the dissolution of Na-feldspar (fraction VI).

4.4 Geochemical Modelling: Process Identification and Estimation of the Resulting Porosity Change

For identification of mineral reactions with influence on the rock porosity and the estimation of the porosity change, the high-pressure bioreactor experiments of the RECOBIO-project (Hoth et al. 2011) were geochemically modelled. Table 4 gives an overview on experimental conditions. Each experiment was performed as microbial test reactor (TR = BIO) with a parallel sterile control reactor (CR = GEO). The experiments were modelled using PHREEQC 3.1 by stepwise

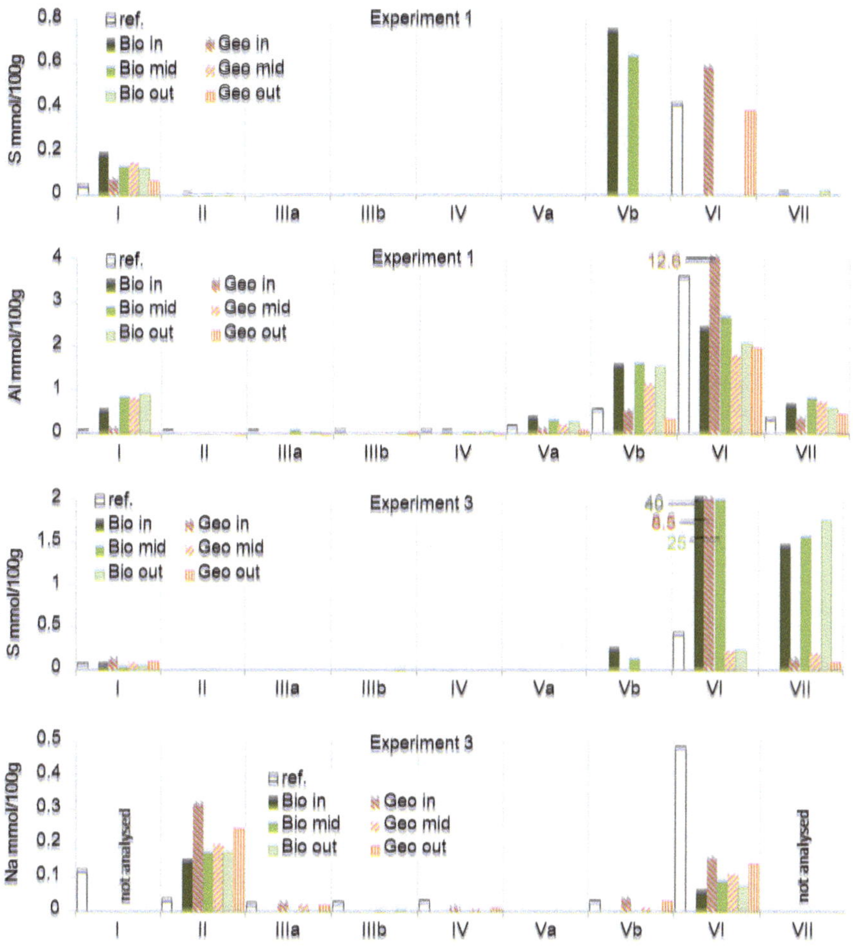

Fig. 9 Extracted amounts of selected elements per extraction step for experiment 1 (*upper part*—S, Al) and experiment 3 (*lower part*—S, Na)—the BIO-cores are marked with *green* colours, the GEO-cores with *red* colours. The initial material is marked *white*

implementing (a) mixing of pore and formation water, (b) cation exchange, (c) equilibrium with gas phase, (d) mineral reactions, (e) microbial reactions (chemolithoautotrophic SO_4^{-2} reduction, autotrophic formic acid formation) for the microbial TR-reactor.

In addition to water and gas analyses during the experiments, the rock materials of reactor 8 were sequentially extracted (procedure see Table 3). Table 5 compares the mineralogical analysis of the unaltered rock sample with the mineral assemblage determined from SE prior and after experiments CR8 and TR8.

With a good reproduction of measured water composition, the modelling of all three experiments identified the following reactions:

Table 4 Experimental conditions of the high-pressure bioreactors

Reactor	R 5	R 6	R 8
Sampling date	27.11.2007	27.05.2008	07.04.2009
Temperature (°C)	40	40	40
Initial pressure (atm)	11.9	12.1	14.5
p CO_2:H_2:N_2 (atm)	4.7:3.6:3.6	6.5:2.8:2.8	6.5:4.0:4.0
Duration: test, control	79 d, 163 d	121 d, 121 d	202 d, 202 d

Always 350 mL freshly sampled formation water from the Schneeren bore hole Z3 was used. Furthermore as solid material always 50 g of milled rock from the bore hole Z3 (depth of 2,800 m) of the drill core archive

Table 5 Composition of the original formation rock from XRF analysis compared with results from sequential extraction of the unaltered rock and material after the experiments CR8, TR8

Original composition from RFA analysis	Related reactive minerals from sequential extraction	Initial	CR8	TR8
Quartz: 78 %	(quartz: inert) pore water pH	7.6	9.3	5.2
Ankerite: 1 %	Ankerite[a] mmol/100 g	1.5	0.15	0.01
Siderite: 1 %	Siderite mmol/100 g	4.16	0.02	<0.01
Calcite: 1 %	Calcite mmol/100 g	2.37	4.73	0.17
	Rhodocrosite mmol/100 g	0.08	0.01	<0.01
Plagioclas: 4 %	Anorthite mmol/100 g	0.52	0.23	0.27
Iron-chlorite: 1 %	Chamosite[b] mmol/100 g	1.16	1.04	1.07
Clay minerals: 14 %	K in silicates mmol/100 g	4.55	49.6	32.5

[a] $CaFe_{0.6}Mg_{0.4}(CO_3)_2$
[b] As representative for iron-chlorite: $Fe^{2+}{}_3Mg_2Fe^{3+}{}_{0.5}Al_{0.5}(Si_3Al)O_{10}(OH)_8$

- dissolution of calcite, magnesite by CO_2 and rhodochrosite at higher p(CO_2),
- precipitation of siderite coupled to a dissolution of iron-silicates by CO_2,
- formation of ankerites $Ca(Fe,Mg)(CO_3)_2$ at lower p(CO_2),
- microbial sulphate reduction causes enhanced dissolution of siderite with precipitation of excess Ca in $CaCO_3$ and formation of sulphides,
- intensified carbonate dissolution after microbial formation of organic acids.

Using molar volumes and an initial porosity of 0.12, the changes in porosity caused by these mineral transformations were calculated for the experiments (Fig. 10). In most cases, dissolution effects balance volume changes, due to mineral precipitation and so the porosity changes are small. For one scenario (microbial TR 5) a significant decrease in porosity was modelled. Thus, CO_2 and CO_2 induced microbial activity may cause mineral transformations that in many cases leave overall porosity unchanged, but in some cases the decrease of rock porosity within CO_2 storage units could be significant.

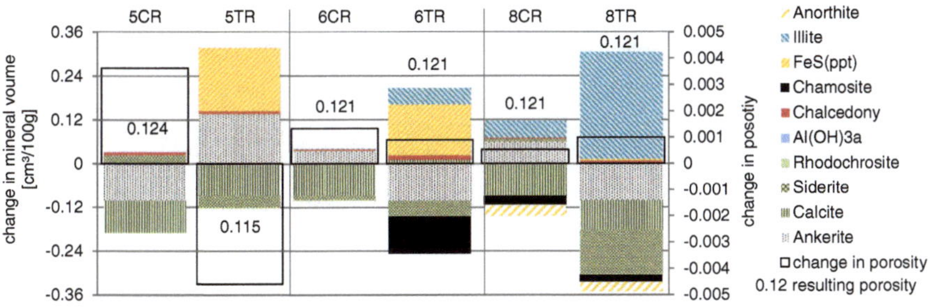

Fig. 10 Changes in mineral volumes calculated from modelling static high pressure bioreactors and resulting change in porosity (values). Initial porosity were assumed to n = 0.12. Different patterns show different mineral phases. Thereby positive values mark precipitation of these phases and negative values marks dissolution

5 Risk Assessment by Application of GoldSim Modelling

5.1 Overview of the Process Model Applied for Dynamic Modelling of Microbial Processes

As shown in the sections before there are biogeochemical reactions, which are relevant for deep geological storage formations. The reaction mechanisms of these were captured by process models and simulated by using the software GoldSim^{TM}. The model concept relies on data from high pressure autoclave experiments of the project RECOBIO. In addition experimental data from the literature were included and suitable boundary conditions were set (Fig. 11). Accordingly, the biogeochemical

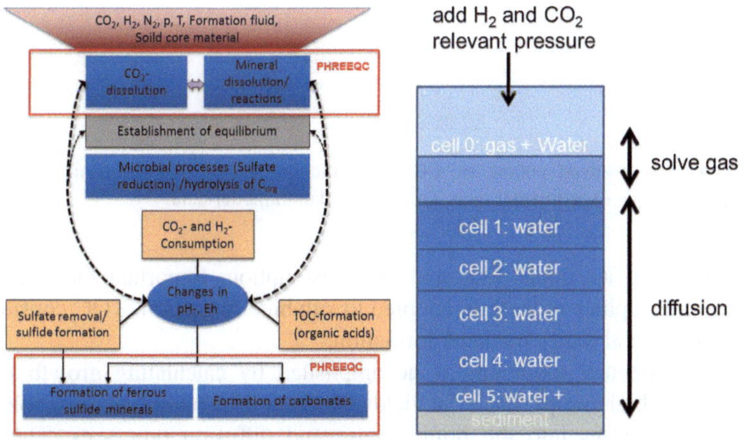

Fig. 11 (*Left*) Model concept for modelling CO_2-affected bio-geochemical processes and (*Right*) partitioning of the experimental reactor for GoldSim modelling

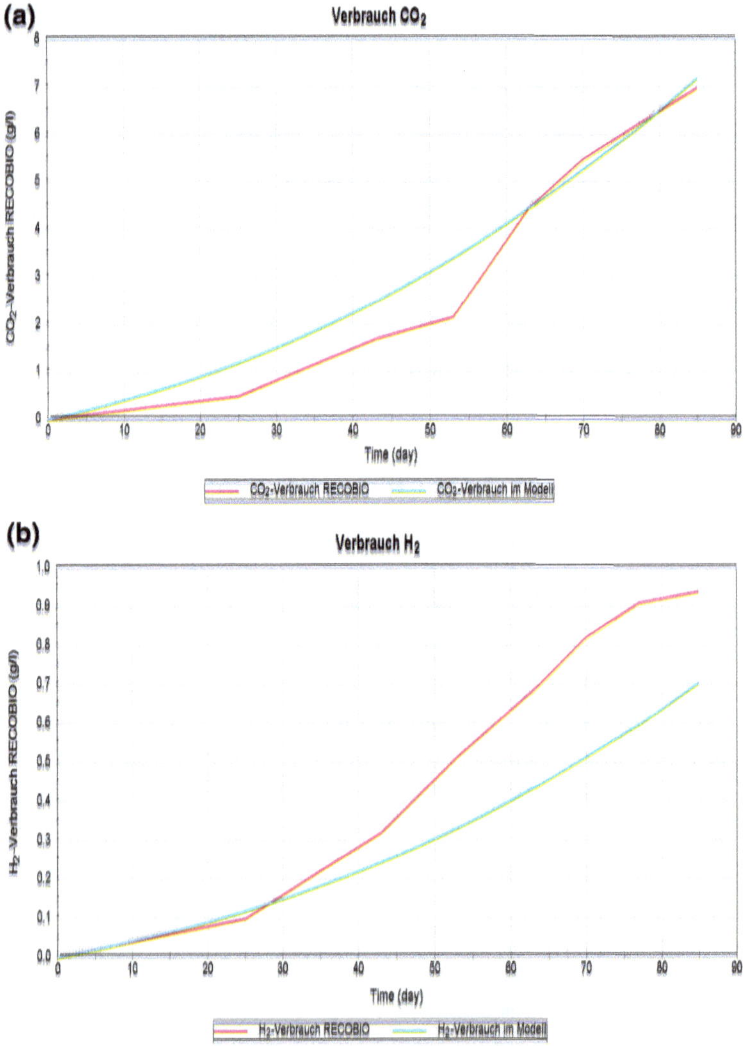

Fig. 12 **a** Time-resolved course of CO_2 consumption and **b** time-resolved course of H_2 consumption—*red*—experimental data, *green*—simulation/model data

reactions take place under CO_2 and H_2 consumption. Important biogeochemical reactions are sulphate reduction, coupled to sulphide formation, TOC consumption (Acetogenese) and partly methane formation.

The experimental modelling is accomplished by calculating growth rates of relevant microbial groups, bacterial yields, biomass concentrations, metabolic rates, dissolution of gases into the liquid phase and diffusion processes. The model concept (Fig. 11) was converted in GoldSim™ by partitioning the reactor in separate cells. The modelling includes diffusion processes and the dissolution of

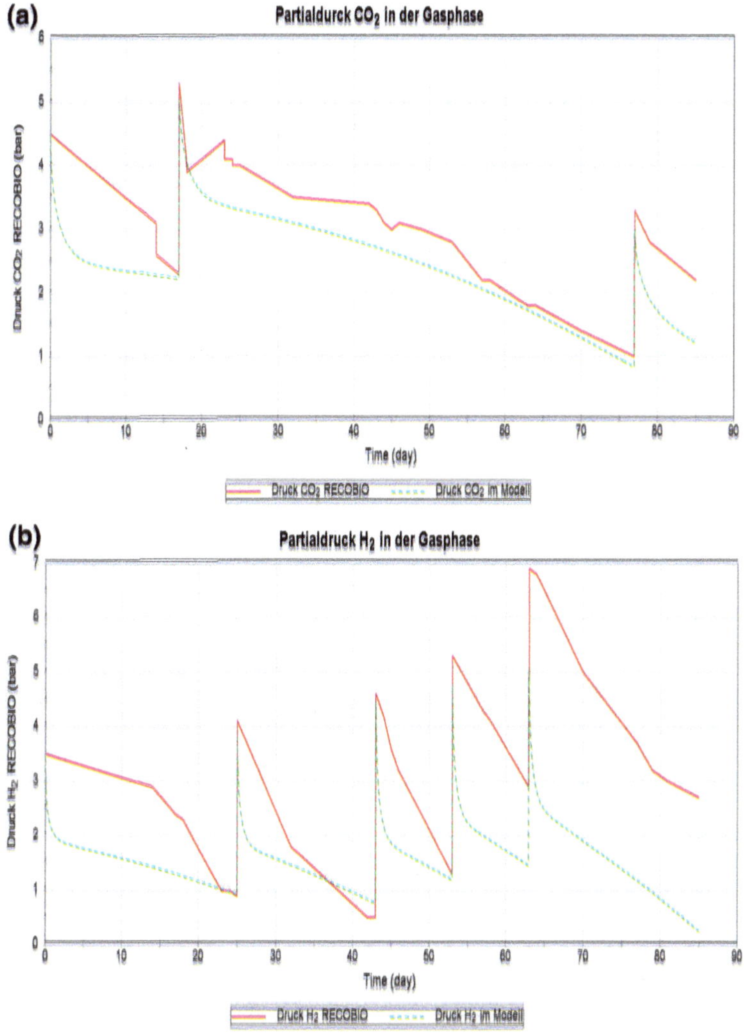

Fig. 13 **a** Time-resolved course of CO_2 partial pressure and **b** time-resolved course of H_2 partial pressure—*red*—experimental data, *green*—simulation/model data

pressurized gases into the liquid phase, microbial conversion reactions and precipitation/dissolution of solid phases.

It was possible to verify the model by the RECOBIO test results and to identify key parameters. In addition, the model enables forecasting conclusions about substrate consumption and precipitation behaviour. Figures 12, 13 and 14 shows selected results of the re-modelling of the RECOBIO experiments.

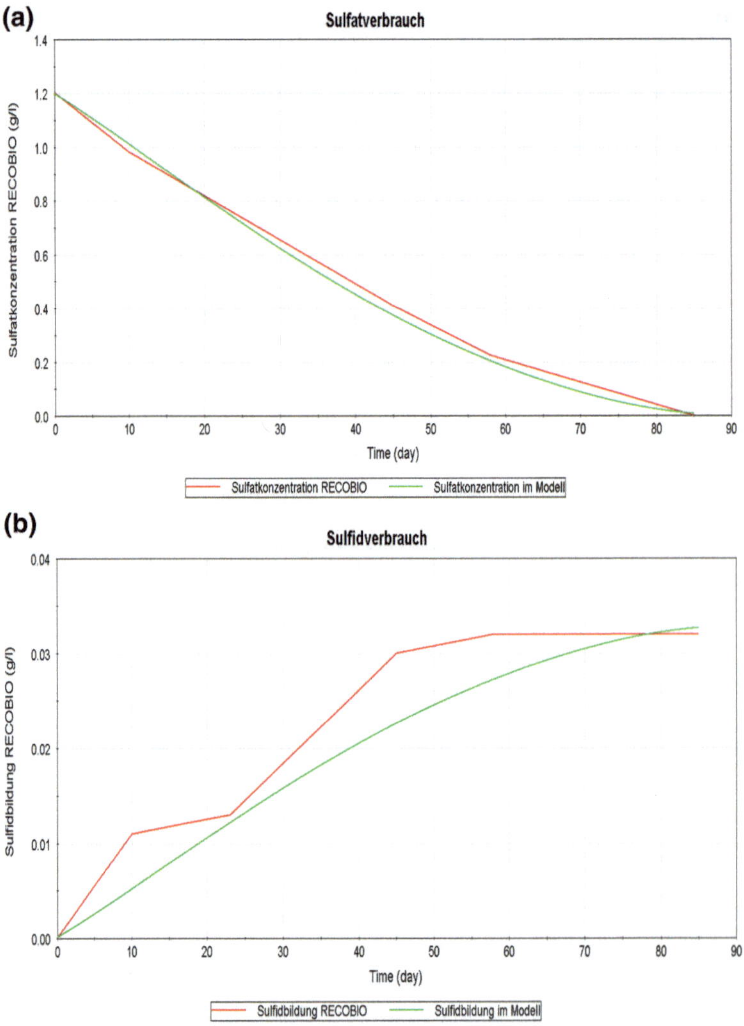

Fig. 14 **a** Time-resolved course of sulphate concentration and **b** time-resolved course of sulphide concentration—*red*—experimental data, *green*—simulation/model data

6 Conclusions

A microbial- or CO_2-depending influence on the permeability behaviour of high permeable rock (10^{-13}–10^{-14} m^2) could not be detected in the flow experiments. However, results of batch experiments during the RECOBIO projects (2005–2011) (see Hoth et al. 2009a, b, 2011; Ehinger et al. 2009) showed that a metabolic active microbial community could produce organic substances such as organic acids from CO_2 that might represent a further CO_2 storage mechanism. On the other hand, the production of organic acids might also cause a pH-drop that can result in a decreased CO_2 solubility.

Although the microbial community sustained the CO_2 stress in the flow experiments, a significant influence on the permeability behaviour of the core material was not detected. Hence, serious variations on the permeability behaviour during the operation of a CO_2 storage site with a high permeable reservoir rock are not expected according to the flow experiments. Nevertheless, beside the geological constitution of a putative CO_2 storage site, the microbial community should always be known in order to predict possible disturbances in the operation of the CO_2 storage.

Acknowledgements We would like to acknowledge the German Federal Ministry of education and Research for funding, Grant numbers 03G0781 A to D and 03G0782.

Furthermore many thanks to the team of Sonja Martens at GFZ Potsdam for providing us core material of the Ketzin site for the discontinuous flow experiments.

References

Chivian D, Brodie EL, Alm EJ, Culley DE, Dehal PS, DeSantis TZ, Gihring TM, Lapidus A, Lin LH, Lowry SR, Moser DP, Richardson PM, Southam G, Wanger G, Pratt LM, Andersen GL, Hazen TC, Brockman FJ, Arkin AP, Onstott TC (2008) Environmental genomics reveals a single-species ecosystem deep within earth. Science 322(5899):275–278

Davey ME, Macgregor BJ, Stahl DA (2001) Genus IV. *Petrotoga.* In: Boone DR, Castenholz RW, Garrity GM (eds) Bergey's manual of systematic bacteriology. The Archaea and the deeply branching and phototrophic Bacteria, vol 1. Springer, New York, pp 382–385

Davidova IA, Gieg LM, Duncan KE, Suflita JM (2007) Anaerobic phenanthrene mineralization by a carboxylating sulfate-reducing bacterial enrichment. ISME J 1(5):436–442

Ehinger S, Seifert J, Kassahun A, Schmalz L, Hoth N, Schlömann M (2009) Predominance of *methanolobus* spp. and *methanoculleus* spp. in the archaeal communities of saline gas field formation fluids. Geomicrobiol J 26:326–338

Foght J (2008) Anaerobic biodegradation of aromatic hydrocarbons: pathways and prospects. J Mol Microbiol Biotechnol 15(2–3):93–120

Frei J et al (1986) "Chemisch und mikrobiologisch bedingte Gasverluste am UGS Ketzin". Report of a research project, Brennstoffinstitut, Freiberg

Gieg LM, Davidova IA, Duncan KE, Suflita JM (2010) Methanogenesis, sulfate reduction and crude oil biodegradation in hot Alaskan oilfields. Environ Microbiol 12(11):3074–3086

Gniese C, Bombach P, Rakoczy J, Hoth N, Schlömann M, Richnow HH, Krüger M (2013) Relevance of deep-subsurface microbiology for underground gas storage and geothermal energy production. Adv Biochem Eng Biotechnol doi:10.1007/10_2013_257

Graupner T, Kassahun A, Rammlmair D, Meima JA, Kock D, Furche M, Fiege A, Schippers A, Melcher F (2007) Formation of sequences of cemented layers and hardpans within sulfide-bearing mine tailings mine district Freiberg, Germany. Appl Geochem 22:2486–2508

Hendry P (2006) Extremophiles: there's more to life. Environ Chem 3:75–76

Hoth N, Kassahun A, Ehinger S, Muschalle T, Seifert J, Schlömann M, Häfner F (2009a) Recycling of sequestrated CO_2 by microbial—biogeochemical transformation in the deep subsurface—RECOBIO. Final report, BMBF-funded research, FKZ 03G0616A and 03G0616C

Hoth N, Kassahun A, Ehinger S, Würdemann H (2009b) Biogeochemische Wechselwirkungen—langfristige mikrobielle Umwandlung des gespeicherten CO_2. Geotechnol Sci Rep 14:58–65

Hoth N, Kassahun A, Krüger M, Gniese C, Frerichs J, Muschalle T, Schlömann M, Reich M (2011) Untersuchung der biogeochemischen transformation von im tiefen Untergrund gespeichertem CO_2—RECOBIO 2. Final report, BMBF-funded research, FKZ 03G0696A, 03G0697A and 03G0616C

L'Haridon S, Reysenbacht A-L, Glenatt P, Prieur D, Jeanthon C (1995) Hot subterranean biosphere in a continental oil reservoir. Nature 377:323–324

Lovley DR, Chapelle FH (1995) Deep subsurface microbial processes. Rev Geophys 33(3):365–381

Maune MW, Tanner RS (2012) Description of *Anaerobaculum hydrogeniformans* sp. nov., an anaerobe that produces hydrogen from glucose, and emended description of the genus *Anaerobaculum*. Int J Syst Evol Microbiol 62(Pt 4):832–838

Rao CRM, Sahuquillo A, Lopez Sanchez JF (2008) A review of the different methods applied in environmental geochemistry for single and sequential extraction of trace elements in soils and related materials. Water Air Soil Poll 189:291–333

Smigan P, Greksak M, Kozankova J, Buzek F, Onderka V, Wolf I (1990) Methanogenic bacteria as a key factor involved in changes of town gas stored in an underground reservoir. FEMS Microbiol Ecol 73:221–224

Vogt C, Kleinsteuber S, Richnow HH (2011) Anaerobic benzene degradation by bacteria. Microb Biotechnol 4(6):710–724

Wagner ID, Zhao W, Zhang CL, Romanek CS, Rohde M, Wiegel J (2008) *Thermoanaerobacter uzonensis* sp. nov., an anaerobic thermophilic bacterium isolated from a hot spring within the Uzon Caldera, Kamchatka, Far East Russia. Int J Syst Evol Microbiol 58(11):2565–2573

Widdel F, Rabus R (2001) Anaerobic biodegradation of saturated and aromatic hydrocarbons. Curr Opin Biotechnol 12(3):259–276

Zeien H (1995) Chemische Extraktionen zur Bestimmung der Bindungsformen von Schwermetallen in Böden. Bonner Bodenkundl. Abh, 17

Seismic and Sub-seismic Deformation Prediction in the Context of Geological Carbon Trapping and Storage

Charlotte M. Krawczyk, David C. Tanner, Andreas Henk, Henning Trappe, Jennifer Ziesch, Thies Beilecke, Chiara M. Aruffo, Bastian Weber, Andrea Lippmann, Uwe-Jens Görke, Lars Bilke and Olaf Kolditz

Abstract In the joint project PROTECT (PRediction Of deformation To Ensure Carbon Traps) we predicted and quantified the distribution and the amount of sub-/ seismic strain in the proximity of the CO_2 reservoir in the Otway Basin. Three approaches fill the sub-seismic space: *seismic multi-attributes* stabilized the interpretation of the 3-D depth model by imaging small lineaments; *retro-deformation* revealed in the seal ca. 3 % as highest strain magnitudes; numerical *forward modelling* shows that the minimum horizontal stress at reservoir is locally overprinted by faults. We calibrated our predictions with new near-surface reflection seismic measurements and used advanced visualization tools. Thus, this seismo-mechanical workflow reveals possible migration pathways, and as such provides a tool for prediction and adapted time-dependent monitoring for subsurface storage in general.

1 Introduction

In saline aquifer and depleted gas fields, storage mechanisms depend, in the first decades of CO_2 injection, to a large extent on the overburden geology of a site, as well as on the reservoir rocks themselves. Thus, the main processes that have to be

C.M. Krawczyk (✉) · D.C. Tanner · J. Ziesch · T. Beilecke
Leibniz Institute for Applied Geophysics (LIAG), Stilleweg 2, 30655 Hannover, Germany
e-mail: lotte@liag-hannover.de

A. Henk · C.M. Aruffo · B. Weber
TU Darmstadt, Institut für Angewandte Geowissenschaften, Schnittspahnstraße 9, 64287 Darmstadt, Germany

H. Trappe · A. Lippmann
TEEC GmbH, Burgwedeler Straße 89, 30916 Isernhagen, Germany

U.-J. Görke · L. Bilke · O. Kolditz
Helmholtz Centre for Environmental Research, UFZ, Permoserstraße 15, 04318 Leipzig, Germany

A. Liebscher and U. Münch (eds.), *Geological Storage of CO_2 – Long Term Security Aspects*, Advanced Technologies in Earth Sciences,
DOI 10.1007/978-3-319-13930-2_5

investigated are structural and stratigraphic trapping. To assess propagation and storage of CO_2 in the subsurface prior to drilling, as well as to detect possible pathways and CO_2-induced alteration, seismic imaging and deformation prediction techniques are essential (Krawczyk and Tanner 2010). If leakage does occur, CO_2 can migrate upwards and dissolve in the groundwater close to the surface, which has been proven for some natural CO_2 and gas reservoirs (Lewicki et al. 2007). Also, repeated reflection seismic measurements in the Sleipner Field prove that seals are not totally leakproof (Zweigel et al. 2004).

A geomechanical assessment is generally undertaken at storage sites, and stress field changes due to gas injection must be determined to estimate fault reactivation (see van Ruth and Rogers 2006; Vidal-Gilbert et al. 2010; Aruffo et al. 2014, for the Otway Basin). The reservoir stress path is one of the key points of any fault stability analysis, but at the same time a major source of uncertainty is the accuracy of the baseline geometrical model (Rogers et al. 2008). The models can be enhanced by the addition of sub-seismic structures, which are defined as structures that are below seismic resolution, but above core/outcrop scale (for more detail and definition of sub-seismic gap, see Krawczyk et al. 2011, 2015).

Filling this general gap in knowledge and working across seismic and sub-seismic scales is the aim of the joint project PROTECT (PRediction Of deformation To Ensure Carbon Traps). The project combines seismo-mechanical behaviour of reservoirs based on both inverse kinematic deformation modelling and forward numerical simulation with in situ observation. Such joint studies for the detection of potential migration paths towards the surface are rare (e.g., Raistrick 2008; Krawczyk et al. 2015). Here, we present the principle components of the workflow, illustrated by a field example from the Otway Basin where the Australian demonstration project is operated by the CO2CRC (see Jenkins et al. 2011).

2 The Seismo-Mechanical Workflow

The workflow developed in PROTECT covers seismic and sub-seismic scales to predict deformation and possible pathways with an enhanced resolution.

As first input for model building the workflow comprises both 3-D seismic volumes on the large scale and well data on the small scale (Fig. 1, left). Using the different modelling procedures outlined in detail below deformation at various seismic and sub-seismic scales is investigated and validated (Fig. 1, middle). The approach ultimately results in the prediction of the complete picture of deformation (Fig. 1, right). These independently-derived results can be inspected using visualization tools, and most importantly, the predicted areas of enhanced deformation can be additionally surveyed by dedicated new reflection seismic measurements (Fig. 1, right).

Applying this workflow at the site of the Australian demonstration project (Fig. 2), we contribute to the monitoring and verification plan of the Australian Otway CO_2 project (Dodds et al. 2009).

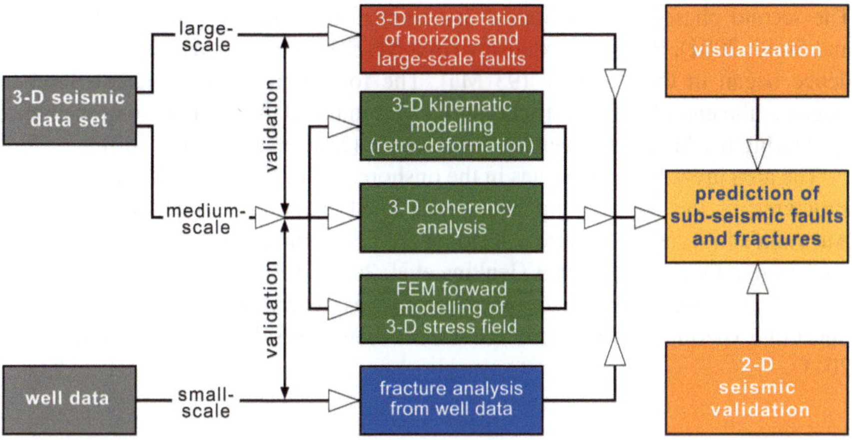

Fig. 1 Workflow of the joint project PROTECT for seismic and sub-seismic deformation prediction. Different scales and wavelengths of deformation are combined and the predictions validated (after Krawczyk et al. 2011)

3 Area of Investigation

The Otway Basin is a NW-striking, passive margin, rift basin (Fig. 2) that formed during the break-up of Gondwana and the Antarctic-Australian separation (Williamson et al. 1990). The first phase of extension began in the Late Jurassic–Early Cretaceous.

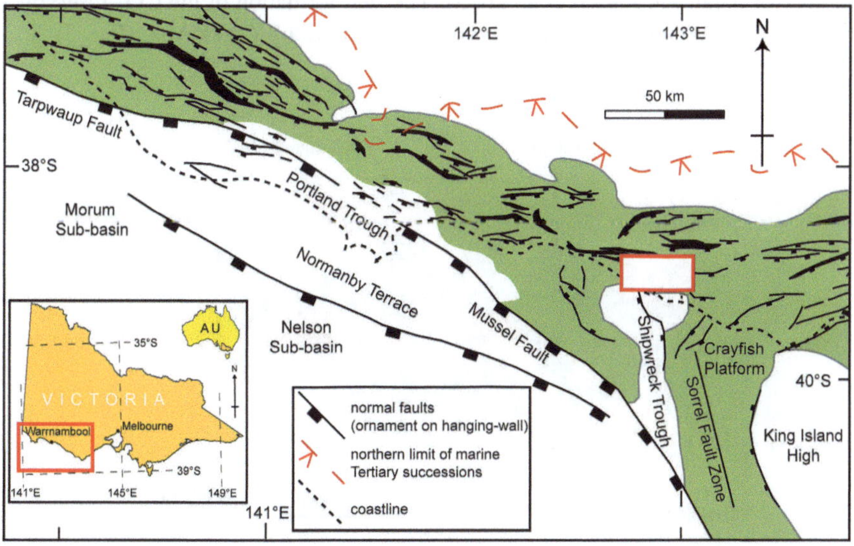

Fig. 2 Map of the investigation area in the Otway Basin (*insert*: location in Australia). The structural map of the Otway Basin (*green*) shows major faults (after Ziesch et al. 2015, see references therein). The large-scale, 3-D seismics used in this study is within the *red rectangle*

The second phase was a period of inversion in the mid-late Cretaceous (Norvick and Smith 2001). Renewed extension (in a NE-SW direction) and rift-related subsidence began in the Turonian (93 Ma). The formation of the incipient Southern Ocean at the end of the Maastrichtian (65 Ma) initiated the third phase of the break-up. During the Miocene, inversion began again (Norvick and Smith 2001).

The area investigated here lies in the onshore part of the Otway Basin. The major structures in the surrounding area are NW-SE striking normal faults (Fig. 2). Australia's first carbon capture and storage (CCS) demonstration project has been in operation at Otway since 2003 (Jenkins et al. 2011). 65.445 t of CO_2-rich gas were injected into the 2,050 m deep, depleted gas reservoir. Subsequently, further injection experiments were carried out on a shallower saline aquifer (Sharma et al. 2011).

4 Three-Dimensional Geological Depth Model

The detailed geological 3-D model of the investigation area includes eight stratigraphic horizons (ranging from Late Creteaceous Top Turonian-Top Waarre Fm., up to Oligocene Top Rupelian-Top Narrawaturk Fm.) and 24 major faults, interpreted at depths between 0.45 to 4.0 km (Fig. 3). The area surrounding injection well CRC-1 is dominated by 60° SW-dipping, NW-SE striking normal faults that have listric character in the SW and planar character in the NE. In addition, secondary, north-dipping, antithetic faults developed due to movement on the major faults (for detail see Ziesch et al. 2015). The seal formation of the CO_2-reservoir is between Top Waarre Fm. (top reservoir) and Top Skull Creek Mst. (Fig. 3).

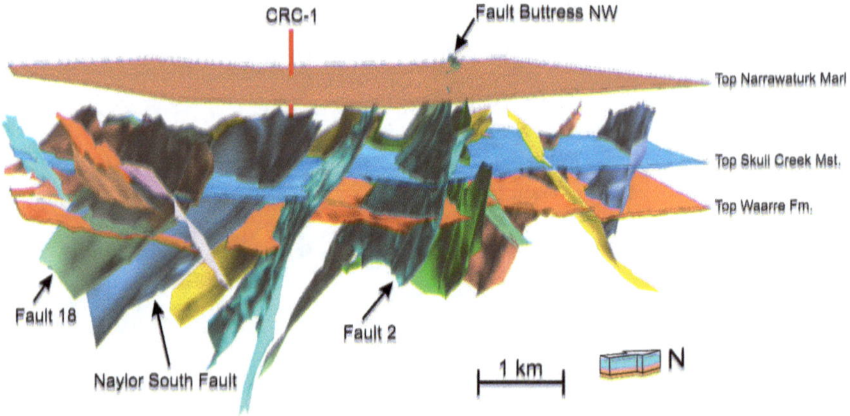

Fig. 3 Geological 3-D depth model between 0.45 and 4.0 km depth (modified from Ziesch et al. 2015). The location of injection well CRC-1 is marked, and faults used for further study are labelled.

The fault offset increases from north to south, to a maximum of 800 m (at the reservoir level-Top Waarre Fm.). The 3-D model shows major sediment accumulation on the hanging-walls of the faults (with respect to the footwall thicknesses), which is typical for passive margins such as the Otway Basin. Most faults only exist below 800 m depth (Late Paleocene Top Thanetian-Top Pebble Point Fm.) and die out above this level. Only Fault Buttress NW and Fault 2 displace all the interpreted stratigraphic horizons (Fig. 3), and appear to tip out in the seismically-unresolved, near-surface area (0–450 m depth). Fault kinematic analyses show that 60 % of the faults have dip-slip movement and 40 % of the faults have a small component of dextral strike-slip movement (Ziesch et al. 2015).

5 Sub-seismic Approaches

In our new seismo-mechanical workflow, sub-seismic scale information and prediction is achieved using three different approaches (c.f., Fig. 1, middle). By kinematically restoring the seismically-acquired volumes, i.e. retro-deforming the faulted strata, we show the spatial distribution of the strain and therefore possible fracture orientation caused by the faulting process. As a complementary approach forward finite element modelling (FEM) based on the same subsurface structure is used. This approach incorporates the mechanical properties of rocks and faults to predict the in situ stress distribution, in particular the local stress perturbations near faults and lithological boundaries. Both these methods are compared critically with coherency analysis of the equivalent seismic volumes, to reveal and categorise the seismic evidence of faulting and fracturing. Coherency estimates are obtained from seismic data as a volume-based measurement. Different coherency measures ensure an optimum resolution of fault systems and associated fault attributes that can be related to sub-seismic deformation.

5.1 Retro-deformation

One of the main challenges is to reconstruct the timing of geological movements within the area and to predict and quantify the amount of seismic and sub-seismic strain between reservoir and surface.

For this purpose we created volumes of tetrahedral elements between all eight interpreted stratigraphic horizons, separated by all major faults. For each stratigraphic horizon, beginning with the youngest and from oldest to youngest fault movement, each fault block is restored to the pre-faulted geometry (retro-deformation; c.f. Ziesch et al. 2014). We thereby image the sub-seismic deformation that occurs in the hanging-wall volume during this process and infer that this sub-seismic strain occurred during the movement on the fault. During restoration each tetrahedral element is attributed with the 3-D strain tensor.

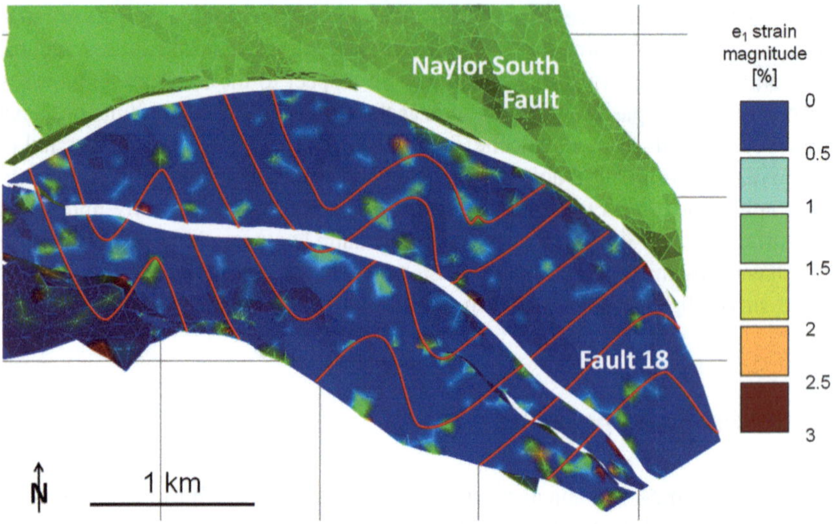

Fig. 4 Map view of the top seal horizon after sequential retro-deformation of Faults 18 and Naylor South (c.f. Fig. 3), colour-coded for e_1 strain magnitude. *Red lines* indicate strike of the e_2-e_3 plane and therefore the strike of sub-seismic fractures

We demonstrate this method on two fault blocks of the seal horizon, Fault 18 and Naylor South Fault (Fig. 3). The hanging-wall volume is colour-coded by e_1-strain magnitude (the major strain eigenvector), which is a direct proxy for the amount of fracturing. In the example shown the strain is diffuse throughout the hanging-walls with a maximum of ~3 % (Fig. 4). Assuming fractures are sub-perpendicular to e_1, i.e. they are parallel to the e_2-e_3 plane, then the strike of the fractures can be shown by plotting the trajectories of the e_2-e_3 plane (Fig. 4).

Analysis of the trajectories of the e_2-e_3 plane indicates that the fracture strike mainly changes in two areas (in the centre and to the west of the faults, Fig. 4). This shows that sub-seismic deformation is dependent on fault morphology, because the changes in fracture orientation correlate with corrugations in the faults (compare Fig. 3).

5.2 Coherency and Curvature Analyses

Highlighting similarities and dissimilarities in the seismic data is the aim of various coherency processing algorithms to improve the structural resolution, and to better identify subtle lineaments (Trappe and Hellmich 2003; Endres et al. 2008; Lohr et al. 2008a). Advanced coherency analysis with IHS (intensity, hue, and saturation are used to combine coherency, dip, and azimuth) and shaded-relief has the potential to reveal small-scale features that are important, especially for reservoir analysis (Gazar et al. 2011).

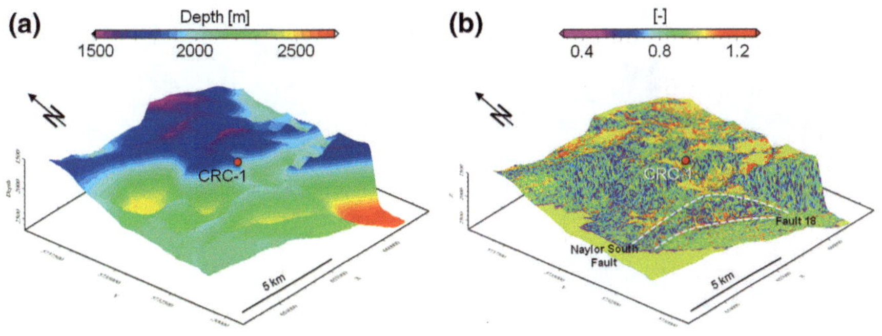

Fig. 5 Three-dimensional attribute images of the top reservoir level. The perspective views of the Top Waarre horizon **a** at depth, and **b** as combined median coherency and minimum curvature attributes (dimensionless summed and scaled) reveal the higher resolution achieved by seismic attribute processing

A subvolume of the Otway 3-D exploration seismic data set (depth domain, 9.5 km × 8.4 km) was used to develop and demonstrate various processing tools. New implementations are curvature analysis (following the principle of Roberts 2001) applied to horizons and faults, the determination of geological fault dip direction, and 3-D attribute mapping and multi-attribute evaluation (Fig. 5). The multi-attribute tool allows the individually calculated attributes, such as curvature and coherency, to be combined with various mathematical operations. Here, the coherency and curvature analyses focused on the main reservoir horizon, Top Waarre (c.f. Fig. 3). The sizes of the analysis windows were 50 m for the coherency and 25 m for the curvature processing.

3-D imaging with multi-attributes enables the detailed analysis of complex fault and lineament systems. While the depth representation of top reservoir (Fig. 5a) yields the general trend and morphology of the horizon, additional small-scale features can be detected in the multi-attribute display, thus complementing the main structures and providing greater fault density (Fig. 5b). For the Top Waarre horizon the combination of median coherency and minimum curvature (Fig. 5b) yields the clearest image (for further examples see Krawczyk et al. 2015). Especially the Naylor South Fault may be interpreted here as a zone of parallel faults rather than a single fault trace (Fig. 5b, transition to fracture zones is indicated by red colour).

5.3 Forward Modelling

Geomechanical models based on numerical methods (e.g. Fischer and Henk 2013) provide an effective tool to study the state of stress in the subsurface and to analyse the changes that may occur in response to CO_2 injection and long-term storage. Two different approaches varying spatially and methodologically have been used for a comprehensive geomechanical analysis in this case study.

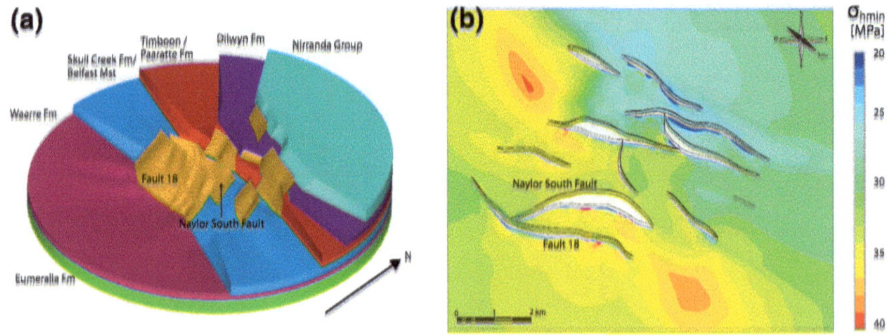

Fig. 6 Geomechanical model setup and stress magnitude for the reservoir level. **a** Cutaway showing setup of steady-state model with lithostratigraphic layers and faults (diameter 8 km, thickness 2.8 km; c.f. Fig. 3). **b** Distribution of minimum horizontal stress magnitudes at mid-reservoir level in map view

A steady-state geomechanical model based on finite element (FE) techniques was used for the Otway data to gain insights into the total stress distribution in the reservoir and entire overburden. The model geometry, including six lithostratigraphic layers and ten faults (Fig. 6a), was adopted from the seismic interpretation (Ziesch et al. 2015). Furthermore, the 3-D seismic velocity distribution was used as a proxy to incorporate both vertical and lateral changes in rock mechanical parameters (e.g. Young's modulus, Poisson Ratio, density). Forward modelling results provide detailed information on the 3-D stress distribution as well as the slip and dilation tendencies of the faults. For example, the minimum horizontal stress shows a dependence with depth overprinted locally by perturbations due to faults, with lowest values in the northern part of the study area (Fig. 6b).

In addition, a dynamic simulation coupling flow and mechanical calculations was performed (Aruffo et al. 2014). Modelling focused on the immediate vicinity of the injection area to estimate the response of the in situ stress field to changes in pore pressure due to CO_2 injection. The main aims of this model were to assess cap-rock integrity and potential fault reactivation during CO_2 injection operation, and to infer the maximum safe injection pressures. The critical pore pressure at which failure occurs in the numerical model is 1.15 times greater than the original pore pressure, while the analytical model predicts values of 1.2 and 1.5 for constant and variable stress fields (Aruffo et al. 2014). This may be due to the fact that numerical models consider much greater complexity that leads to a more conservative estimation.

6 Calibration

The deformation predictions derived above determined areas of enhanced interest regarding the possible extent of deeper faults towards surface, as well as the sub-seismic strain contained in different layers. To compare our results (c.f. Fig. 1, right), visualization tools and new field measurements were realized.

6.1 Advanced Visualization

Three-dimensional visualization and interaction techniques are crucial to gain a better understanding of complex and heterogeneous data sets, especially in the field of geological reservoir characterization, as well as to support decision making processes. For the Otway Basin data set, we generated an integrated, interactive visualization that combines original geophysical data (i.e. seismic data, coherency and borehole data), the geological 3-D model as well as modeling results (from static and dynamic reservoir simulations) into an immersive presentation that can be shown on virtual reality (VR) display systems such as the UFZ's TESSIN VISLab, via mobile 3-D projectors or head-mounted displays (HMDs).

The agreed software protocol for data exchange between all partners is Petrel. To enhance the visualization data converting tools were developed and used to import data sets such as grids (Park et al. 2014) and borehole information from Petrel, 3-D seismics (Bilke 2014) from standard SEG-Y format, and geological horizons from GOCAD into the general-purpose scientific visualization software ParaView as well as the OpenGeoSys Data Explorer (Rink et al. 2013). Both of these are built using the graphics library VTK (The Visualization Toolkit) which provides a wide range of visualization algorithms and data analysis tools.

As a main achievement of the visualization analyses in the PROTECT project, the complete procedure for the integrated visualization of different types of data were developed including software components necessary for the interaction of the used software systems (for details see Zehner 2012). These tools are used in a virtual reality environment that allows interactive exploration, selection of interesting subsets, and modification of visualization parameters. The example for the visualization of geophysical data of the Otway site (Fig. 7) used the transparency tool. Thereby, a consistent tracing of elements for vertical, horizontal, and arbitrary slices enables the consolidation of the interpretation (c.f. sub-chapters 3, 4).

6.2 Reflection Seismic Shear-Wave Measurements

High-resolution validation of interpretation and predictions was gained in November 2013 by a near-surface 2-D shear wave (SH) reflection seismic survey. Using LIAG's equipment for SH-wave seismic acquisition (hydraulic vibrator MHV4S with source point spacing of 4 m, and SH-geophones mounted on a 240-m-long land streamer with 1 m spacing; see Krawczyk et al. 2013) five reflection seismic profiles were acquired (for details of seismic acquisition and processing see Beilecke et al. 2013, 2014a).

Here, we present profile PROTECT2 because it spans across the Buttress NW fault zone that can be traced in the large-scale, 3-D data set from reservoir up to ca. 350 m depth but not further upwards (c.f. Fig. 3). Data processing of this profile comprised geometry setup, elevation statics, surface-wave noise suppression via

Fig. 7 Interactive 3-D seismic visualization in the UFZ visualization centre: the transparent *grey-scaled* cross-sections allow a fast exploration, while *coloured contours and isolines* help to easily identify major seismic features

fk-filtering, velocity determination after DMO-correction, stacking, and post-stack depth migration (Beilecke et al. 2014a).

The ca. 1.5 km long profile PROTECT2 achieved good imaging quality down to ca. 750 m depth confidence level (Fig. 8), which is unexpectedly deep for SH-wave surveying.

The post-stack depth-migrated section shows several continuous reflectors from 10 m depth downwards and with variable amplitude at depth. From both the northern and the southern ends of the profile, these horizontal layers can be traced for ca. 450 m distance towards the profile centre. However, all of them are cut abruptly in the central part of the section (Fig. 8, 650–900 m distance, transparent trapezoid). This southern boundary of the weakly-imaged central part of the section is presumably the near-surface expression of the Buttress NW fault zone that can be imaged here with a vertical resolution of ca. 8 m in the upper part and ca. 10 m at 350 m depth (Fig. 8, red line). The northern limit of the less-resolved central part of the section is characterized by smaller, north-dipping events that are also faintly indicated in the variance cube of the 3-D data set. Thus, the new 2-D data ideally complement the 3-D exploration volume that has a resolution limit of ca. 20 m at 350 m depth.

The combined analysis of 3-D data variance and 2-D shear wave structure further suggests to interpret the Buttress NW fault zone as the southern boundary of a ca. 500 m wide zone with hard-linked splay faults that indicate some dextral movement along the fault.

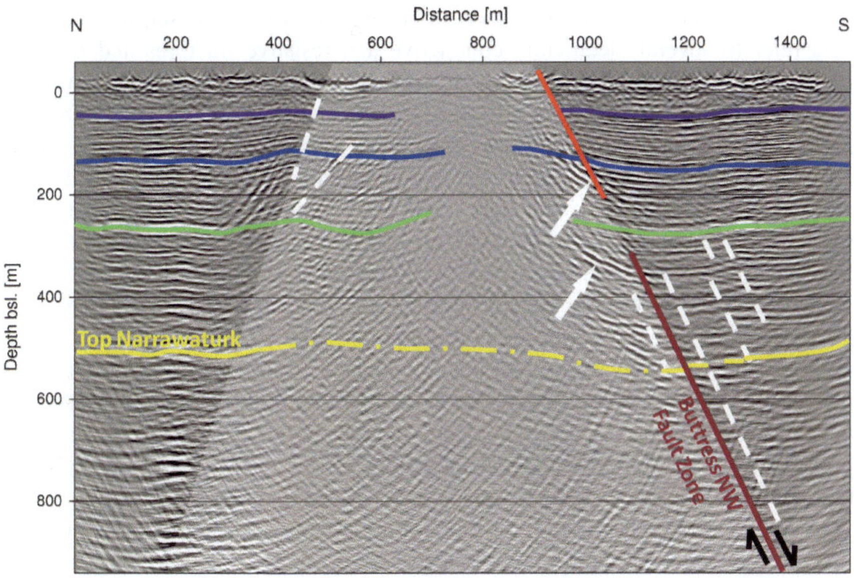

Fig. 8 Interpreted depth-migrated section of shear-wave reflection seismic profile PROTECT2 crossing the Buttress NW fault zone. The Top Narrawaturk reflector (*yellow*) is calibrated by the reflector interpreted in the 3-D seismic data. The trace of the almost vertical boundary of the Buttress NW fault is seen itself in the near-surface data (*red line*), and it continues the fault trace transferred from the exploration volume in the deeper part of the section (*dark red line*). *White dashed lines*-minor north- and south-dipping faults, *white arrows*-out-of plane features, *transparent area*-area of distorted reflectivity

7 Discussion

Tectonic features are important for reservoir characterization, since fractures and faults may represent predominant pathways for migration and possible leakage out of the reservoir. These features range from small-scale fractures to large-scale faults that are imaged either by 3-D seismic exploration data or in wells. The different sub-seismic approaches and steps constituting our workflow aim to decrease the sub-seismic resolution gap, and have shown for the Otway Basin case study that this is possible.

Having worked from the large towards the smaller scales, we finally compare our findings with the results from FMI data in the injection well CRC-1 (see Fig. 3 for location). From borehole breakouts, the present-day maximum horizontal stress direction was determined to be 142° (van Ruth and Nicol 2007). Together with nearby mini-frac and leak-off tests (van Ruth 2007; Tenthorey et al. 2010) it is confirmed that the Naylor Fault is presently in an extensional regime. This correlates with the results from our study.

In general, more features are observed in the 3-D seismic data set, while the well data refer to several electrically conductive and resistive fractures and only few faults (Fig. 9). However, a consistent trend is seen in all cases; they all show faults and fractures that dip between 50° and 60° SW and strike ESE–WNW. Only the 3-D seismics exhibits an additional component, which are antithetic faults that dip NE by 60°. This may be caused by the one-dimensional direction of the well profile and the fact that the FMI data only come from depths between 600 to 2,000 m (M. Lawrence, pers.com. 2014), while the seismics covers the 450–3,000 m depth interval. Another possible effect may be simple undersampling of well data.

Research at the onshore Ketzin pilot site in Germany (see also Martens et al. 2015) showed that the application of simple seismic attributes helps to identify more complex geological features than anticipated from the first interpretation of the seismic volume (Kling 2011), and suggested that multi-attribute processing as developed in this case study, would provide constraints for the geological interpretation and subsequent numerical modelling of, for instance, CO_2 arrival times (e.g. Kempka and Kühn 2013).

A hydrocarbon-related study in the North German Basin provided comparable first steps towards the workflow presented here (Lohr et al. 2008b), but validation with newly-acquired data for calibration was not possible. Thus, with the case study presented here, we can evidence the first time that the approaches of the workflow enable an enhanced understanding of a faulted area, and that it is capable of calibrating the predictions made.

The structural component provided by the new reflection seismic shear-wave measurements not only supplements the large-scale exploration data towards surface and provides a high-resolution near-surface image. In addition, the specific setup using the horizontally-polarized shear-wave component for source and receivers allows the derivation of the shear-wave velocity (Vs) from surface

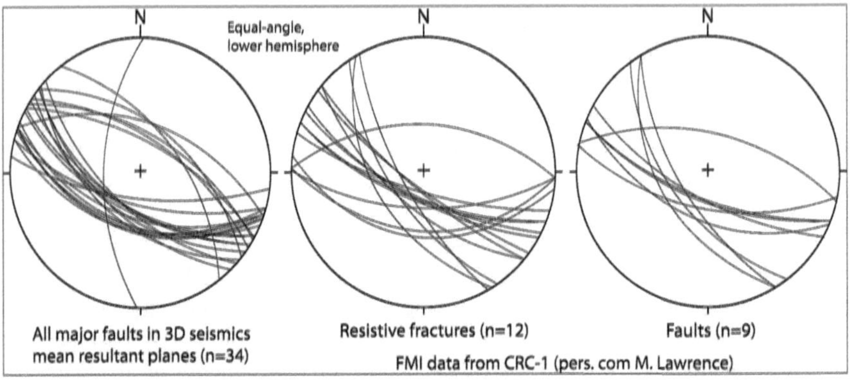

Fig. 9 Comparison of fault and fracture analyses based on 3-D seismic interpretation (c.f., Ziesch et al. 2015) and nearby well data (provided by M. Lawrence, pers.com. 2014; see Figs. 3 and 5 for location of well CRC-1); equal-area stereographic projection of planar data as great circles in the lower hemisphere

measurements and along profiles. Combining this with P-wave velocity (Vp) information determined from former surveys (e.g. Pevzner et al. 2011), we can provide petrophysical parameters away from the well, and also use it to populate geomechanical or other models. Exemplified for the PROTECT3 profile, slightly variable Vp/Vs (Fig. 10a) and the Poisson's Ratio (Fig. 10b) down to ca. 350 m confidence depth are defined. As expected, they follow in the first order the sedimentary layering with rather homogeneous values in vertical direction. The extremely high values in the uppermost layer (Fig. 10; −45 to −35 m) are due to the road construction and the weathering layer. The laterally-varying values will be analysed further for their robustness, and will contribute to better constrained geomechanical predictions in the future.

Since the shear-wave seismic survey was 2-D, sideswipe energy related to near-surface splay-fault reflections may have been projected onto the 2-D line, and thus distorting the image. However, the survey has locally greatly improved resolution, showed its capability as seismic validation tool at the CO2CRC project site, and forms the reference for future local high-resolution near-surface 3-D seismic

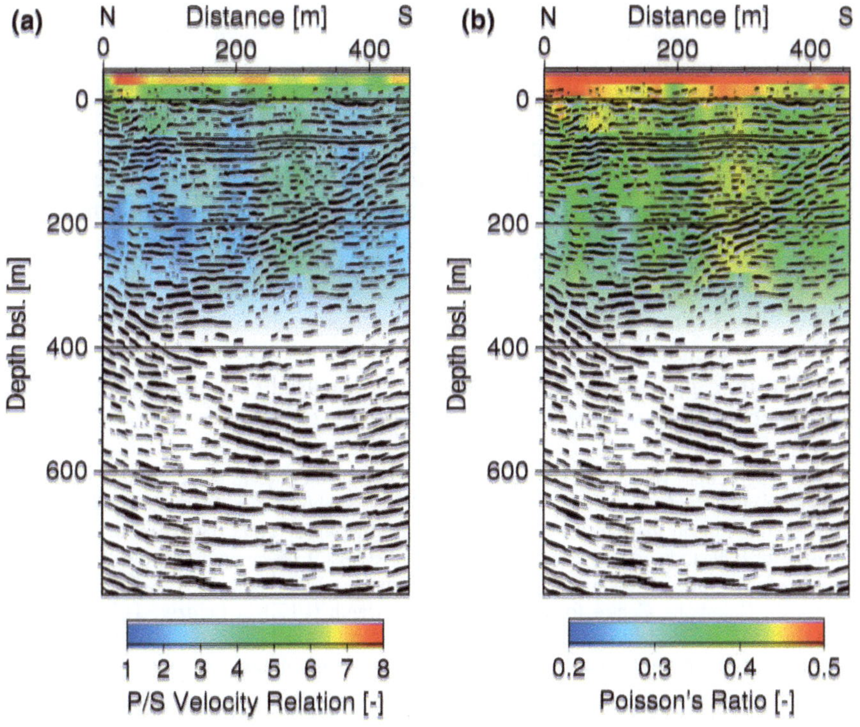

Fig. 10 Elastic parameters, derived from available and newly-gained seismic data, are *colour-coded* and overlain on the shear-wave seismic structural image of profile PROTECT3. The shear-wave velocities (Beilecke et al. 2014b) are combined with P-wave information (from Pevzner et al. 2011) to image **a** the velocity ratio, and **b** the Poisson's Ratio

structure mapping. This may be further supported by high-resolution elevation mapping for the detection of possible indicators for blind near-surface structures.

The independent sub-seismic approaches have been proved capable of narrowing the sub-seismic gap. Especially the high-resolution 3-D coherency and curvature processing including multi-attribute evaluation is a tool applicable also to large depths where structural resolution decreases. If combined iteratively with the structural interpretation, backward and forward modelling results are also stabilized.

8 Summary and Conclusion

The seismo-mechanical workflow of PROTECT considers both the physical data base for processing and modeling of seismic data, as well as the interpreted geological 3-D depth model for backward and forward modeling approaches. The independently-derived predictions of areas with enhanced sub-seismic strain were investigated by additional seismic measurements and advanced visualization tools. To this end the near-surface structural inventory is imaged at a higher resolution, physical properties for model population are provided from surface seismic measurements, and areas for monitoring are better defined which additionally supplements safety measures.

Our recommendation for a safe storage and monitoring of CO_2 thus comprises:

(a) deformation quantification using large-scale reconnaissance seismic surveys,
(b) characterisation of in situ tectonic stress distribution and magnitude,
(c) identification of potential areas with a high sub-seismic deformation risk,
(d) surveying in more detail using high-resolution seismics or other methods,
(e) setup of a both spatially and temporally variable monitoring plan.

We consider the whole approach to be a very powerful tool to categorise, validate and demonstrate the potential of existing pathways in the overburden of a reservoir. It is applicable to many tasks, and can be used to detail management and monitoring plans.

Acknowledgments This work was sponsored in part by the Australian Commonwealth Government through the Cooperative Research Centre for Greenhouse Gas Technologies (CO2CRC). PROTECT is funded through the Geotechnologien Programme (grant 03G0797) of the German Ministry for Education and Research (BMBF). The PROTECT research group consists of Leibniz Institute for Applied Geophysics (LIAG) in Hannover (coordinator), Technical University Darmstadt, Helmholtz Centre for Environmental Research (UFZ) in Leipzig, TEEC GmbH in Isernhagen (all in Germany) and Curtin University in Perth, Australia. We thank Midland Valley Exploration Ltd for their software suite Move2012 and for discussion.

References

Aruffo CM, Rodriguez Herrera A, Tenthorey E, Krzikalla F, Minton J, Henk A (2014) Geomechanical modelling to assess fault integrity at the CO2CRC Otway Project, Australia. Aust J Earth Sci 61:987–1000. doi: 10.1080/08120099.2014.958876

Beilecke T, Tanner D, Singh R (2013) Operations closeout report LIAG survey. CO2CRC Publication RPT13-4785, pp 27

Beilecke T, Krawczyk CM, Tanner DC (2014a) Seismic program closeout report—shallow shear wave data acquisition survey. CO2CRC Publication RPT14-4950, pp 73

Beilecke T, Krawczyk CM, Tanner DC, Ziesch J, PROTECT Research Group (2014b) Poisson's ratio model derived from P-and S-wave reflection seismic data at the CO2CRC Otway Project pilot site, Australia. In: Proceedings EGU conference Vienna/Austria 16:EGU2014-5721

Bilke L (2014) SimpleSeismicReader—a para view reader plugin for simple seismic files exported from OpendTect, doi:10.5281/zenodo.10509

Dodds K, Daley T, Freifeld B, Urosevic M, Kepic A, Sharma S (2009) Developing a monitoring and verification plan with reference to the Australian Otway CO_2 pilot project. TLE 28(7):812–818. doi:10.1190/1.3167783

Endres H, Lohr T, Trappe H, Samiee R, Thierer PO, Krawczyk CM, Tanner DC, Oncken O, Kukla PA (2008) Quantitative fracture prediction from seismic data. Pet Geosci 14(4):369–377. doi:10.1144/1354-079308-751

Fischer K, Henk A (2013) A workflow for building and calibrating 3-D geomechanical models—a case study for a gas reservoir in the North German Basin. Solid Earth 4:1–9

Gazar AH, Javaherian A, Sabeti H (2011) Analysis of effective parameters for semblance-based coherency attributes to detect micro-faults and fractures. J Seismic Explor 20:23–44

Jenkins CR, Cook PJ, Ennis-King J, Undershultz J, Boreham C, Dance T, de Caritat P, Etheridge DM, Freifeld BM, Hortle A, Kirste D, Paterson L, Pevzner R, Schacht U, Sharma S, Stalker L, Urosevic M (2011) Safe storage and effective monitoring of CO_2 in depleted gas fields. Proc Natl Acad Sci 109(2):E35–E41

Kempka T, Kühn M (2013) Numerical simulations of CO_2 arrival times and reservoir pressure coincide with observations from the Ketzin pilot site, Germany. Environ Earth Sci 70(8):3675–3685. doi:10.1007/s12665-013-2614-6

Kling C (2011) Structural interpretation and application of spectral decomposition for facies analysis of three-dimensional reflection seismic data at the Ketzin CO_2 storage site. MSc thesis, Technische Universität, Berlin, p 123

Krawczyk CM, Tanner DC (2010) Subseismic deformation analysis—a prediction tool for a safe CO_2-reservoir management. In: Proceedings EAGE CO_2 workshop, Berlin/Germany. http://www.earthdoc.org/detail.php?pubid=38428

Krawczyk CM, Tanner DC, Henk A, Trappe H, Urosevic M (2011) Sub-/seismic deformation prediction—development of a new seismo-mechanical workflow in the Otway Basin. In: Proceedings SES conference. Valencia, Spain, pp 1–5. www.earthdoc.org

Krawczyk CM, Polom U, Beilecke T (2013) Shear-wave reflection seismics as valuable tool for near-surface urban applications. Lead Edge 32(3):256–263. doi:10.1190/tle32030256.1

Krawczyk CM, Henk A, Tanner DC, Gurevich B, PROTECT Research Group (2015) Prediction of deformation to ensure carbon traps (PROTECT)—a new seismo-mechanical workflow to predict seismic and sub-seismic deformation. Int J Greenhouse Gas Control subm

Lewicki JL, Birkholzer J, Tsang CF (2007) Natural and industrial analogues for leakage of CO_2 from storage reservoirs: identification of features, events, and processes and lessons learned. Environ Geol 52:457–467

Lohr T, Krawczyk CM, Oncken O, Tanner DC (2008a) Evolution of a fault surface from 3D attribute analysis and displacement measurements. J Struct Geol 30(6):690–700. http://dx.doi.org/10.1016/j.jsg.2008.02.009

Lohr T, Krawczyk CM, Tanner DC, Samiee R, Endres H, Thierer PO, Oncken O, Trappe H, Bachmann R, Kukla PA (2008b) Prediction of sub-seismic faults and fractures-integration of 3D seismic data, 3D retrodeformation, and well data on an example of deformation around an inverted fault. AAPG Bull 92(4):473–485. http://dx.doi.org/10.1306/11260707046

Martens S, Conze R, De Lucia M, Henninges J, Kempka T, Liebscher A, Lüth S, Möller F, Norden B, Prevedel B, Schmidt-Hattenberger C, Szizybalski A, Vieth-Hillebrand A, Würdemann H, Zemke K, Zimmer M (2015) Joint research project CO_2MAN (CO_2MAN reservoir management): continuation of research and development work for CO_2 storage at the Ketzin pilot site. This volume

Norvick M, Smith MA (2001) Mapping the plate tectonic reconstructions of southern and southeastern Australia and implications for petroleum system. APPEA J 41:15–35

Park C-H, Shinn YJ, Park Y-C, Huh D-G, Lee SK (2014) PET2OGS: algorithms to link the static model of petrel with the dynamic model of OpenGeoSys. Comput Geosci 62:95–102. doi:10.1016/j.cageo.2013.09.014

Pevzner R, Shulakova V, Kepic A, Urosevic M (2011) Repeatability analysis of land time-lapse seismic data: CO2CRC Otway pilot project case study. Geophys Prospect 59(1):1–12. doi:10.1111/j.1365-2478.2010.00907.x

Raistrick M (2008) Carbon capture and storage projects to challenge governments, scientists and engineers. First Break 26(1):35–36

Rink K, Fischer T, Selle B, Kolditz O (2013) A data exploration framework for validation and setup of hydrological models. Environ Earth Sci 69(2):469–477

Roberts A (2001) Curvature attributes and their application to 3D interpreted horizon. First Break 19(2):85–100

Rogers C, van Ruth PJ, Hillis RR (2008) Fault reactivation in the Port Campbell Embayment with respect to carbon dioxide sequestration, Otway Basin, Australia. Geol Soc London 306:201–214. doi:10.1144/SP306.10

Sharma S, Cook P, Jenkins C, Steeper T, Lees M, Ranasinghe N (2011) The CO2CRC Otway project: leveraging experience and exploiting new opportunities at Australia's first CCS project site. Energy Procedia 4:5447–5454

Tenthorey E, John Z, Nguyen D (2010) CRC-2 extended leak-off and mini-frac tests: results and implications. CO2CRC Report RPT10-2228, pp 32

Trappe H, Hellmich C (2003) Seismic volume attributes for fracture analysis. Geol Soc London Spec Publ 209:65–75

van Ruth P, Rogers C (2006) Geomechanical analysis of the Naylor structure, Otway Basin, Australia. CO2CRC Report RPT06-0039, pp 27

van Ruth P (2007) CRC-1 extended leak-off test report. CO2CRC Report RPT07-0608, p 8

van Ruth P, Nicol A (2007) Structural interpretation of the CRC-1 resistivity image log. CO2CRC Report RPT07-0962, pp 34

Vidal-Gilbert S, Tenthorey E, Dewhurst D, Ennis-King J, van Ruth P, Hillis R (2010) Geomechanical analysis of the Naylor Field, Otway Basin, Australia: implications for CO_2 injection and storage. Int J Greenhouse Gas Control 4(5):827–839

Willamson PE, Swift MG, O'Brien GW, Falvey DA (1990) Two-stage early Cretaceous rifting of the Otway Basin margin of southeastern Australia: implications for rifting of the Australian southern margin. Geology 18(1):75–78

Zehner B (2012) Geometric modelling, gridding and visualization. In: Kolditz O, Görke U-J, Shao H (eds) Benchmarks and examples for thermo-hydro-mechanical/chemical processes in porous media. Springer, Berlin

Ziesch J, Tanner DC, Krawczyk CM (2014) Strain associated with the fault-parallel flow algorithm during kinematic fault displacement. Math Geosci 46(1):59–73. doi:10.1007/s11004-013-9464-3

Ziesch J, Aruffo CM, Tanner DC, Beilecke T, Dance T, Henk A, Weber B, Tenthorey E, Lippmann A, Krawczyk CM (2015) Geological structure of the CO2CRC Otway project pilot site, Australia: fault kinematics based on quantitative 3D seismic interpretation. Basin Research (in review)

Zweigel P, Arts R, Lothe AE, Lindeberg EBG (2004) Reservoir geology of the Utsira Formation at the first industrial-scale underground CO_2 storage site (Sleipner area, North Sea). Geol Soc Spec Publ 233:43–58

Long-Term Safety of Well Abandonment: First Results from Large Scale Laboratory Experiments (COBRA)

Frank R. Schilling, Andreas Bieberstein, Jörg-Detlef Eckhardt, Michael Haist, Astrid Hirsch, Steffen Klumbach, Marco Kromer, Josephin Mühlbach, Birgit I.R. Müller, Harald S. Müller, Thomas Neumann, Stefan Schläger and Theodoros Triantafyllidis

Abstract A long-term safe and reliable abandonment of wells is a crucial prerequisite for a secure abandonment of underground storage sites, while considering the long-lasting environmental impact, e.g. leakage through wells. To study the impact of the cementation on the tightness of wells, both, the cementation of a well during completion and abandonment are investigated in Full-scale laboratory experiments. Different autoclaves from small to Full-scale have been developed to test cementations under various conditions and to perform long-term tests under in situ conditions of a CO_2 storage site. The experiments show that the surface-texture (e.g. roughness) of the drilled well has a significant influence on the formation of mud-channels—for rough surfaces, up to 75 % of the serrations consist of non-displaced mud and only 25 % are well hardened cement, creating possible leakage pathways. Even under idealized cementation conditions using a cement recipe characterized by a very low shrinkage, micro-annuli are formed. These micro-annuli are connected throughout the whole oil-field casing in the Full-scale experiments for which the widths of the micro-annuli in the order of 10–20 μm could be deduced. Along the micro-annuli the cement is carbonated due to the flow of CO_2-bearing fluid. The fluid flow could also be verified by a Spatial Time-Domain-Reflectometry (TDR) setup embedded in the cementation, which was successfully tested as an in situ monitoring system. In the Full-scale experiments, chemical reactions in the system casing—cement—rock—fluid were examined. The geochemical analyses during and after the experiment show variations in pH, conductivity and chemical composition of the brines, which are well described by an interplay of corrosive processes and precipitations of carbonate minerals.

F.R. Schilling (✉) · A. Bieberstein · J.-D. Eckhardt · M. Haist · A. Hirsch · S. Klumbach · M. Kromer · J. Mühlbach · B.I.R. Müller · H.S. Müller · T. Neumann · S. Schläger · T. Triantafyllidis
Karlsruhe Institute of Technology (KIT), Kaiserstr. 12, 76131 Karlsruhe, Germany
e-mail: frank.schilling@kit.edu

A. Liebscher and U. Münch (eds.), *Geological Storage of CO_2 – Long Term Security Aspects*, Advanced Technologies in Earth Sciences,
DOI 10.1007/978-3-319-13930-2_6

1 Introduction

The development of strategies for a sustainable and safe reduction of greenhouse gas emissions to the atmosphere is one of the major challenges of this century. Geological CO_2 storage has been identified as one of the most promising technologies to effectively reduce anthropogenic greenhouse gas emissions to the atmosphere (IPCC 2005). However, the sustainability of the technology strongly depends on the quality of the leak tightness of a storage site—especially after the abandonment of the site. Migration of CO_2 along wells has been identified as one of the major leakage pathways for CO_2 storage sites (e.g. Koorneef et al. 2012). Ensuring a highly durable, long lasting well abandonment is therefore considered as an important task for a long-term safe storage of CO_2. According to Kühn et al. (2013) a proper well abandonment has to prevent

- all physical hazards potentially induced by the well,
- any migration of contaminants between various formations and,
- the possibility of hydrologic communication between originally separated aquifer systems.

In this context the quality of the cemented well bore is of particular interest. For both, prior to CO_2 injection (Nelson and Guillot 2006) and as a part of the abandonment (Randhol et al. 2007), a cement slurry is pumped through the casing downward and displaces the drill-mud and further fluids present in the borehole. The durability of a hydraulic sealing strongly depends on the success of this displacement (Jamot 1974). Different cementation flaws may arise in the displacing process, especially sections with remaining mud, so called mud channels, may significantly impede the leak tightness of the well (Brice and Holmes 1964). The effects of interactions between cement and mud during the displacement process, especially the combination of rheology and density differences between cement and mud are not yet sufficiently understood (Abdu et al. 2012; Jakobsen et al. 1991; Lajeunesse et al. 1999; Nguyen et al. 1992). Micro-annuli or gaps, which form due to fracturing of the cement, can form further effective pathways for gas migration and emission (Fig. 1).

Besides the test of abandonment strategies at field test sites such as at Ketzin, various laboratory experiments have been performed to test the suitability of different materials for abandonment (e.g. Hirsch et al. 2013; Kamali et al. 2008; Lesti et al. 2013; Nasvi et al. 2014). In this context, the degradation of various cement-recipes has been tested in small-scale experiments. Other studies were carried out either in small scale autoclaves (Marbler et al. 2012) or under non-realistic pressure and temperature conditions for storage sites (Druckenmiller and Maroto-Valer 2005). However, the long-term safe abandonment for CO_2 storage sites does not only depend on the long-term resistance of the single materials (rock, cement, steel), but also—and possibly much more pronounced—on the physical, chemical and mechanical interaction of the different materials of the seal in the vicinity of a CO_2-rich brine. With our real scale experiments, the time dependent interaction of

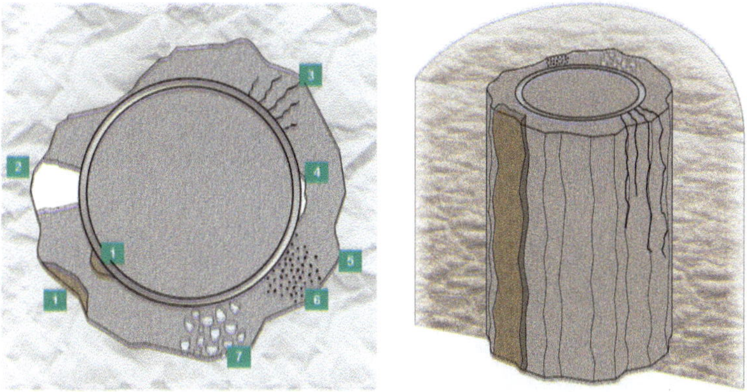

Fig. 1 Backfilled borehole with casing and cementation flaws in cross-section (*left*) and in longitudinal section (*right*) with various defects: *1* and *2* channels filled with and without mud, which can form during cementation, *3* cracks, *4* gap, *5* micro-annuli, *6* higher permeable area and *7* invaded gas (Kromer et al. 2014)

rock, cement and casing in contact with CO_2-rich brine can be studied in Full-scale under quasi in situ conditions of a reservoir.

Within the COBRA project (Full-scale borehole—simulator) different experiments have been performed in order to assess abandonment strategies. As the processes leading to a leakage of the borehole may act on different spatial scales, experiments in small to Full-scale autoclaves are performed to quantify the transport and transformation processes in abandoned wells. The thus obtained results allow quantifying of different leakage pathways of storage sites.

2 Experimental

2.1 Small-Scale Autoclave

The Small-scale autoclave system consists of four separate small autoclaves, each 0.99 dm^3 of volume with a pressure generator and a control panel (Fig. 2). The autoclaves are operated independently and can be heated up to 150 °C. Samples of hardened cement or rock are placed on polytetrafluoroethylene (PTFE) racks within the pressure vessels (Fig. 2 right). Brine (up to 230 g/dm^3 NaCl) or pure water were filled into the vessels. The autoclaves can be operated at pressures up to 300 bar (30 MPa), pressure is built up with CO_2 with a constant rate of up to 40 bar/h. During the experiments, continuous temperature and pressure control is ensured. The experimental conditions as well as the CO_2 supply are recorded digitally. The key advantage of this autoclave system lies in the possibility to investigate the chemical processes inside the autoclaves during the experiments by fluid sample extraction and the option to replace the attack medium during tests under nearly

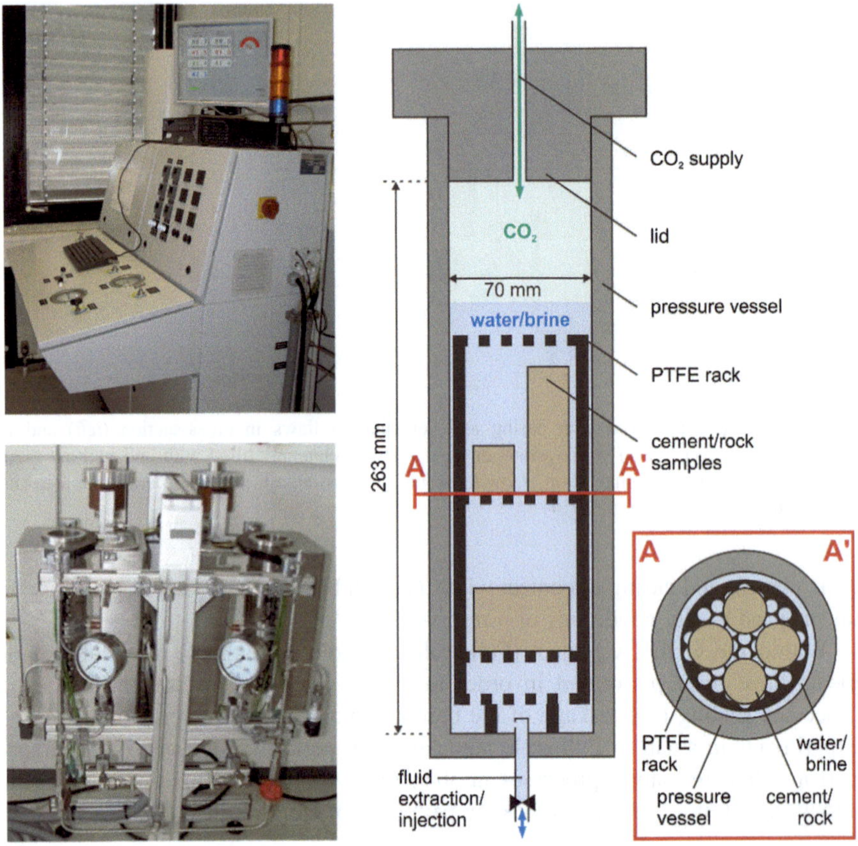

Fig. 2 Small autoclave system; pressure generator with control panel (*top left*); frame with two separate autoclaves (*bottom left*); schematic sectional view of a filled autoclave (*right*)

unchanged pressures and temperature conditions (pressure variations <10 bar, temperature variations <5 K).

Different cement mixtures, all based on ClassG well cement, have been studied with regard to their corrosion behaviour in the autoclave system. An overview of the main constituents is given in Table 1. Details regarding the investigated raw materials, mixtures and experimental techniques can be found elsewhere (e.g. Hirsch et al. 2013).

2.2 Full-Scale Autoclave

To investigate the long-term integrity of well bore abandonments at Full-scale and under reservoir conditions a large-scale borehole simulator has been developed. This facility is designed to test standard oilfield casings with a diameter of 5.5″

Table 1 Composition of cement mixtures

Mixtures	Raw materials						
	ClassG well cement	Water	Salt	Quartz powder	Black coal fly ash	Super-plasticizer	Latex dispersion
ClassG	+	+	−	−	−	−	−
ClassG+Salt	+	+	+	−	−	−	−
SaltCem24	+	+	+	−	−	+	+
SaltCem36	+	+	+	−	−	+	+
MWC-FA	+	+	−	+	+	+	+
MWC-1	+	+	−	+	−	+	+
MWC-2	+	+	−	+	−	+	+
MWC-01-A	+	+	−	+	−	+	+

+ incorporated in the mix; − not incorporated in the mixture

(140 mm) and a length of up to 10 m. Experiments can be performed at pressures and temperatures up to 100 bar and 100 °C, respectively. First experiments were carried out with a well casing of 2.5 m in length.

The Full-scale autoclave system (Fig. 3) comprises of three parts: (i) a lower reaction chamber (lower vessel) to investigate the alteration processes within the cement/steel/rock/fluid system, (ii) the connecting (cemented) oilfield casing, and (iii) an upper reaction chamber (upper vessel) to study processes of cementation and transport within the well casing. The pressure in the lower and upper vessel can be controlled independently, allowing for the establishment of a defined pressure gradient through the cemented oilfield casing.

The lower vessel has a volume of 30 dm^3 and can be filled with water, brine, CO_2, rock, and cement samples or a combination of the prior mentioned media. The fluid in the lower vessel can be circulated through a sampling chamber with a flow rate of 0–10 l/h. The sampling chamber has a volume of 250 cm^3 and can be removed from the circulation system using a bypass. The set-up enables for an in situ fluid sampling during the experiment.

The metal parts in the autoclave system are protected from corrosion by a polyetheretherketone (PEEK) coating, except for the materials with are intentionally to be corroded during the experiment, such as the well casing. The lower reaction chamber is connected to the upper vessel through a cemented standard oilfield casing (5.5″). The upper reaction chamber is partially filled with tab-water, to prevent dehydration of the cement plug, which can create additional cracks. The upper vessel can be pressurized with gas, in order to stabilize the plug and adjust a defined pressure gradient between the lower and upper vessel.

Potential upward migration processes of carbon dioxide through leakage pathways in the cemented borehole casing is determined by recording the temporal development of the pressure in the upper vessel. The pressure in the upper vessel was not controlled whereas the pressure in the lower vessel was kept constant. Rising carbon dioxide concentrations within the upper vessel are monitored by gas

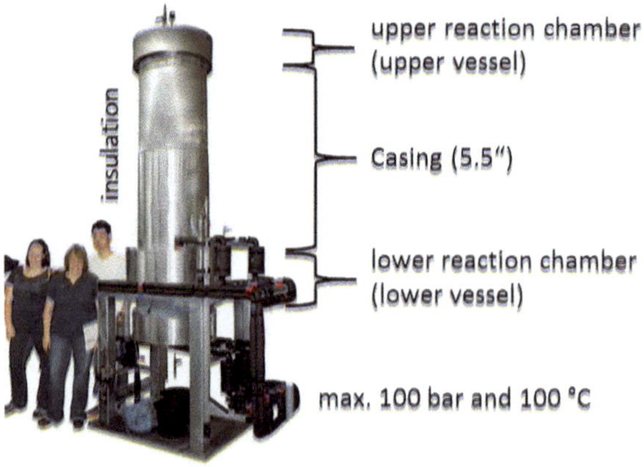

Fig. 3 Full-scale autoclave consisting of two parts: A lower vessel to investigate the alteration processes within the cement/steel/rock/fluid system, and an upper vessel to study processes of cementation and transport within the well casing. The setup can be operated up to 100 °C and 100 bar

sampling from the upper vessel. Besides thermocouples and pressure gauges, a Time-Domain-Reflection-Monitoring (TDR) system is implemented into the autoclave to monitor the migration of CO_2 (see below).

2.3 Borehole Cementation Simulator

In order to investigate formation mechanisms of cementation flaws, the borehole geometry as well as the topography of the borehole surface and the rheology of the fluids and cement suspension were varied systematically. The borehole-cementation-simulator (BCS) developed within the COBRA project is a replica of a bottom 11 m section of a borehole in scale 1 to 1. The simulator consists of cylindrical segments (cap-rock segments) with a height of 1.20 m and a diameter of 500 mm each, which can be stacked to a total height of approximately 11 m (Fig. 4). The inner void in the segments represent the borehole. The inner surface of the segments is designed to represent the surface topography (especially its roughness) of the geological formation, in which a borehole is drilled. In order to ensure a defined roughness and repeatable geometry and topography of the borehole, artificial cap-rock segments made out of a low thermal strain concrete are used (Fig. 5). A polypropylene matrix is inserted in a formwork for casting artificial cap-rock segments. After hardening of this artificial cap-rock, the matrix is removed, providing a model-borehole with a defined and reproducible geometry.

Within this study, the mean inner diameter of the cap-rock—i.e. the diameter of the artificial borehole—is fixed to $7^5/_8$″ (200 mm). The artificial borehole is contoured with circumferential serrations with a defined depth d and a defined height

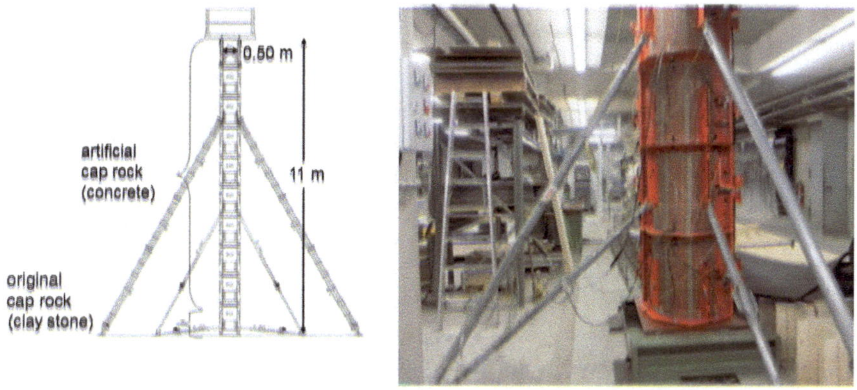

Fig. 4 Sketch of the borehole-cementation-simulator (*left*) and the two segments at the bottom without insulation (*right*)

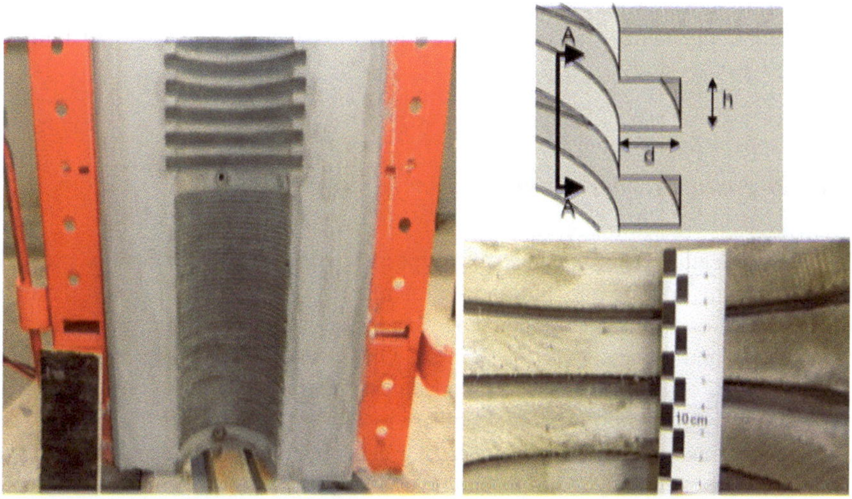

Fig. 5 One semi-ring of the borehole-cementation-simulator (Fig. 4) with a borehole diameter of 200 mm ($7^5/_8''$), circumferential serrations with a depth and height of 5 mm in the lower and 20 mm in the upper part, respectively (*left*). Sketch of serrations with height *h* and depth *d* (*right* at *top*) and magnified section of the surface in the longitudinal section A–A (*right* at the *bottom*)

h (Fig. 5). In the first step *h/d* ratios between 5 mm/5 mm and 40 mm/40 mm are chosen, corresponding to typical borehole conditions (Lux et al. 2012). As all cap-rock rings are dividable into semi-rings along their longitudinal axis, the rings are not able to take up radial forces. Hence, the rings are surrounded by a 5 mm thick steel formwork to adsorb forces and guarantee sufficient sealing. Furthermore, the formwork can also be equipped with a drilled core of formation rock.

Each artificial cap-rock segment is equipped with up to eight temperature and four pressure sensors to monitor the cementation process. The sensors are positioned in the artificial cap-rock, with the sensor surface being in line with the surface of the borehole, allowing a direct contact to the fluids. To simulate in situ conditions, each segment can be heated up to a temperature of max. 100 °C. The applied eight heating and control circuits allow the simulation of constant temperatures or heat gradients over the height of the borehole. The lowest part of the borehole-cementation-simulator can be placed on a Full-scale autoclave (under construction).

In the unfilled artificial borehole, different kinds of monitoring instruments such as pressure p or temperature T-sensors can be installed. After this, the cementation of the annulus between casing and the formation rock is simulated (scenario 1). In a second set-up, the borehole is completely filled and sealed with a cement suspension, simulating abandonment (scenario 2).

The simulator is limited to a maximum pressure of up to 2.5 bar and therefore able to take the hydraulic pressure of the cement suspension during cementing. As the fluid dynamic of the drill-mud and cement-suspension is mainly affected by temperature, the much smaller isostatic pressure effect on the flow behavior is neglected in this study.

2.4 Corrosion Experiments

The experimental set-up includes four types of potential reservoir rocks and one seal rock from Triassic and Jurassic sediments. These rocks are a Lower Triassic Bunter Sandstone, three different Keuper Sandstones (Carnian, Norian and Rhaetian Sandstones) and as seal rock the Jurassic Opalinus-Clay. The reason to use the different types of rocks is to evaluate specific alteration effects caused by different mineralogical compositions and to compare results with other working groups, e.g. CO_2—brine interactions studied by Marbler et al. (2012) and Fischer et al. (2013). The rocks are cut to a cylindrical shape of 10 cm length and 5 cm in diameter and are used for petrophysical investigations. Before starting the experiments, all samples were mineralogically, geochemically and petrophysically characterised.

Fluids were analysed immediately after sampling during the experiments (in situ sampling—for details see above) for various hydrochemical parameters (temperature, pH, electrical conductivity and alkalinity). Small portions of the fluid are extracted for major cations (inductively coupled plasma optical emission spectrometry—ICP-OES) major anions (ion chromatography—IC) and H- and O-isotope (cavity ring-down spectroscopy—CRDS) measurements. To study alteration processes, rock samples were taken from the reaction chamber at the end of the experiments to analyse their mineralogical and chemical composition by X-ray diffraction (XRD) and X-ray fluorescence (XRF), respectively. Optical microscopy on thin sections in combination with μXRF as well as by using powders from a micromill preparation make it possible to observe geochemical processes on a μm-scale.

In the first experiments, the Bunter Sandstone and the Norian Sandstone were considered as reservoir rocks. Together with PTFE sticks (to protect the coating of the reaction chamber), 15 cylinders of one type of the rocks were placed into the lower vessel of the Full-scale autoclave which was flooded with brine prior to the pressurizing with CO_2. The synthetic brine was mixed from tab water and food quality sodium chloride with a concentration of 50 g/l NaCl and initial concentrations of 75 mg/l Ca^{2+}, 40 mg/l K^+, and 47 mg/l Mg^{2+}.

Two experiments have been performed. Experiment 1 is subdivided in three phases. In the *first phase* the heating system was tested to maintain constant temperature conditions within the simulator of 55 °C at ambient pressure (without CO_2) within the entire autoclave. During this experiment, the lower reaction chamber was filled with ca. 20 l of brine and 15 cylindrical rock samples of the Norian Sandstone. The fluid was sampled daily over 13 days. In the *second phase* (30 days), the upper vessel was pressurized in three steps [2 bar (2 days), 4 bar (3 days) and 6 bar (15 days)] with compressed air to perform a permeability experiment. For the *third phase* of experiment 1 the brine was exchanged. The pressure was increased to 60 bar by CO_2 injection into the lower vessel at a rate of 1–2 bar/h. Simultaneously, the pressure in the upper vessel was increased by nitrogen injection to maintain a constant pressure in the upper and lower vessel. During the third phase, fluids were sampled over a period of 17 days from the autoclave. About 180 ml fluid are required for the analyses. Water was refilled to keep the water table constant. Two additional fluid samples were extracted after finishing the experiment and the rock samples were removed from the reaction chamber.

The second experiment was carried out at slightly higher temperature (70 °C) and pressure (80 bar). Bunter Sandstone was used as reservoir rock. The brine composition was the same as in the first experiment. Eight fluid samples were extracted during the experiments directly from the autoclave. The pressure of the upper vessel was systematically varied to perform permeability experiments.

3 Time Domain Reflectometry (TDR) for Monitoring the Integrity of Cemented Casing

Spatial TDR has been developed originally to determine soil moisture profiles using variations of dielectric properties of involved components (soil, water, air; Schlaeger 2005). Within this study TDR is tested as a monitoring technology for the surveillance of deep wells, e.g. abandoned wells for carbon dioxide storage sites. The focus is to quantitatively observe and detect both, defects in the cemented plug and the migration of CO_2 through the cemented casing. Therefore, different laboratory tests were performed in order to identify dielectric properties of relevant cements and to investigate the ability of TDR in detection of imperfections in a cementitious vicinity.

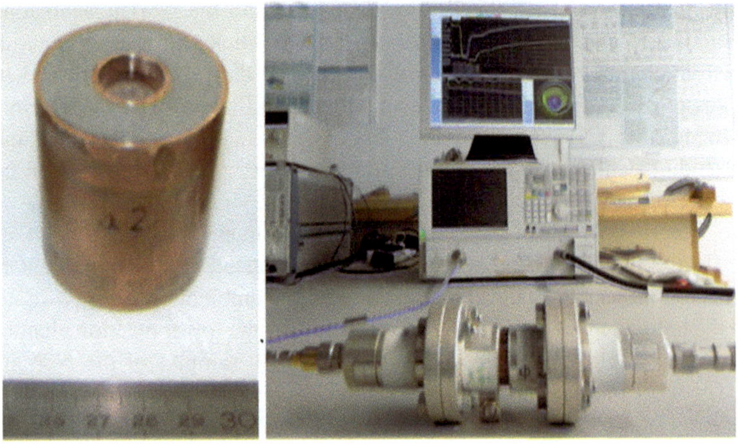

Fig. 6 Measurement of the dielectric properties at various frequencies of relevant cements by means of small coaxial samples (*left*) with a network analyser (*right*)

3.1 Dielectric Material Properties of Various Cements

The complex dielectric permittivity $\varepsilon = \varepsilon' + i\,\varepsilon''$ of the investigated carbonated and non-carbonated cements was examined in the frequency range between 1 MHz and 5 GHz at room temperature and under atmospheric pressure with a network analyzer (VNA, Agilent E5061B ENA). Prior to the dielectric measurements, a mechanical calibration kit (Agilent 85032F) is used to calibrate the VNA measurement port(s) to effectively eliminate systematic errors and imperfections from the experimental setup. For the accurate determine of dielectric properties further calibration and algorithms are used to calibrate the sample cell and the adapters (Chen et al. 2014); setup shown in Fig. 6.

3.2 Investigations for Testing the Ability of Spatial TDR in Detection of Imperfections in a Cementitious Vicinity

Based on the different types of potential defects in backfilled boreholes with casing (Fig. 1) by means of large-scale laboratory tests, the sensitivity of Spatial TDR for defect detection has been investigated. For this purpose, artificial defects were placed in cemented pipes with a diameter of 150 mm and length between 1.2 and 5.5 m (cf. Fig. 7). The tests were performed at ambient temperature and pressure conditions with analogue materials (e.g. air, styrofoam, glass beads, glass-beads-water mixtures—Fig. 7) to simulate defects with dielectric properties similar to the high *p*-*T* conditions of reservoirs.

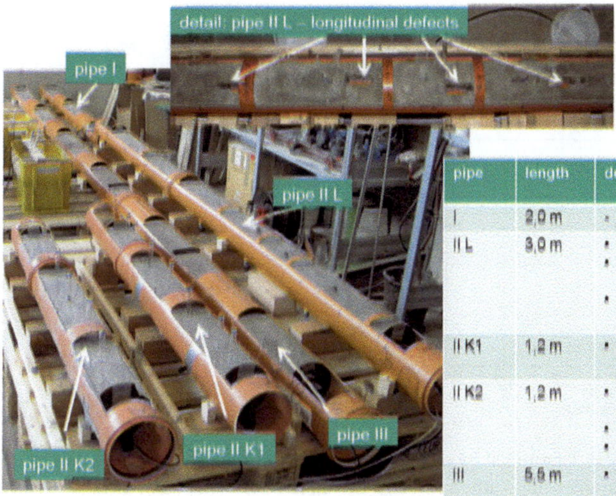

pipe	length	defect	simulation purpose
I	2,0 m	-	concrete
II L	3,0 m	• styrofoam • longitudinal defects • test tubes	• air in void • material inclusions • specific inclusions
II K1	1,2 m	• test tubes	• specific inclusions
II K2	1,2 m	• glass-spheres • test tubes • drills	• gravel bed • specific inclusions
III	5,5 m	• lateral defects	• material inclusions

Fig. 7 Large scale test setup for defect determination with spatial TDR

4 Results and Discussion

4.1 Cement Alteration

The cement alteration studies are based on prior studies (Hirsch et al. 2013; Kromer et al. 2014); out of eight types of cement suspensions (see Table 1), results for three are presented here. Mixtures ClassG and MWC-2 were examined in the small-scale autoclaves. ClassG contains an API ClassG HSR Well Cement at a water/cement-ratio of 0.4 (reference material). In MWC-2 (MWC stands for "modified well cement") a total of 46 wt% of API ClassG cement was replaced by quartz powders with defined particle size gradation curves at a water/cement-ratio of 0.48. The MWC-01-A-cement that has been used in the Full-scale autoclave experiments is a development based on MWC-2. MWC-01-A and MWC-2 differ in the amount of additives as well as in the used quartz powders and thereby in their particle size gradation curves. Details regarding the composition of the mixes can be found in Hirsch et al. (2013) and Kromer et al. (2014).

The experimental conditions in the small scale-autoclaves (samples submerged in carbonic acid at 40 °C and 90 bar for 28 days) led to a strong cement alteration in the contact zone between hardened cement and carbonic acid (named carbonation in the following) resulting in e.g. changes of petrography, pore space distribution or pH-value of the pore fluid (Fig. 8). After the experiment, X-ray powder diffraction shows considerable amounts of newly formed calcium carbonates (Calcite, Aragonite and Vaterite) especially in the ClassG samples.

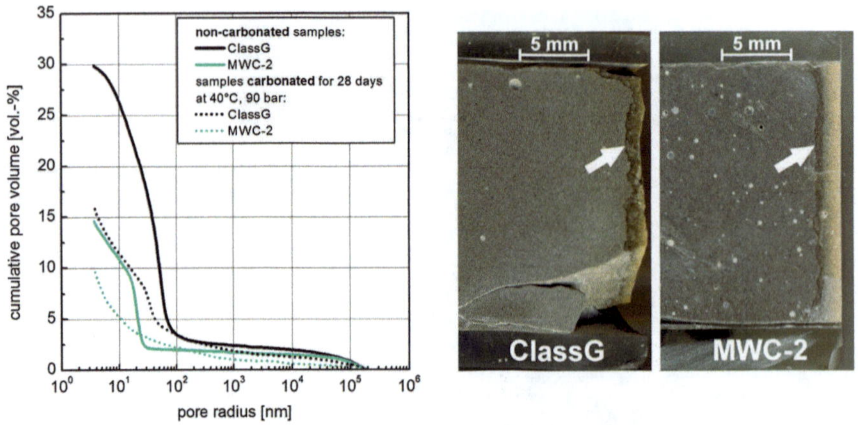

Fig. 8 Pore space distribution determined by mercury intrusion porosimetry (*MIP*) of ClassG- and MWC-2-cement samples; samples non-carbonated and carbonated for 28 days at 40 °C and 90 bar (*left*); scan of ClassG- and MWC-2-cement samples carbonated for 28 days at 40 °C and 90 bar, carbonation front indicated by *white arrows* (*right*)

The porosity of a non-carbonated ClassG-sample is comparatively high (30 vol. %) whereas that of a MWC-2-sample is only 15 vol.%. The 28 days carbonation experiment reduces both porosities, that of ClassG to only 15 vol.%, that of MWC-2 to 10 vol.% (Fig. 8 left).

The right side of Fig. 8 shows carbonated samples of hardened ClassG-cement and MWC-2-cement. The carbonation front is clearly visible (white arrows). In both samples the front has reached a depth of 1–2 mm after 28 days. In contrast to ClassG, MWC-2 shows a more uniform front and a slightly higher carbonation depth.

Complementary to the Small-scale autoclave tests, first investigations were carried out on the carbonation of a cement core (MWC-01-A) in the Full-scale Autoclave. The top of Fig. 9 shows a cut of the open casing and the cement core after experiment 1 (56 days at a max. temperature of 70 °C and pressure of 80 bar). The bottom of the casing and the cement core (Fig. 9, bottom left) were in contact with CO_2-saturated brine throughout the experiment, whereas the top was covered with water and N_2 at the beginning. Figure 9 (bottom right) illustrates that the bottom of the cement core is distinctly carbonated after the experiment (carbonation depth approx. 5 mm). However, a much thinner but yet present carbonated layer is detectable on the whole lateral surface and on the top of the cement core.

As detected by XRD, carbonated samples of the two presented cements from the small-scale autoclave tests (ClassG and MWC-2) show different contents of calcium carbonates after the experiment. The formation of calcium carbonates is directly related to the amount and type of calcium source that is present in the non-carbonated cements. The calcium mainly originates from the dissolution of calcium hydroxide and CSH-phases (Wigand et al. 2009). As the non-carbonated ClassG-cement

Fig. 9 Cement core after autoclave experiment of 56 days at 70 °C and 80 bar; whole cement core and casing cut open (*top*); bottom part of cement core and casing (*bottom left*); bottom part of split cement core wetted with phenolphthalein, *yellow/brown* areas are carbonated, *pink areas* are non-carbonated (*bottom right*)

is richer in calcium hydroxide than the quartz-powder rich MWC-2-cement, the alteration of the ClassG sample is stronger than that of the MWC-2 sample. Therefore, a mixture very similar to the MWC-2-cement was used for the large-scale experiments.

The carbonation of the cement has direct consequences on the porosity of the materials. The observed reduction of porosities in both cement types (ClassG: from 30 to 15 vol.%, MWC-2: from 15 to 10 vol.%—Fig. 9) is explained by the precipitation of calcium carbonates in the pore space of the hardened cement. In addition, calcium carbonates possess a larger volume than the main precursor mineral calcium hydroxide, which further reduces the pore space. As the non-carbonated ClassG-cement contains more calcium hydroxide than the MWC-2-cement, the porosity reduction is much higher in ClassG than in MWC-2.

The slightly lower carbonation depth for cement ClassG shows that pronounced calcium carbonate formation in the cement's pore space might lead to a "clogging" of the pore system and thus a slowing down effect of cement carbonation, at least for short durations (28 days).

The carbonation depth at the bottom of the cement core in Full-scale experiments after experiment 1 (56 days at 70 °C and 80 bar) is in accordance with the results of the small samples. The observed carbonation of the lateral surface and the top of the cement core proves that the boundary surface between cement and casing (steel) forms a micro-annulus acting as a pathway for CO_2 migration.

4.2 Cementation Experiments

To investigate the influence of the surface roughness on the formation of cementation flaws, first tests were carried out with a borehole replica of 3 m height (Borehole-Cementation Simulator—Fig. 4). Before as well as during the cementation process, the rheological properties [such as the viscosity, e.g. flow spread (DIN EN 1015-3 2007)] and specific weights of the cement slurry and bentonite suspension were determined. The resulting change of the properties (e.g. viscosity, density) by the pumping process was smaller than 2.0 % compared with the properties prior to pumping.

In Fig. 10 sections of the hardened cement, i.e. the filled borehole, are shown. Obviously, the cement slurry did not completely displace the borehole fluid in the vicinity of the serrations and various cementation flaws developed. For a complete displacement of the borehole fluid by the cement suspension, the serrations of the hardened cement in the annulus should not fall below a height of 20 mm and a depth of 20 mm (Fig. 10a, b) or a height of 5 mm and a depth of 5 mm (Fig. 10c, d), respectively.

During casting, the casing was replaced by a pipe of transparent acryl, to observe the non-displaced borehole fluid. Further cementation flaws are shown in Fig. 11.

The cementation flaws are quantified by measuring the volume of the empty borehole before the cementation and the volume of the hardened cement afterwards. Figure 12 shows the cementation flaws, i.e. the volume of the non-displaced borehole fluid divided by the net volume of the serration. A cementation flaw ratio of 0.75 indicates that 75 vol.% of the serration consists of non-displaced borehole fluid and only 25 vol.% of the serration is formed by hardened cement.

The formation mechanisms of the cementation flaws strongly depend on the borehole geometry and especially the borehole surface (Figs. 10 and 12). The results show that very rough borehole topography with large serrations of h/d = 20 mm/20 mm have a higher flaw ratio than the smaller serrations with h/d = 5 mm/5 mm. Further studies are required to define constitutive laws for the prediction of the formation mechanism of cementation flaws in dependence of the rheological behavior of borehole fluids and cement suspensions as a function of the borehole surface.

In addition to the large-scale cementation experiments with the borehole-cementation-simulator, the used cement suspension was evaluated regarding its durability. The porosity was determined by mercury intrusion porosimetry. The used cement (MWC-01-A) has a reduced porosity, compared to the reference cement. Gas migration via the pore space of the cement is unlikely. Therefore, the tightness of the cementation will be dominated by cementation flaws (Kromer et al. 2014).

Furthermore, the autogeneous shrinkage was investigated on prisms based on DIN 52450 (1985) and Acker et al. (1998). The shrinkage deformations corresponded to 0.182 mm/m after 28 days and were smaller than the deformation measured on the reference cement (w/c = 0.44, according to DIN EN ISO 10426-1

Fig. 10 Filled borehole strand with cap-rock geometry *h/d* = 20 mm/20 mm (*left*) and with *h/d* = 5 mm/5 mm (*right*) in longitudinal section (Fig. 6—A–A); *top row* lighter sections are areas of non-displaced borehole fluid; *bottom row* identical areas after cleaning with water

2010), which showed a deformation of 0.309 mm/m. After 28 days, the used cement exhibited a compressive strength of 64 N/mm^2 and a bending tension strength of 9.5 N/mm^2 determined according to DIN EN 196-1 (2005).

4.3 Development of Fluid-Chemistry in the Vicinity of Cemented Bore-hole Casing

The mineralogical composition of potential reservoir rocks has been determined by X-ray diffraction. Whereas the Bunter Sandstone has an initial mineralogical composition of quartz, feldspars, hematite and accessory mica, clay minerals,

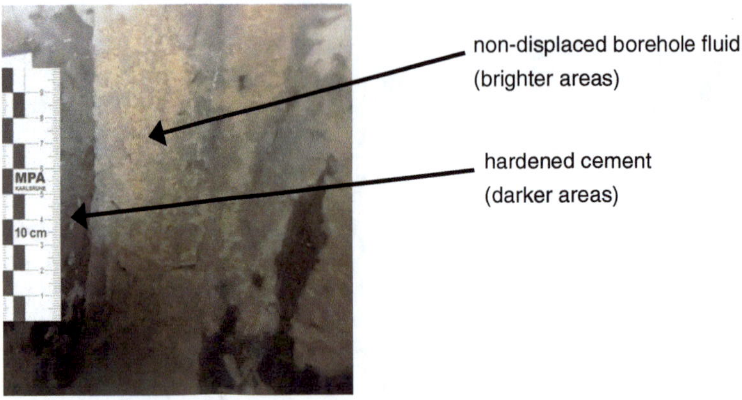

Fig. 11 Areas of non-displaced borehole fluid (*brighter areas*) at the inner side of the annulus, which is typical for a channel or micro-annulus

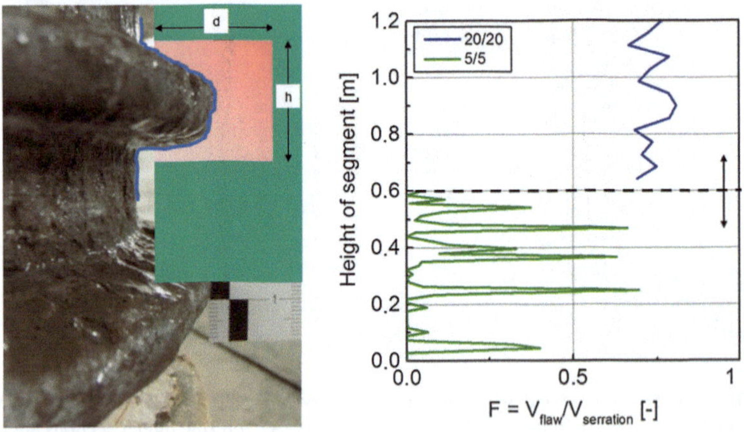

Fig. 12 Definition of the cementation flaw ratio (*red area* in the *left picture*) and the volume of the cementation flaws referred to the volume of the net serration dependent on the cap-rock geometry

carbonates (dolomite and calcite) and goethite, the Norian Sandstone shows a mineral assemblage of quartz, kaolinite, carbonates (calcite and dolomite) feldspars and accessory mica/clay minerals and iron hydroxides.

During the first alteration experiments the evolution of the composition of the CO_2-rich brine was determined. Variations in dissolved concentrations of major cations (Ca^{2+}, K^+, Mg^{2+}, Na^+ and Si^{4+}) during experiment 1—phase 3 are presented in Fig. 13. Three events can be determined that affect the fluid composition within the reaction chamber significantly. The first CO_2 injection into the system after 136 h led to an increase of the dissolved Ca^{2+}-concentration up to approx. 200 mg/l. The maximum Ca^{2+} concentration of 266 mg/l was reached after 284 h. The same

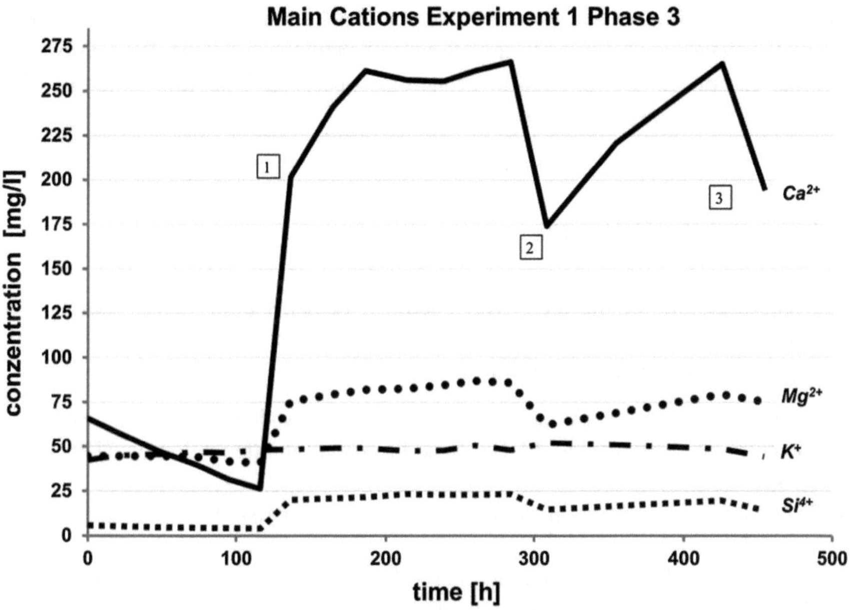

Fig. 13 Main cations of experiment 1 phase 3. *1* start of CO_2 injection, *2* refilling of the reaction chamber with 5 l fresh brine, *3* end of daily sampling

trend but smaller in extent was observed for Mg^{2+} (75 mg/l) and Si^{4+} (20 mg/l). After the drop in concentration during refill, the concentrations of Ca^{2+}, Mg^{2+} and Si^{4+} increased again to the same maximum concentrations. The K^+ concentration was increasing continuously up to the refill from 42 to 52 mg/l with small fluctuations. Prior to the CO_2-injection (event 1), the quench pH-value slightly increased from 8.3 to 8.5 due to the contact with the saline brine. After the start of injection, the expected rapid decrease to a constant value of pH 6.0 was observed.

Prior to the CO_2-injection, Ca^{2+} shows a decrease in concentration. This is interpreted as precipitation effects due to the observed increasing pH-value and the high Ca^{2+}-concentration of the initial brine. The change in the pH-value can be interpreted through the contact of the brine with the cemented casing.

The CO_2-injection led to a strong decrease in pH-value and thus the dissolution of minerals, which raised the concentration of dissolved ions in the fluid—especially of Ca^{2+} cations, but also of Mg^{2+} and Si^{4+}. Only K^+ seems nearly unaffected by the CO_2 alteration during the time of the experiment (Fig. 13).

The decreasing of the concentrations during refill can be explained as a dilution effect by refilling of 5 l fresh brine [Fig. 13, point (2) and (3)]. Our first data show a complex interaction of solution and precipitation processes which seems to be dominated by the pH value. The contact of the brine with the cement results in an increase of pH value. While increasing the pressure through CO_2 injection carbonic acid is formed and the pH-value decreases which results in an increase of the solubility of carbonate and silicate minerals. Calcite could be identified as one of the major phases contributing to the observed behaviour, however, the involvement of other carbonates are required to explain the observed behaviour which involves besides Ca^{2+} cations, Mg^{2+} and Si^{4+}.

4.4 Spatial TDR Results

The measurements of the dielectric spectra of cements under investigation (non-carbonated and carbonated) are shown in Fig. 14. The results of these measurements can be summarized as follows:

- Analysed cements show unusually high frequency dependence (dispersion) in dielectric properties up to 2 GHz. In addition, a high attenuation on TDR-signals is observed.
- Increasing salt content in the composition of the cements increases frequency dependence.
- Increasing additive content (e.g. quartz, fly ash, polymers) in the composition of the cements decreases the dispersion.
- Carbonation of the cements decreases frequency dependence.

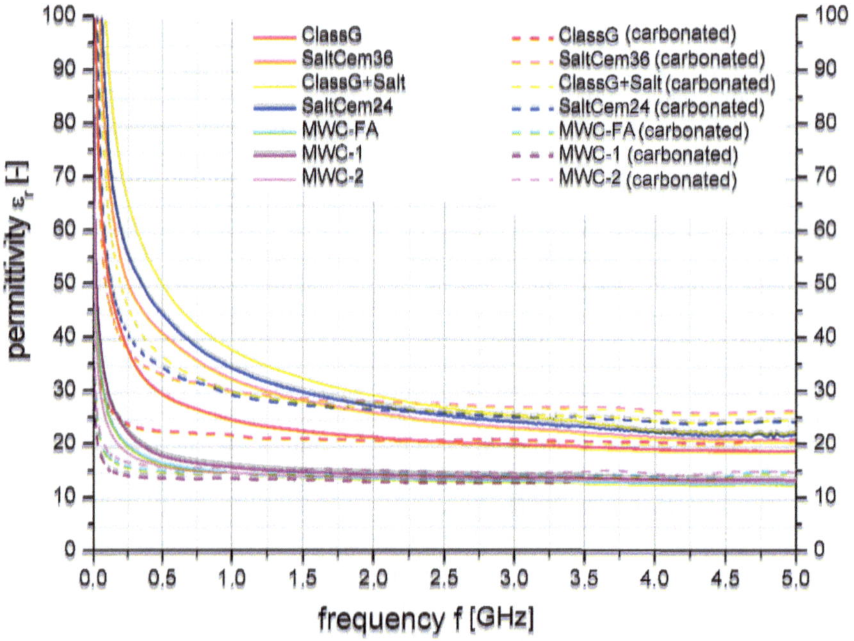

Fig. 14 Dielectric spectra (real part) of cements under investigation (several types of modified well cement (MWC) and cement with additional salt content)

- Cement mixtures with salt additions are most suitable to differentiate between non-carbonated and carbonated material.

The ability of the TDR method to detect imperfections in the cement has been analyzed using large-scale tests (Fig. 7). One aspect of this investigation was to determine the maximum possible length of the sensor in order to identify given imperfections or gaps. Imperfections with a size of 5 cm could be securely determined in sensors of up to 4 m length, whereas imperfections with a size of 2 cm required a reduction of the sensor length to 2 m. The minimum imperfection size, which could be determined using a 1.2 m sensor, was for enclosures of at least 1 cm in size. Smaller enclosures, such as mud conglomerates could not be detected. Therefore, it was decided to use a maximum sensor length of 90 cm in the Full-scale Autoclave experiments with carbon dioxide. The frequency dependence of the dielectric permittivity of the cements resulting in an enhanced attenuation of the TDR-signal makes it difficult to use longer sensors if small imperfections should be detected. As one main result a new TDR-sensor has been developed to fulfill the requirements of resolution, precision and practicability for future application.

The results of first TDR-measurements in the standard oilfield well with a cement plug in the Full-scale Autoclave show that pressure variations in the lower and upper vessel of the Full-scale Autoclave can be observed as significant sensor signals. Variations of the chemical composition of the fluids could be identified from the sensor signal. Different processes seem to be responsible for the observed behavior in the experiments:

- pressure effects—resulting in a very fast change of the sensor signal,
- Variation of the chemical composition of the fluid (e.g. water, gas, CO_2)—resulting in a fast change of the sensor signal,
- chemical reactions of the cements—probably resulting in a comparably slow change of the sensor signal.

To fully deconvolute the sensor signal, new algorithms need to be implemented which include the strong frequency dependency of the complex dielectric permittivity. As these different processes influence the sensor-signal in a complex manner and with different time constants, further studies are required to separate the different processes quantitatively.

4.5 Fluid Flow Through a Cemented Well

While systematically varying the pressure difference between the lower and upper vessel of the Full-scale Autoclave (Fig. 3) a fluid flow through the cemented casing is forced. This fluid flow could be detected by TDR (see above) and by the pressure equilibration in the vessels. For the data shown in Fig. 15, the applied pressure in the lower vessel was kept constant (automatically controlled) whereas the pressure

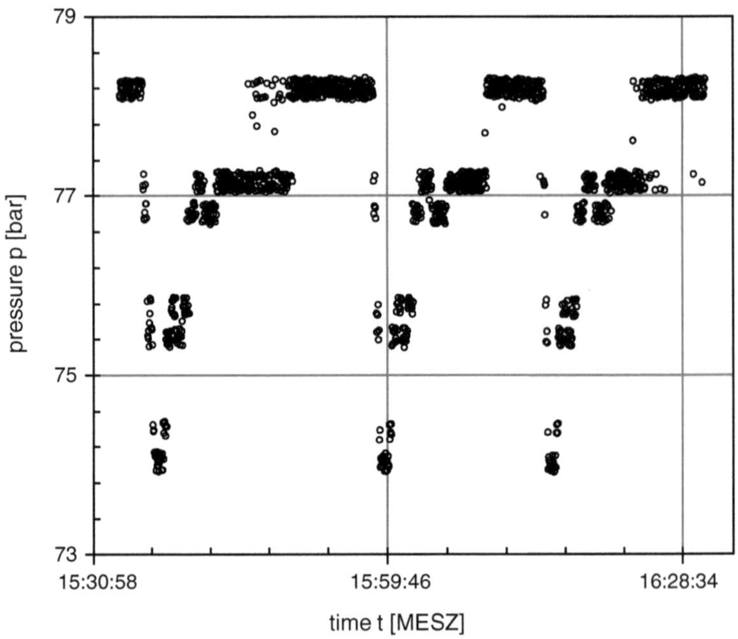

Fig. 15 Three times pressure reduction of the upper vessel by about 4 bar and reequilibration due to CO_2 migration during experiment 2. The steps in the pressure readings represent the accuracy of the upper pressure sensor

of the upper vessel was reduced by about 4 bar for three times. The resultant pressure increase due to a fluid flow through the cemented casing is shown in Fig. 15.

The flow of CO_2 through a cemented casing may have different reasons:

1. Macroscopic fluid/gas flow via flaws, mud channels etc.
2. Diffusion and/or permeation through the cement matrix
3. Transport through cracks or micro-annuli

For the cementation of the casing, idealized conditions have been applied; no drill mud or spacer has been used prior to the cementation of the oil field casing. Therefore, mud channels or flaws due to a complex flow pattern of poly-phase flow are eliminated. In a more realistic environment, these additional fluid-pathways would have to be taken into account (Figs. 10 and 12, see above, experiments with Borehole-Cementation Simulator).

A fluid flow through the cement matrix could be an effective pathway for highly permeable cements. To distinguish between different pathways, the permeability of the cement matrix was studied in Small-scale experiments, where the reaction front (carbonization) was determined to reach a few millimetres (Fig. 8). The permeability amounted to 0.01–0.02 mD. For the Full-scale experiments, the same penetration depth of the reaction front was observed (Fig. 9). The additional determination of the pore structure (Fig. 8 left) shows that the average pore space is rather small—which results in a very small bulk permeability (Kromer et al. 2014). As a consequence, the observed (Fig. 15) rapid pressure equilibration in the upper vessel cannot be the result of permeation through the cement matrix.

As shown in Fig. 9, the interface between steel and cement shows a significant carbonation. As the contact surface of the cement is not entirely carbonated, one can roughly assume that only half of the interface forms an annulus capable of a fluid flow.

No cracks reaching from the lower vessel to the upper vessel are observed after dismantling the cemented well. However, some cracks—mainly perpendicular to the rod axis—are identified as carbonated. As a consequence, the observed flow can be mainly linked to the formed micro-annulus.

To model the fluid flow through a micro-annulus an in-house Finite-Difference-program was used, assuming an annulus, which covers half of the steel-cement interface. Mainly CO_2 or CO_2-rich fluids seem to reach the upper vessel. Therefore, for the applied p, T-conditions the dynamic viscosity and compressibility of CO_2 are set to 0.02 10^{-3} Pa s and 9.3 10^{-8} Pa^{-1}, respectively (IPCC 2005). The width of the annulus was varied to match the observed pressure-time relation. The flow rate was calculated using a Hagen-Poiseuille relation

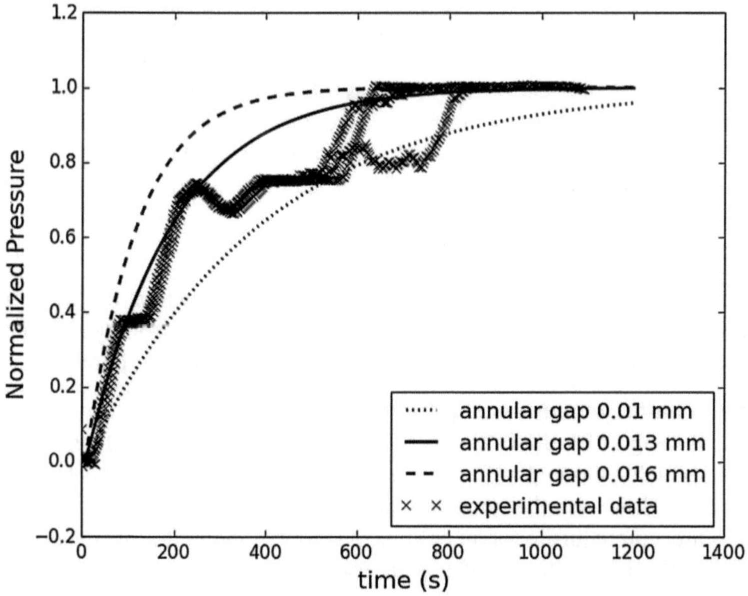

Fig. 16 Averaged and smoothed data for the three permeability experiments (x) (Fig. 15) and modelled data (*lines*). A good approximation while using an FD-algorithm is obtained for a semi-circular micro-annulus between casing and cement with an annular gap of ~13 μm

$$\dot{Q} = \frac{Kh^3 b}{12\eta} \frac{\Delta p}{\Delta l}$$

where K is a constant with $K = 1 - \sum_{n=1}^{\infty} \frac{192}{(2\pi n - \pi)^5} \frac{h}{b} \tan h\left((2n-1)\frac{\pi b}{2h}\right)$ and close to 1 for the applied dimensions, h is the width of the annulus, b is half the perimeter of the casing, η the dynamic viscosity and $\Delta p/\Delta l$ the pressure gradient. As h is very small with respect to b, the curvature of the annulus was not taken into account.

The approximation (Fig. 16) leads to a width of the annulus in the order of 10–20 μm (best approximation 13 μm), which is in good agreement to published data (Nelson and Guillot 2006). Consequently, even for an idealized cementation under optimal conditions with minimal shrinkage (see above) a micro-annulus in the order of 10–20 μm will develop.

5 Conclusion

The interpretation of the suite of results from Full-scale experiments enables to test and improve abandonment strategies, and to evaluate monitoring techniques. The first results of the Full-scale experiments indicate that in a chemical complex system, which includes rock–cement–steel–brine, different solution and precipitation processes interact.

The reaction controlled processes in and diffusion through the solid matrices of rocks, cement and steel can be neglected as efficient pathways for fluid migration. However, the experimentally observed mud-channels and micro-annuli have a considerable higher leakage potential. The leakage could be monitored using Spatial TDR and by pressure equilibration in the first suite of experiments. Further studies are required

- to better quantify the observed processes,
- to develop and verify constitutive laws for the prediction of the formation mechanism of cementation flaws in dependence of the rheological behavior of borehole fluids and cement suspensions as a function of the borehole surface,

in order to further improve and qualify abandonment strategies.

The advantage of Full-scale experiments is their potential to quantify the processes, especially those that are not assessable through smaller scale laboratory experiments. However, the smaller scale laboratory experiments and model calculations are a prerequisite for an optimized layout of the Full-scale experiments, as well as for the interpretation of the observations on the larger scale in terms of process understanding and scenario development.

Acknowledgments Financial support through the research projects COSTEC (BMBF-GEOTECHNOLOGIEN grant number 03G0702A) and COBRA (BMBF-GEOTECHNOLOGIEN grant number 03G0763A) are gratefully acknowledged. We like to thank our colleagues from University of Weimar (MPA) for their support during TDR calibration.

References

Abdu A, Naccache MF, De Souza Mendes PR (2012) Effect of rheology on oil well plugging process. In: Proceedings of the ASME 2012 international mechanical engineering congress and exposition, IMECE 2012

Acker P, Bažant ZP, Chem JC, Huet C, Wittmann FH (1998) Recommendation for measurement of time-dependent strains of concrete. Mater Struct 31:507–512

Brice JW, Holmes BC (1964) Engineered casing cementing programs utilizing turbulent flow techniques. J Pet Technol 16:503–508

Chen Z, Schwing M, Karlovšek J, Wagner W, Scheuermann A (2014) Broadband dielectric measurement methods for soft geomaterials: coaxial transmission line cell and open-ended coaxial probe. Int J Eng Technol 6:373–380

DIN 52450 (1985) Bestimmung des Schwindens und Quellens an kleinen Probekörpern. Beuth Verlag, Berlin

DIN EN 196-1 (2005) Prüfverfahren für Zement Teil 1: Bestimmung der Festigkeit. Beuth Verlag, Berlin

DIN EN 1015-3 (2007) Prüfverfahren für Mörtel und Mauerwerk - Teil 3: Bestimmung der Konsistenz von Frischmörtel (mit Ausbreittisch). Beuth Verlag, Berlin

DIN EN ISO 10426-1 (2010) Erdöl- und Erdgasindustrie - Zemente und Materialien für die Zementation von Tieflochbohrungen - Teil 1: Anforderungen (ISO 10426-1:2009 + Cor.1:2010). Beuth Verlag, Berlin

Druckenmiller ML, Maroto-Valer MM (2005) Carbon sequestration using brine of adjusted pH to form mineral carbonates. Fuel Process Technol 86:1599–1614

Fischer S, Liebscher A, De Lucia M, Hecht L, the Ketzin Team (2013) Reactivity of sandstone and siltstone samples from the Ketzin pilot CO_2 storage site—laboratory experiments and reactive geochemical modeling. Environ Earth Sci 70:3687–3708

Hirsch A, Haist M, Müller HS (2013) Durability of borehole cements used in carbon dioxide capture and storage. In: Sakai K (ed) Proceedings of the first international conference on concrete sustainability, ICCS13, Japan Concrete Institute, Tokyo, 27–29 May 2013

IPCC (2005) IPCC special report on carbon dioxide capture and storage. Prepared by working group III of the intergovernmental panel on climate change

Jakobsen J, Sterri N, Saasen A, Aas B, Kjosnes I, Vigen A (1991) Displacements in eccentric annuli during primary cementing in deviated wells. In: Proceedings of the SPE production operations symposium, 7–9 April 1991, Society of Petroleum Engineers, Oklahoma City

Jamot A (1974) Deplacement de la boue par le laitier de ciment dans l'espace annulaire tubage-paroi d'un puits. Revue de l'Association Francais Technique Petroliére 224:27–37

Kamali S, Moranville M, Leclercq S (2008) Material and environmental parameter effects on the leaching of cement pastes: experiments and modeling. Cem Concr Res 38:575–585

Koorneef J, Ramírez A, Turkenburg W, Faaij A (2012) The environmental impact and risk assessment of CO_2 capture, transport and storage: an evaluation of the knowledge base. Progr Energy Combust Sci 38:62–86

Kromer M, Haist M, Müller HS (2014) Formation mechanisms of cementation flaws in well cementations under consideration of paste rheology. In: Bastien J, Rouleau N, Fiset M, Thomassin M (ed) Proceedings of the 10th fib international PhD symposium in civil engineering, Université Laval, Quebec

Kühn M, Wipki M, Durucan S, Korre A, Deflandre J-P, Boulharts H, Lüth S, Frykman P, Wollenweber J, Kronimus A, Chadwick A, Böhm G and CO_2CARE Group (2013) Key site abandonment steps in CO_2 storage. Energ Proc 37:4731–4740

Lajeunesse E, Martin J, Rakotomalala N, Salin D, Yortsos YC (1999) Miscible displacement in a Hele Shaw cell at high angles. J Fluid Mech 398:299–319

Lesti M, Tiemeyer C, Plank J (2013) CO_2 stability of Portland cement based well cementing systems for use on carbon capture & storage (CCS) wells. Cem Concr Res 45:45–54

Lux K-H, Czaikowski O, Rutenberg M, Seeska R (2012) Untersuchungen zur Validierung von Modellansätzen für Tongestein anhand von Feldexperimenten am Standort Tournemire (F) im Rahmen DECOVALEX-THMC. Abschlussbericht zum BMWi-Forschungsvorhaben mit dem Förderkennzeichen 02E10427, Clausthal-Zellerfeld

Marbler H, Erickson KP, Schmidt M, Lempp C, Pöllmann H (2012) Geomechanical and geochemical effects on sandstones caused by the reaction with supercritical CO_2: an experimental approach to in situ conditions in deep geological reservoirs. Environ Earth Sci 69:1981–1998

Nasvi MCM, Ranjith PG, Sanjayan J (2014) Effect of different mix compositions on apparent carbon dioxide (CO_2) permeability of geopolymer: suitability as well cement for CO_2 sequestration wells. Appl Energ 114:939–948

Nelson EB, Guillot D (2006) Well cementing. Schlumberger, Sugar Land

Nguyen D, Kagan M, Rahman SS (1992) Evaluation of drilling fluid removal by cement slurry from horizontal wells with the use of an accurate mathematical model. J Pet Sci Eng 8:191–204

Randhol P, Valencia K, Taghipour A, Akervoll I, Carlsen IM (2007) Ensuring well integrity in connection with CO_2 injection. SINTEF petroleum research report, Trondheim

Schlaeger S (2005) A fast TDR-inversion technique for the reconstruction of spatial soil moisture content. Hydrol Earth Syst Sci 9:481–492

Wigand M, Kaszuba JP, Carey JW, Hollis WK (2009) Geochemical effects of CO_2 sequestration on fractured wellbore cement at the cement/caprock interface. Chem Geol 265:122–133

“CO_2RINA”—CO_2 Storage Risk Integrated Analysis

René Kahnt, Alexander Kutzke, Mirko Martin, Michael Eckart, Ralph Schlüter, Thomas Kempka, Elena Tillner, Alexandra Hildenbrand, Bernhard M. Krooss, Yves Gensterblum, Markus Adams, Martin Feinendegen, Stefan Klebingat and Christoph Neukum

Abstract While risk assessment for CO_2-storage often has been conducted by using a lot of simplifications and conservatisms, our approach developed in the CO_2RINA-research project is based on the integration of all models, existing at a moment in time. These models will be coupled by the so called transfer function approach which has been proven to be very powerful in the risk assessment for low level radioactive waste. This concept ensures, that the risk assessment is always consistent with the state of the models existing for a specific site. It can be immediately improved if the existing models will be improved over time. The approach was verified by comparison of the direct coupling of different process models in an overall model with a coupling by transfer functions by conducting a wide range of test calculations showing very good accuracy of the approach. The new coupling approach allows the incorporation of a variety of additional effects which are difficult to handle in an overall model. Examples for such processes are complex chemical and microbiological interactions, geomechanical feedback loops and migration of CO_2 in the atmosphere. In the project there have been developed specified models for a generic site with parameters similar to the Ketzin site. These models

R. Kahnt (✉) · A. Kutzke · M. Martin
G.E.O.S. Ingenieurgesellschaft mbH, Schwarze Kiefern 2, 09633 Halsbrücke, Germany
e-mail: r.kahnt@geosfreiberg.de

M. Eckart · R. Schlüter
DMT & Co KG, Essen, Germany

T. Kempka · E. Tillner
Helmholtz-Zentrum Potsdam Deutsches GeoForschungsZentrum GFZ, Potsdam, Germany

A. Hildenbrand · B.M. Krooss · Y. Gensterblum
Geology, Geochemistry und Deposits for Oil and Coal (LEK), RWTH Aachen, Aachen, Germany

M. Adams · M. Feinendegen
Institute for Geotechnics in Construction Engineering (GIB), RWTH Aachen, Aachen, Germany

S. Klebingat · C. Neukum
Engineering Geology and Hydrogeology (LIH), RWTH Aachen, Aachen, Germany

A. Liebscher and U. Münch (eds.), *Geological Storage of CO_2 – Long Term Security Aspects*, Advanced Technologies in Earth Sciences,
DOI 10.1007/978-3-319-13930-2_7

include a reservoir model, a model for the alteration of the cementation of a mature well, a fault model, a model describing advection and diffusion through the cap rock, a complex model for the migration in a Quaternary aquifer including complex chemical interactions and a geomechanical model. Additionally there was shown the way of integration of microbiological processes which have been modelled in detail in the CO_2BIOPERM project. The new approach is ready to be adapted to a specific CO_2-storage site.

1 Background and Purpose

Risk assessment for CO_2 storage projects is mostly conducted using expert knowledge to evaluate specific scenarios. Features, events and processes (FEPs) are investigated to assess their impacts on operational and environmental safety. Oldenburg (2007) suggest an approach similar to a calculation sheet, whereas the interacting user defines relative parameter weight and processes as well as reliability of the available information by qualitative input and sorting methods. The drawback of such methods is the subjectivity related to decisions made by the interacting user. Oladyshkin et al. (2009) suggest the utilization of stochastic methods using a model reduction for specific scenarios to allow a reasonable application of Monte-Carlo simulations. However, efficient application of this method has to be verified for CO_2 storage projects acting at large spatial scales.

Numerical simulations are an established tool for prediction of thermal-hydraulic processes in geothermal reservoirs, oil and gas depotsits as well as CO_2 storage formations at different time scales (Bielinski et al. 2008; Flemisch et al. 2007; Pruess and Spycher 2007; Pruess 2004; Pruess et al. 2003; Helmig 1997, u.v.a.) and were successfully applied within the scope of the EU-Project CO_2SINK to provide reliable predictions for the CO_2 arrival in the observation well CO_2 Ktzi 200/2007 (Kempka et al. 2010a).

The thermal-hydraulic-mechanical simulations scheduled for prediction and evaluation of stress field changes in the vicinity of the storage reservoir were already successfully applied by parameterization and verification using InSAR data resulting from the CO_2 enhanced gas recovery (EGR) at the In Salah site in Algeria (Krechba gas field). The simulation results showed very good agreement between InSAR data and the time-specific modeling results of the coupled numerical tools for different areas in the near field of the injection wells of a storage formation located at about 1,800 m depth (Rutqvist et al. 2009a). Further studies conducted by Rutqvist and Tsang (2002), Rutqvist et al. (2002, 2007, 2008, 2009b), Wang et al. (2009), Watanabe et al. (2010), McDermott et al. (2010), Nowak et al. (2010) and Park et al. (2010) demonstrate the suitability of the coupled numerical models intended for application with the proposed project to determine stress field changes in a CO_2 storage formation to predict and evaluate failure criteria of relevant parameters in the storage formation and its cap rocks as well as potentials for reactivation of faults.

It can be concluded that the extensive national and international efforts in the research field of CO_2 storage in geological formations provided a basic conceptual understanding of the relevant processes in the CO_2 storage formation as well as the different migration pathways into the overburden and surrounding rocks (CSLF-T 2009; Deel 2007; Metz et al. 2005; Walton 2005; Nordbotten et al. 2005; DOE 2009). Currently, these processes can be reliably described assuming well-known geological boundary conditions and input parameters (Zhang et al. 2006; Zhou et al. 2005; Celia et al. 2005).

Very different approaches exist to integrate the specific processes relevant for the risk assessment of long-term geological storage of CO_2 (Jagger 2009; Espie 2005; Cugini et al. 2010; Saripalli et al. 2003; Wildenborg et al. 2005; Pawar et al. 2006; Condor and Unatrakarn 2010). Suggested methodologies integrate approaches such as systematic cause and effect analysis (SWIFT) and Monte-Carlo based methods (RISQUE). In addition, methods based on evidence (TESLA) not considering the classic Boolean logic, but taking into account that an assumption may be wrong, correct or even involve an uncertainty. The evidence based method was originally developed for application in nuclear waste storage. Further approaches for risk assessment of CO_2 storage projects were derived from the evaluation of nuclear waste storage sites (North 1999). Regarding nuclear waste storage, the methods for risk assessment were extensively addressed by Ho (1992) and North (1999) and updated by Maul et al. (2007).

Within the scope of the proposed study we aim to implement and apply an approach comparable to those developed for nuclear waste storage considering its methodology (Stauffer et al. 2005, 2006, 2007, 2009; Viswanathan et al. 2005, 2008). However, the targeted approach strongly deviates in the implementation of the model coupling required for the overall risk assessment. The simulation tool intended to be applied to couple the specific process models will be the GoldSim software package as already addressed by Stauffer et al. (2007) within the scope of the CO_2-PENS methodology. This probabilistic simulation tool was originally developed for the assessment of the Yucca Mountain nuclear waste storage site and provides high flexibility regarding the simulation of arbitrary and highly dynamic systems. Kahnt (2006, 2008), Kahnt and Paul (2008) developed a methodology for the assessment of longterm storage of low radioactive waste materials based on the GoldSim software package (CSM-Approach). The CSM-Approach is based on the coupling of different dynamical processes. While process models such as FEHM (Zyvoloski et al. 1997) and TOUGH2 (Zhang et al. 2006) as well as (semi-) analytical methods (Nordbotten et al. 2005) are directly operated by the GoldSim software package in the CO_2-PENS project, the CSM-Approach provides generic flexible interfaces. Thus, arbitrary process models can be exchanged and extended independently. These flexible interfaces have been used for coupling the Box Model (Module C) applied for coupled 3D-simulations of geochemical processes to the GoldSim system simulation model (Kahnt 2006, 2008; Kahnt and Paul 2008).

The main goal of the conducted work was the adaptation of the rist assessment methodology developed for low level radioactive waste for the risk assessment of CO_2 storage in geological media, its numerical realization and specific validation based on data from natural analogue studies and a present CO_2 storage pilot site. The newly developed methodology is mainly based on the international risk assessment concepts described before.

For this purpose, the geological system involving its technical elements is simulated over a specific timespan in order to assess the probability of CO_2 release with regard to spatial and temporal parameter dependence.

The challenge of risk assessment is the fact that different processes and their combinations influence CO_2 migration into shallow geological formations or the atmosphere. Here, conceptual uncertainties (which processes will dominate?) as well as parameter uncertainties (specific process parameters and their variability cannot be exactly determined due to geological heterogeneities) and parameter variability have to be considered. Even though a deterministic methodology is required for the assessment of the multiple specific processes and their interaction, a probabilistic approach has to be applied at the cross-system level to represent parameter variability and parameter uncertainties in the quality required for risk assessment.

To allow for high transparency for the evaluation and selection of potential storage sites as well as decision support system tools for regulatory authorities, a common and site-unspecific methodology flexible enough to be adapted to any specific site location was developed.

The developed approach provides an improved methodology in comparison to the tools available on the international marked, since it is set up consequently modular taking into account clearly defined interfaces. By this, existing site-specific models can be consistently implemented into the risk assessment model and coupled with each other. The single modules can be further developed independently and be flexibly exchanged.

The methodological solution is based on segmentation of multiple coupled general problem simulation into single defined problem simulation modules. Communicating via expert interfaces a combination of sub-solutions from these single modules can be integrated to solve the main problem module simulation.

An general overview of the integration of the different modules is given in Fig. 1.

Before the description of single modules is presented in detail, major aspects of the procedure will be explained here to ensure the understanding of their relevance in the overall solution approach:

- The project targets the development of a consistent methodology for risk assessment based on modular description of a main system. At the core of this work the method of probabilistic modelling and associated correlation analysis allows the identification of processes and parameters regarding their site-specific contribution strength to the overall risk, which is an essential step to optimize objectivity in the assessment process. Processes and parameters from different fields of work are further investigated regarding the influence of potential cross-communicating sub-processes by directly simulating the influence of diverse counteractive measures on different single units and the main system itself.

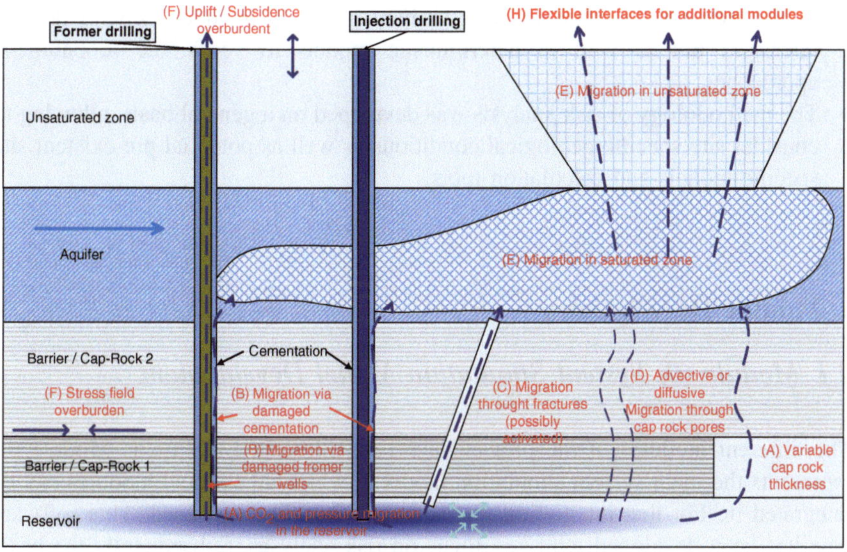

Fig. 1 Schematic overview about CO_2-release pathways and related single processes covered by the different modules (for processes *red text* was used, while for structures *black text* has been used). The letters *A–H* have been used for the enumeration of the different modules

This vice versa allows the assessment of effectivity and quality of each counteractive measure to reduce the overall risk. In a possible additional step the costs of the different counteractive measures can be compared to the effectivity (quality) factor of each counteractive measure, giving the basis for optimized risk assessment solutions. The methodology thus directly reclines to the concept of specific site conditions that has been developed for clean up decisions in the field of low radioactive mining wastes.

- The identification of applicable interfaces between the single modules is of major importance. The basic target behind this approach is the idea to guarantee the possibility of single module replacement. Following this idea, different conceptual approaches as well as different software tools are applicable for the sub-processes simultaneously without communication conflicts. A main focus was kept on the possibility to further develop the single modules independently.
- In order to demonstrate the applicability of the solution approach it was necessary to test real problem scenarios and use site-specific data streams. This has been done using generic data from the EC funded project CO_2SINK at the Ketzin-site together with associated modelling tools to verify the applicability of the single modules. An important aspect of the concept-verification is the possibility to resort to much more quantitative and qualitative data at the Ketzin-site, as in comparison to theoretical validation.
- Besides the development of probabilistic risk solutions, the solutions themselves have been validated for several sub processes using deterministic approaches.

In this context the coupling was verified not only in modules with defined interfaces but as well in deterministic models to verify the probabilistic approaches.

- The methodology of risk analysis was developed on a general basis, allowing to consider site specific geological conditions as well as potential pre-existent site specific models and calculation tools.

2 Module Description

2.1 Module (0) System Simulation Model Development

The different modules of the project have been illustrated in Fig. 1. Module (0) represents the main system simulation model in which all residual modules can be integrated in. For that integration, the software GoldSim was used. This software type has been developed with key focus on risk analyses. Subsequently the basic conceptual idea of the main system simulation model is presented.

Each single module consists of either partly already available and/or new methodologies which have been developed for the sub-processes. Using the complete spectrum of these methodologies the interrelationship between module input parameters and target functions can be calculated. Using various simulation scenarios the total number of target functions in dependency of sensitive module input parameters will be determined. The resulting interrelationship data sets will be stored in multidimensional look-up-tables. These tables are used in the main system simulation model in order to consider all interrelationships of underlying single processes. One of the key challenges was to figure out the interface parameters and the types of interfaces/look up tables.

Within the main system simulations model all single processes are interconnected in a way that, all former single parameter dependencies among each other are maintained. The total spectrum of key parameters is determined together with the related uncertainties and integrated into the main system simulation model by using distribution functions.

Resorting to Monte-Carlo simulations the influence of parameter uncertainties on target parameters was determined. The target parameters of each module as well as the interface between the modules have been specified.

Due to the legal foundations, CO_2 storage operations have to consider the risks for the subjects of protection air, water and soil. Especially there have been calculated concentrations of CO_2 and other species at different points in the geosphere as well as mass fluxes. Additionally geomechanical deformations have been evaluated.

The main simulation model was developed under consideration and awareness of the possibility to be adapted to different site-specific conditions.

The presented solution approach offers the possibility to identify the parameter uncertainties with greatest effect on target parameters using sensitivity analyses. Subsequent investigations can thus be focussed on the resulting key parameters. Operating this way, the procedure enforces the reduction of investigation demand and resulting costs. The structure of the methodological solution into single modules reflecting the process chain is shown in Fig. 1.

The single modules are interconnected via multi-dimensional look-up tables, delivering input–output inter-relationships. The definition of appropriate input and output modes was a major challenge for the development of the methodology, because the coupling has to ensure consistency.

2.2 Module (A): Reservoir Simulations/Numerical System Simulation Model

A 3D numerical system simulation model including the CO_2 storage reservoir, the hydraulically conductive fault and the shallow aquifer has been implemented to serve as reference and for verification of the process and model coupling approaches. Numerical simulation results such as pore pressure and mass flow across specific interfaces serve as main input for the GoldSim system model in module 0, and thus for all models coupled to it. Figure 2 shows the employed generic numerical model grid with an areal extent of 10 km × 10 km and a total thickness of 700 m. Three main areas are distinguished: the reservoir being used for CO_2 injection, the leaky fault and the shallow Quaternary groundwater aquifer.

Applied generic porosities and permeabilities are listed in Table 1, while relative permeabilities were derived from Kempka et al. (2010) and Kempka and Kühn

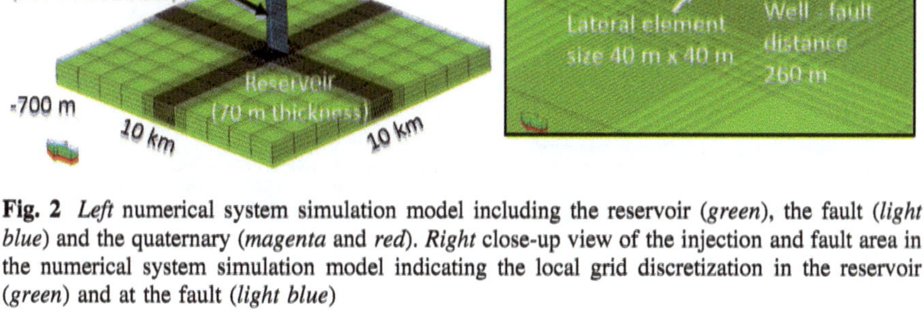

Fig. 2 *Left* numerical system simulation model including the reservoir (*green*), the fault (*light blue*) and the quaternary (*magenta* and *red*). *Right* close-up view of the injection and fault area in the numerical system simulation model indicating the local grid discretization in the reservoir (*green*) and at the fault (*light blue*)

Table 1 Porosity and permeability applied in the numerical system simulation model

Model unit	Porosity (%)	Permeability in x- and y-direction (m^2)	Permeability in z-direction (m^2)
Quaternary	25	10^{-10}	3.33^{-10}
Reservoir	25	10^{-13}	3.33^{-13}
Fault	1	10^{-12}	10^{-12}

(2013). Reservoir salinity is set to 25 % by weight (Klein et al. 2013) to represent realistic conditions for density, Quaternary salinity is zero. This density difference has major impacts on reservoir pressure which is required for brine flux through the fault. 1 kg CO_2/s is injected for 30 years via the injection well plotted in Fig. 3 (left) resulting in a total injected CO_2 mass of about 0.95 Mt at the end of the simulation. A post-injection phase of 20 years is applied thereafter, so that the total simulation time amounts to 50 years. The numerical simulator TOUGH-MP/ECO$_2$N (Pruess 2005a, b; Zhang et al. 2008) was applied for the simulation.

Simulation results indicate that about 73 % of the injected CO_2 have been dissolved in the formation fluid until the end of the simulation time (50 years) as plotted in Fig. 3 (left). As expected, CO_2 dissolution is proceeding even after the injection stop at 30 years of simulation. Figure 3 (right) illustrates that CO_2 and brine arrive after about 2 years of injection in the shallow Quaternary aquifer, while water (no mineralization) being present in the residual fault pore volume is migrating into the Quaternary aquifer from the beginning of CO_2 injection. With the CO_2 arrival in the Quaternary aquifer (after about 2 years) the H_2O mass flow through the fault decreases until the saline formation fluid arrives at the Quaternary aquifer in simulation year 9. This is caused by the increasing density in the fault. Thereafter, CO_2, H_2O and NaCl displacement into the Quaternary aquifer increases to a maximum until the stop of injection in year 30. From then on, a significant reduction of CO_2 and NaCl as well as H_2O flux rates can be observed. H_2O starts to flow back into the fault and reservoir after year 32 at a low rate of about 0.05 kg/s. In total, about 65 % of the injected CO_2 is displaced into the Quaternary accompanied by 0.7 Mt NaCl originating from the reservoir. Hereby, the maximum reservoir pressure elevation amounts to about 1.4 MPa at the injection well top. The

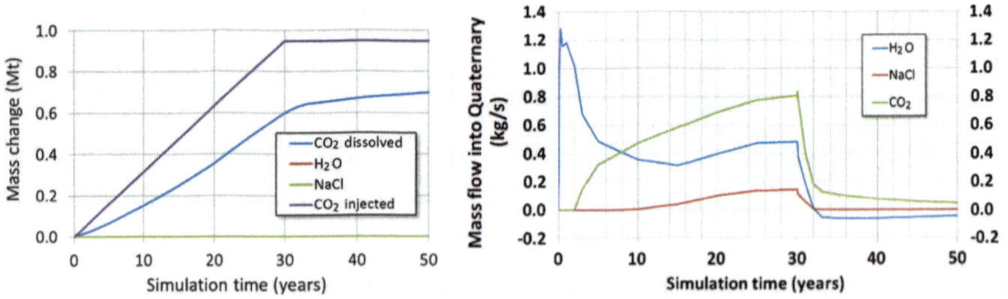

Fig. 3 *Left* numerical system model mass balance change during the entire simulation time of 50 years. *Right* component mass flow of CO_2, H_2O and NaCl into the shallow quaternary aquifer

huge part of CO_2 displaced into the Quaternary is caused by the boundary conditions of the system which have been chosen such, that during the simulation time significant leakage effects will show up.

The introduced reference model with the given parameterization is employed in a simulation framework that allows for automated massive parallel simulations. Consequently, Monte-Carlo simulations were applied to validate the system model based on the GoldSim software package based on varying parameter realizations. Furthermore, the coupled process simulations were implemented for a specific validation of hydro-mechanical processes related to effective stress dependent fault aperture as a function of pore pressure.

2.3 Module (B): Migration Through an Abandoned Well

The model for an abandoned well was initially developed in a separate GoldSimTM model. In this model there is described the transient development of the degradation of the cementation of the well and the resulting widening of an initially existing micro fracture. The basic conceptual model is shown in Fig. 4.

Because of that, the module (B) could be used for initial benchmark calculations of the overall coupling concept, because the GoldSim model can be used separately and the effects can be incorporated as transfer functions/look up tables on the one hand side and the model can be incorporated into the overall GoldSim model directly. By comparison of the results of both modelling concepts, the coupling approach was verified. The comparison is shown for one example calculation in Fig. 5. The result demonstrates that the transfer function coupling approach seems to work very well.

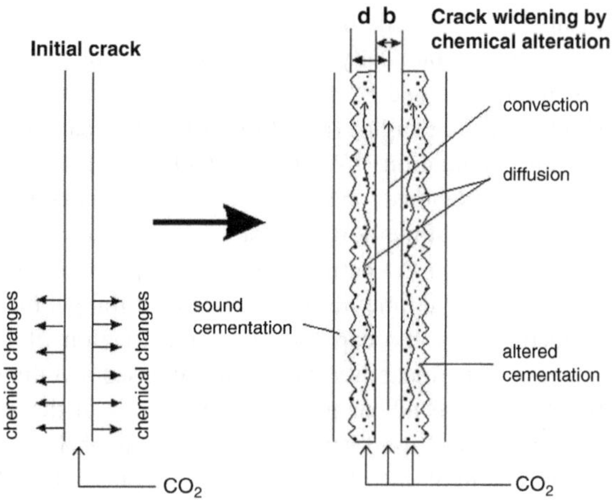

Fig. 4 Conceptual model for the chemical widening of an initial microfracture in the cementation *b* is the opening of the fracture and *d* is the carbonization depth

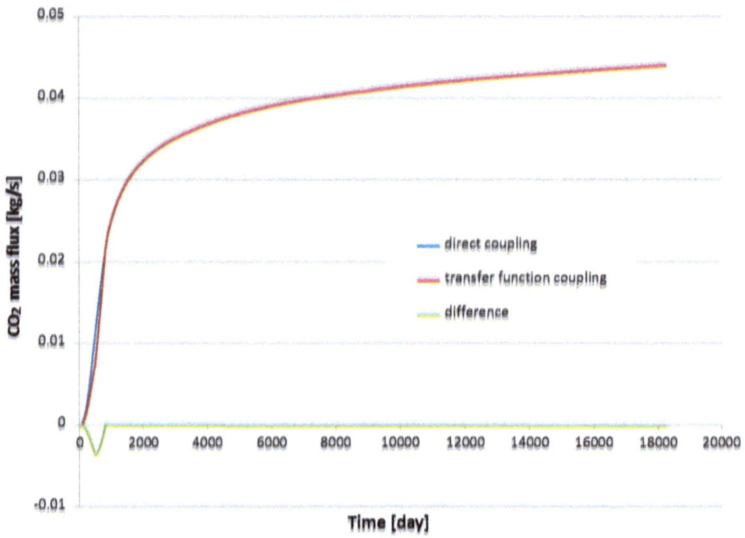

Fig. 5 Comparison of the CO_2-mass flux through the cementation of an abandoned well calculated by direct model coupling and alternatively by transfer functions according to module (B). The *green curve* is the difference between the two calculations

Using this model for the abandoned well coupled with the reservoir model, there has been calculated the probability distribution of the CO_2 mass flux through the well as a function of time taking into account all parameter uncertainties. The result is shown in Fig. 6.

2.4 Module (C): Migration Through Fractured Zones

Faults and fractured zones are of major importance for a safe long-term storage as they may hydraulically connect various geological units. The width of the fault/fractured zone and its permeability are determining factors for the CO_2-storage. Faults can show high permeabilities but they can also act as flow barriers. Changing the stress conditions by injecting CO_2 and, thus, increasing the pore pressure can reactivate previously inactive zones of weakness. Chemical reactions between CO_2 and minerals of the host rock or of precipitated minerals in the fault zone can have an influence on the permeability—both positive and negative.

In geological time frames, induced natural geomechanical processes may lead to a fracturing of the cap rock or a reactivation of fault zones associated with an increase in permeability. Here, an increased swelling capacity of the cap rock will be of particular advantage for the tightness of the storage complex.

In the model used here with a thickness of the cap rock barrier of 120 m with a displacement of only 20–30 m, a sealing effect is likely to occur. Based on diagrams

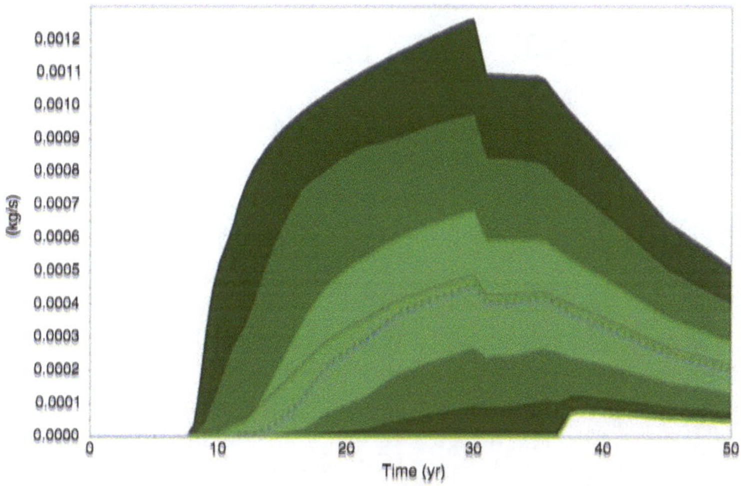

Fig. 6 Distribution function for the time dependent CO_2 mass flux through the cementation of an abandoned well taking into account different parameter uncertainties, like initial opening of the fracture, reaction rate and initial transmissivity of the fracture. The *dotted blue line* is the average value, the *green line* is the median and the different green areas give the percentiles (5, 10, 25, 75, 90 and 95 %)

showing typical ratios the width of the (typically tight) fault core should amount to a maximum of 3 m, the (more permeable) damage zone up to 100 m. It has to be noted, however, that the values can vary in the range of two orders of magnitude!

The modeling in the CO_2RINA-frame-project is concentrated to these fault zones, especially to consider following processes:

- Turbulent flow considering variable gradients with rough-laminar, rough-turbulent, smooth-laminar, smooth-turbulent flow conditions
- Time depended fracture width (coupled with an geomechanical model)
- Two phase flow CO_2–H_2O
- Reactive multi-migrant-transport with precipitation and dissolution of minerals in the fault
- Density driven flow

In the first step the 2-phase-flow was adapted into Reacflow3D (an existing multi-migrant-model) according to the concept of the "fractional flow formulation" [Niessner] and thoroughly tested in comparison with the modeling systems [MUFTE-UG] and [Dumux]. Furthermore some plausibility checks carried out, especially for the control of the density effects in comparison with the modeling system. [Tough2].

After developing an interface for transferring time depending fractures-widths and implement turbulent flow laws, the hydraulic situation in the fault was investigated with the help of the expanded model.

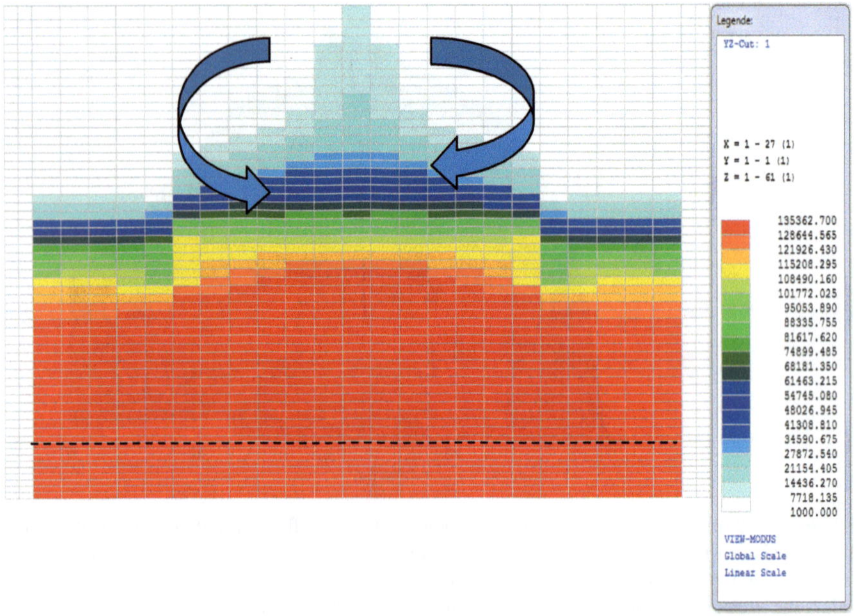

Fig. 7 Distribution of chloride concentration in the fault zone (concentration on a *vertical* plane inside the fault zone) 2 years after start of CO_2–injection in the reservoir. *Red colors* indicate concentration of about 135 g/l representing the initial concentration of the reservoir brine, while the concentration on *top* of the fault has been diluted (see concentration legend). The reservoir below the fault zone is represented by the lowest 8 grid rows (below *dotted black line*) while the quaternary aquifer is located above the *uppermost* grid row

Apart of the new developments in Reacflow3D the emphasis is the investigation, how several types of models (injection zone, fault, groundwater and geochemistry) can interact with each other to get a joint result. It is known from series of real examples that the models of several hydraulic horizons are becoming increasingly complex, so that the overall model is limited in the numbers of considered processes or by the calculation time. As a result a proper coupling procedure is necessary. This means: "decoupling of the complex problem".

Figure 7 gives an impression about the complex flow process in the fault zone caused by density effects. In the middle part of the fault zone in a short rectangular distance to the injection point, the chloride-front is reaching after nearly 10 years the upper quaternary groundwater layer. A few meters away the pressure in the injection horizon is not high enough to lift the heavy salt-column up to the groundwater layer. A circulation of salt water occurs from the middle column to the sides. It illustrates the sensitivity of the flow field caused by density effects.

Figure 8 simplifies this problem into one dimension. The pressure gradient is approximately in (very sensitive) equilibrium with the gravity and the pressure in the injections horizon is increasing up to the level, where the salt water is raising to the quaternary horizon:

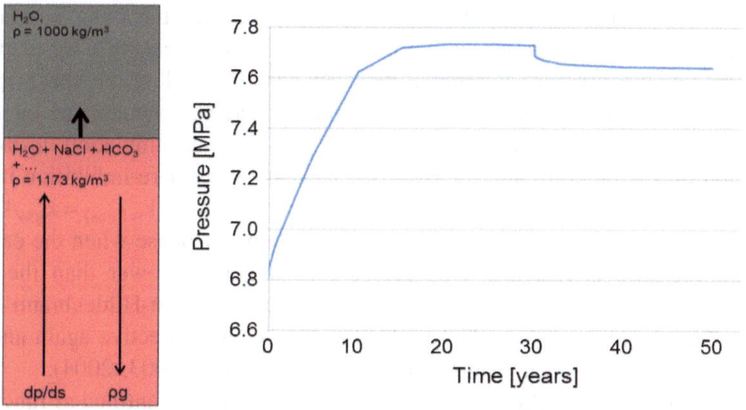

Fig. 8 *Left* force balance between gravity term and pressure gradient. *Right* temporal development of the pressure at the *bottom* of the fault for the central element of below the fault

The original concept to couple the model of the injection horizon with the fault model by transfer of the boundary-variables "pressure" and "CO_2-Saturation" failed, because little differences in some model internal formulas (e.g. density-concentration) led to very big differences in flux of CO_2-gas and raising salt water flow, although in principle all models worked correct.

The consequence of this effect was that the fault zone has to be incorporated into the reservoir model as a boundary condition to maintain an exact overall mass-balance. Additional processes of higher order can be incorporated separately.

From this point of view there was developed a concept, were the Reacflow3D-Model is modeling the change of hydraulic condition by different effects, e.g. fracture width, chemical reactions, and the results were transferred to GoldSim.

GoldSim assigns the different variants with the overall mass balance using the mass-flow of salt water and the hydraulic gradient.

2.5 Modul (D): Advective and Diffusive Flux Through the Cap Rock

Two processes have to be taken into account when considering the safe and long-term sealing capacity of an intact and water-saturated seal (cap rock) that is in direct contact with the free CO_2 phase of the storage layer: (1) Capillary pressure-controlled viscous flow of CO_2 through the seal and (2) molecular diffusion of CO_2 dissolved in the pore water of the cap rock.

The first process depends on the critical capillary threshold pressure (p_{brthr}), which must be overcome for CO_2 to displace the water and enter the interconnected pore system of the cap rock. Whenever this pressure is exceeded viscous flow of CO_2 into the cap rock will occur which can lead to a leakage flow in the long term.

In the *worst-case scenario* of capillary gas breakthrough the water saturation of the cap rock pore system successively decreases with increasing CO_2 (capillary) pressure, i.e. the pressure difference between gas (p_1) and water phase (p_2). In consequence the conductivity for the gas phase increases, leading to increasing leakage rates (Q). Successive depletion of the CO_2 accumulation leads to a reduction of gas pressure and the (reverse) process of water re-imbibition into the cap rock pores. This will reduce the relative gas permeability ($k_{rel(gas)} = k_{gas}/k_{intrinic}$) until the pore system is completely blocked by the water phase when the capillary snap-off pressure is reached. This pressure is somewhat lower than the initial breakthrough pressure ($p_{brthr} > p_{snap\text{-}off}$) (Busch and Amann-Hildenbrand 2013). Due to this phenomenon capillary sealing would become effective again and CO_2 leakage from the reservoir would stop (Hildenbrand et al. 2003, 2004).

The gas leakage rate after capillary breakthrough can be quantified as function of the capillary pressure. Based on measured capillary pressure versus relative gas permeability data (Amann-Hildenbrand et al. 2013) and assuming a log-normal pore size distribution we have estimated the leakage rates from a CO_2 reservoir using Darcy's law for compressible media (Fig. 9).

In contrast to capillary controlled transport, diffusion is considered a continuous and ubiquitous process with low transport capacity. At the reservoir/cap rock interface CO_2 will dissolve in and equilibrate with the pore water. The dissolved CO_2 will diffuse upward, following the concentration (chemical potential) gradient. This process is described by Fick's laws of diffusion (Crank 1975).

The process is approximated by assuming diffusion through a homogeneous plane sheet (cap rock) of given thickness ($x = l$) (Crank 1975; Krooss et al. 1992a, b; Schlömer 1998). The bulk CO_2 concentration at the *base* of the cap rock ($c_1 = c_{pore\ water} \cdot \phi$) is a function of temperature, pressure and salinity (Duan and Sun 2003) and porosity. Because the variations are small this concentration is assumed constant throughout the process. The concentration at the *top* of the cap rock (c_2) is assumed

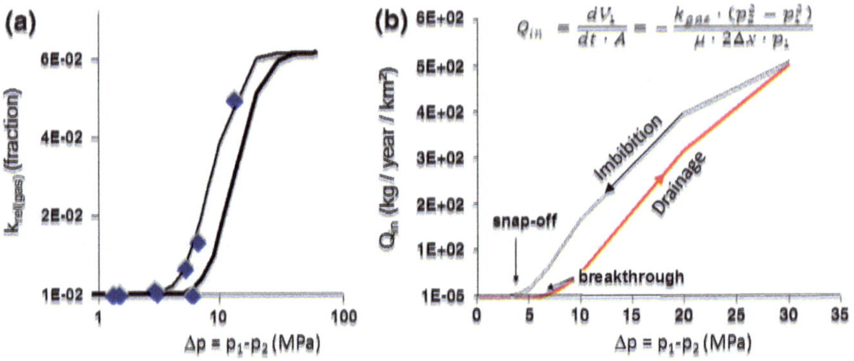

Fig. 9 **a** Relative gas permeability as function of capillary pressure based on data from Amann-Hildenbrand et al. (2013). **b** Leakage rate out of the reservoir into the sealing lithology (Q_{in}) by immiscible viscous gas flow

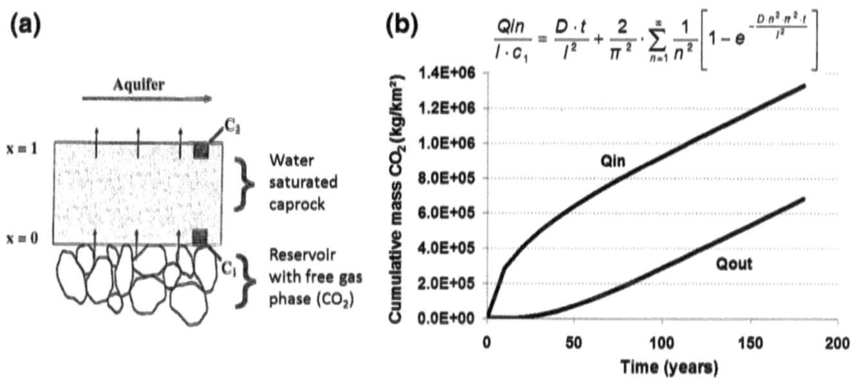

Fig. 10 **a** Basic model for diffusion through a cap rock with thickness l, c1 the concentration on the reservoir/cap rock interface and c^2 at the *top* of the cap rock sequence ($c^2 = 0 =$ const.) (Schlömer 1998). **b** Cumulative mass of CO_2 diffused into (Q_{in}) and out of (Q_{out}) 1 m^2 of a cap rock layer of 160 m thickness

to be constantly zero. Figure 10 shows the calculated cumulative masses of CO_2 entering (at the bottom) and leaving (at the top) one square metre of a 160 m thick cap rock layer as a function of time. Input parameters were: $l = 160$ m, $\phi = 12$ %, T = 34 °C, $p_l = 6.3$ MPa, salinity 1.1 mol NaCl/kg water. The effective diffusion coefficient assumed for this calculation (5×10^{-10} m^2/s) is at the upper boundary of values reported in literature, ranging from 10^{-10} to 10^{-12} m^2/s for shales, clays and carbonates (Busch et al. 2008; Song and Zhang 2013).

The contribution of the described processes to the overall CO_2-leakage risk has been assessed extensively. These processes only contribute in the long term, i.e. on time scales of some thousand years to the leakage and therefore can be neglected if processes during the injection phase become dominant in their risk contribution. This approach is the consequence of the top-down approach underlying the whole CO_2RINA concept.

2.6 Module (E): Fluid Migration in the Unsaturated and Saturated Zone

Quantified prognoses for atmospheric CO_2 outgassing via the unsaturated zone and accompanied groundwater chemistry changes like acidification and salinization are key challenges for reliable integrated long-term CO_2 storage risk assessment tools. In view of numerical solutions a first starting point may thus be set to CO_2 phase partitioning behaviour between liquid and gas phase within the saturated zone. Referring to exemplary brine and groundwater chemistry data sets near the Ketzin pilot site (e.g. LUA 2007; Seis et al. 2013) in this context currently various scenarios are analysed using the software code boxmodel ReacFlow3D (Eckart 2011).

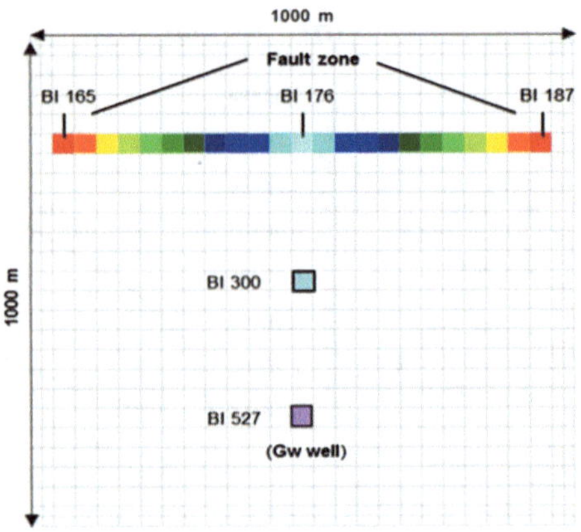

Fig. 11 Model aquifer (*top view*) with different observation points (BI) at which the concentration histories have been evaluated. The different observation points are marked with different colours. Same colours have been used for symmetric points which show the same concentrations

With regard to the schematic conceptual overview and main aquifer boundary conditions in Fig. 11 and Table 2, Figs. 12, 13, 14 and 15 provide first rough long-time saturated zone snapshots of expected CO_2 liquid/gas phase partitioning, chloride- as well as related density-levels, assuming a leaking reservoir brine at a mass flow rate of 1 kg/s with a CO_2/H_2O mass-ratio of 0.08.

With focus on main trends under the selected boundary conditions seen so far commonly >90 % of CO_2 remains dissolved in the liquid phase and only marginal amounts are available in the gas phase for atmospheric outgassing (cf. Figs. 12 and 13). CO_2 liquid phase transport is merely controlled by density changes related to variable brine chemistry Cl- levels, as seen in quite similar spatial distributions. Simultaneously many scenarios indicate Cl- is likely to exceed the drinking water limit of 250 mg/l.

Table 2 Aquifer boundary conditions

Length × width × height	(m)	1,000 × 1,000 × 50
h-potential well BI 527	(m)	25
h-potential residual aquifer	(m)	30
Bottom level well BI 527	(m)	−16
Start up groundwater background levels at t = 0		
CO_2 dissolved liquid phase	(mg/l)	290
CO_2 gas phase	(kg/s)	0
Cl^- dissolved liquid phase	(mg/l)	70
Density	(kg/m^3)	1,000

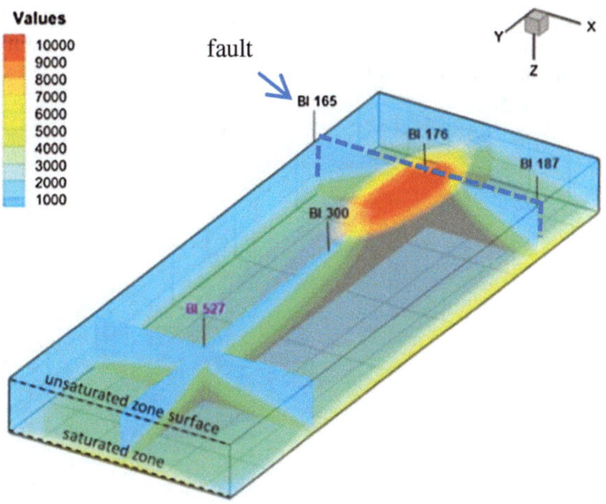

Fig. 12 CO_2 dissolved in aqueous phase (mg/l) distribution after t = 50a

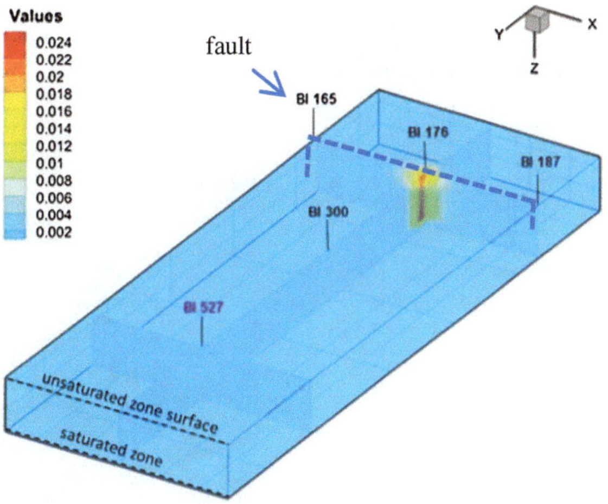

Fig. 13 CO_2 gas phase flux rate (kg/s) distribution after t = 50a

Within the aquifer overall minor mixed fluid phase upward migration is observed, in accordance with higher brine related density levels (cf. Figs. 14 and 15).

2.7 Module (F): Geomechanical Modelling of Stress/Strain State of the Cap Rock and Reactivation of Fault Zones

Module F deals with the geomechanical modelling of an onshore CO_2 storage site. For this the cap rock integrity, the deformation of the ground surface and the hydraulic reactivation of existing fault zones are investigated. Based on the synthetic model of the CO_2RINA project (see module 0), the injection of CO_2 into a

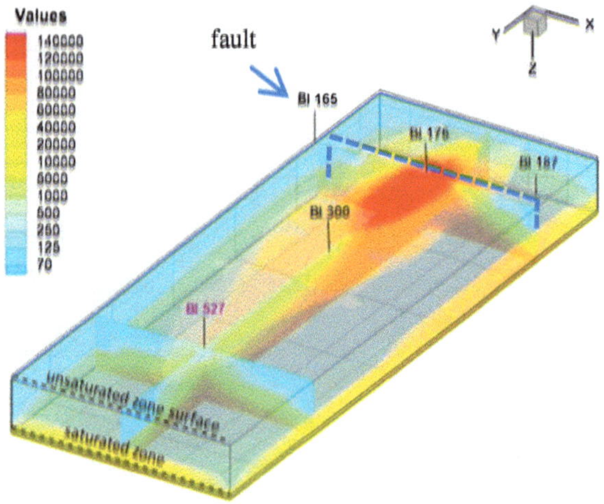

Fig. 14 Chloride concentration distribution in quaternary aquifer (mg/l) after t = 50a

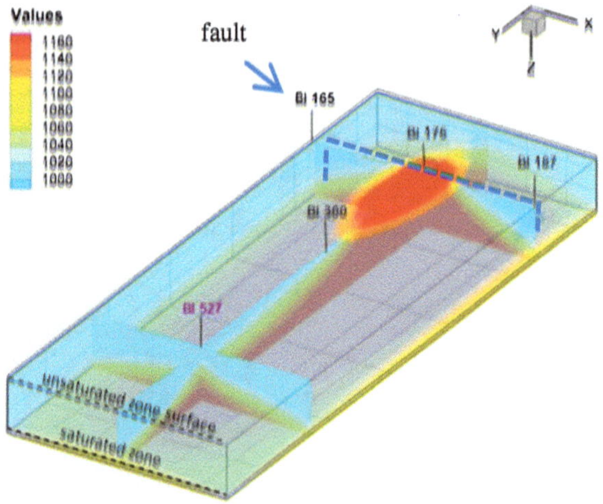

Fig. 15 Fluid density distribution in quaternary aquifer (kg/m^3) after t = 50a

saline aquifer was simulated for 25 scenarios (see module A). For this purpose, the intrinsic permeability of the reservoir was varied between 10^{-14} and 10^{-12} m^2 and that of the fault zone between 5×10^{-13} and 10^{-10} m^2. The initial porosity was 25 %. The pore pressure increases, as determined by module A, were imported into the geo-mechanical model in form of a one-way hydro-mechanical coupling as a time function. A maximum pore pressure increase of approximately 0.25 MPa at the lower end of the fault zone was reached.

As a first approximation, a 2D plane strain geo-mechanical simulation was used. To prevent influences of the boundaries, the synthetic model was extended to a width of 30 km. For the simulations 4-node plane strain elements were used. Ideal elastic, ideal plastic material behavior and the Mohr-Coulomb failure criterion were assumed. The 40 m wide fault zone was modelled with solid finite thickness elements as

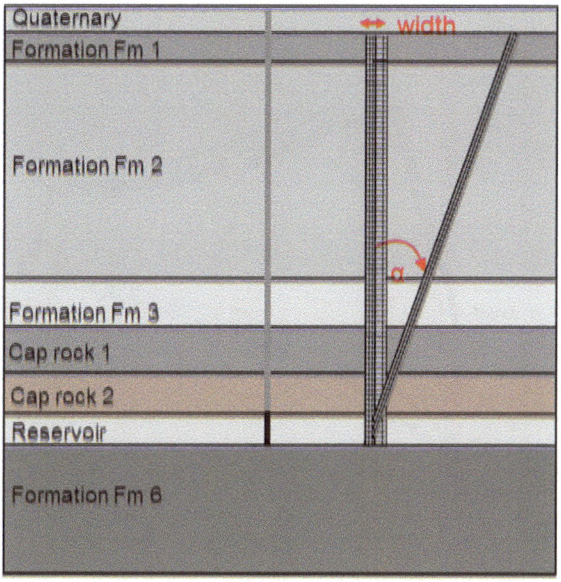

Fig. 16 Synthetic model applied in the present study (personal communication Adams, 2014). The fault width (Δb) and the inclination angle α have been marked in *red* in the figure

recommended by Cappa and Rutquist (2011), Gudmundsson et al. (2010) and Faulkner et al. (2010). Detailed information of the boundary conditions used the geometry and the rock parameters of the model are given by **personal communication Adams 2014**. A cross section of the synthetic model is shown in Fig. 16.

Exemplary, the changes of the opening width of the fault zone, also called hydraulic aperture Δb, as calculated for the 25 scenarios, are shown in Fig. 17. The results were generated at a calculation node in cap rock 2 as a function of time. Due to the high transmissivity of the fault zone, the maximum aperture was determined for scenario S1 to only about 3.5 mm. As the fault zone was modelled as porous media, this aperture equates to a very small volumetric strain of 0.009 which in fact doesn't result in a significant increase of the intrinsic permeability. Therefore, a fully hydro-mechanical coupling was not necessary. Furthermore, only elastic deformations of the fault zone and the reservoir occurred.

Volumetric strains as a function of time at certain depths of the fault zone were delivered to module 0 as input data for the risk assessment software.

Furthermore, a parametric study with hydro-mechanical single phase flow simulations was executed in the frame of module F to determine the conditions for which the volumetric strains of the fault zone are relevant regarding a fully hydro-mechanical coupling. Therefore, the transmissivity of the fault zone was reduced and the inclination was varied between 0° and 20°. The reduction of the transmissivity by 50 times leads to a maximum pore pressure increase of approximately 1 MPa at the foot of the fault zone, which is four times higher than in the 25 scenarios discussed above. In combination with an inclined fault zone, this pore pressure increase leads to a plastification of the fault zone, which causes volumetric

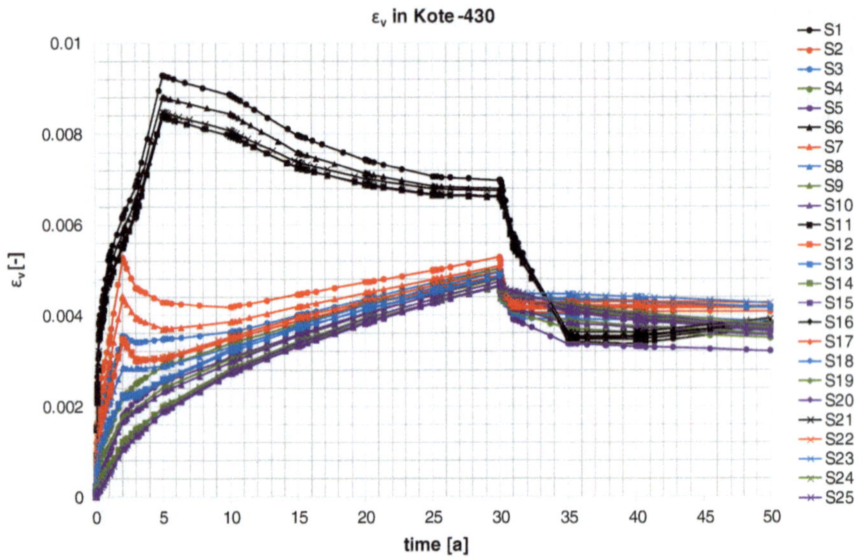

Fig. 17 Hydraulic aperture Δb of the fault at 430 m depth (personal communication Adams 2014), The different parameter combinations of reservoir permeability and initial fault permeability represent the scenarios S1–S25

strains up to 0.037. This results in an increase of the intrinsic permeability in about one order of magnitude.

Taking the results of the parametric study into account (**personal communication Adams 2014**) it can be summarized that significant volumetric strains, which require fully coupled hydro-mechanical simulations, only occur when plastification in the fault takes place.

2.8 Module (H) Flexible Interfaces

Because the general goal of the CO_2RINA-project was the development of a generalized methodology, the incorporated processes are not complete. To be able to incorporate additional processes, there should be the possibility to incorporate additional processes or models. This can be done in two ways: Internal coupling/refinement of process models or external coupling of models for additional risk pathways.

External coupling means, that the results of a simulation, i.e. concentrations or mass fluxes can be used as primary input for an additional model. One example is the CO_2 which leaks to the surface can be used as input for any kind of near surface atmospheric transport model.

Internal coupling means, that process parameters may change due to the incorporation of additional processes. This kind of coupling was demonstrated by incorporation of the effect of microbiological processes onto the porosity and permeability of the reservoir rocks which has been extensively investigated in the project CO_2BIOPERM (Hoth et al. this volume). The example is described there more extensively.

For the reservoir a simple form of the transfer function reads:

$$T\{\dot{m}_{inj}(t), K_{res}, K_{fault} \longrightarrow \text{transfer } \dot{m}_i(t), p_i(t), S_i(t)\}$$

with the abbreviations:

$\dot{m}_{inj}(t)$ injection rate
K permeability
θ porosity
$\dot{m}_i$ mass flux through leakage i
$p_i(t)$ pressure at point i
$S_i(t)$ saturation at point i

On the other hand side, for the microbiological effects there can be derived a transfer function which changes the average porosities and permeabilities due to the effect of microbiological processes. This transfer function reads:

The transfer function for K and θ to incorporate microbiological effects can be written:

$$T\left\{K, \theta, [c_\alpha(t)] \longrightarrow transfer\ \tilde{K}(t), \tilde{\theta}(t)\right\}$$

with the abbreviations:

$[c_\alpha(t)]$ vector of concentrations
$\tilde{K}(t)$ resulting average permeability
$\tilde{\theta}(t)$ resulting average porosity

The combination of both functions results in a new transfer function, in which microbial processes are incorporated:

$$T\{\dot{m}_{inj}(t), \tilde{K}_{res}(t), \tilde{K}_{fault}(t) \longrightarrow \text{transfer } \dot{m}_i(t), p_i(t), S_i(t)\}$$

By using this approach, the results of the CO_2BIOPERM project can be integrated into the CO_2RINA methodology.

3 Results and Conclusions

The selection of appropriate kinds of independent variables and transfer functions (=interfaces between the modules) was a major challenge. This has to be done for each coupling specifically.

Example for the transport in an alluvial aquifer: Transfer from inflowing mass flux at point r_1 to concentration at point r_2 by using convolution integral (aquifer—module E):

$$c(t) = \int_{\tau=0}^{\tau=t} T\left\{t - \tau, \dot{m}(\tau) \overset{transfer}{\longrightarrow} c\right\} * \dot{m}(\tau) * d\tau$$

Transfer function for constant mass fluxes:

The concentration has to be calculated for a set of constant mass fluxes in the related module (aquifer model in this example) which are the base for interpolation.

$$T(X_1, X_2, \ldots, X_N \overset{transfer}{\longrightarrow} Y_1, Y_2, \ldots, Y_N)$$

The approach is valid for several independent variables $X_1,\ldots,$ and dependent variables $Y_1,\ldots,Y_M$ by interpolation in a multi-dimensional space.

$$\dot{m}_i{:}T\left\{\tau, \dot{m}_i(\tau) \overset{transfer}{\longrightarrow} c\right\} = \frac{1}{\dot{m}_i} \frac{d}{d\tau} c(\tau, \dot{m}_i)$$

The validation of the approach by comparison of the overall model which was developed in module (A) with the coupling concept has been demonstrated at different examples. In Fig. 18 there is shown a benchmark calculation with the overall model for the reservoir, the fault and the alluvial aquifer in one model on the one hand side and the models coupled with transfer function on the other hand side for two different parameter scenarios. Figure 19 shows preliminary results regarding the incorporation of geomechanical effects.

The following conclusions can be derived from the work:

- The methodology was implemented and extensively tested
- Most of the interfaces work properly
- Some instabilities for coupling reservoir model and fault model via pressure and saturation
- A new coupling concept was developed and has to be completely implemented and tested
- The iterative coupling of reservoir model and geomechanical model is difficult to be handled
- A new coupling concept for geomechanical effects has been developed and has to be implemented and tested as well
- Implications to coupling of other modules have to be evaluated

The developed methodology is very general and a universal base for risk assessment which has to be adapted to a specific site. It can be adapted to each level of investigation and continuously improved.

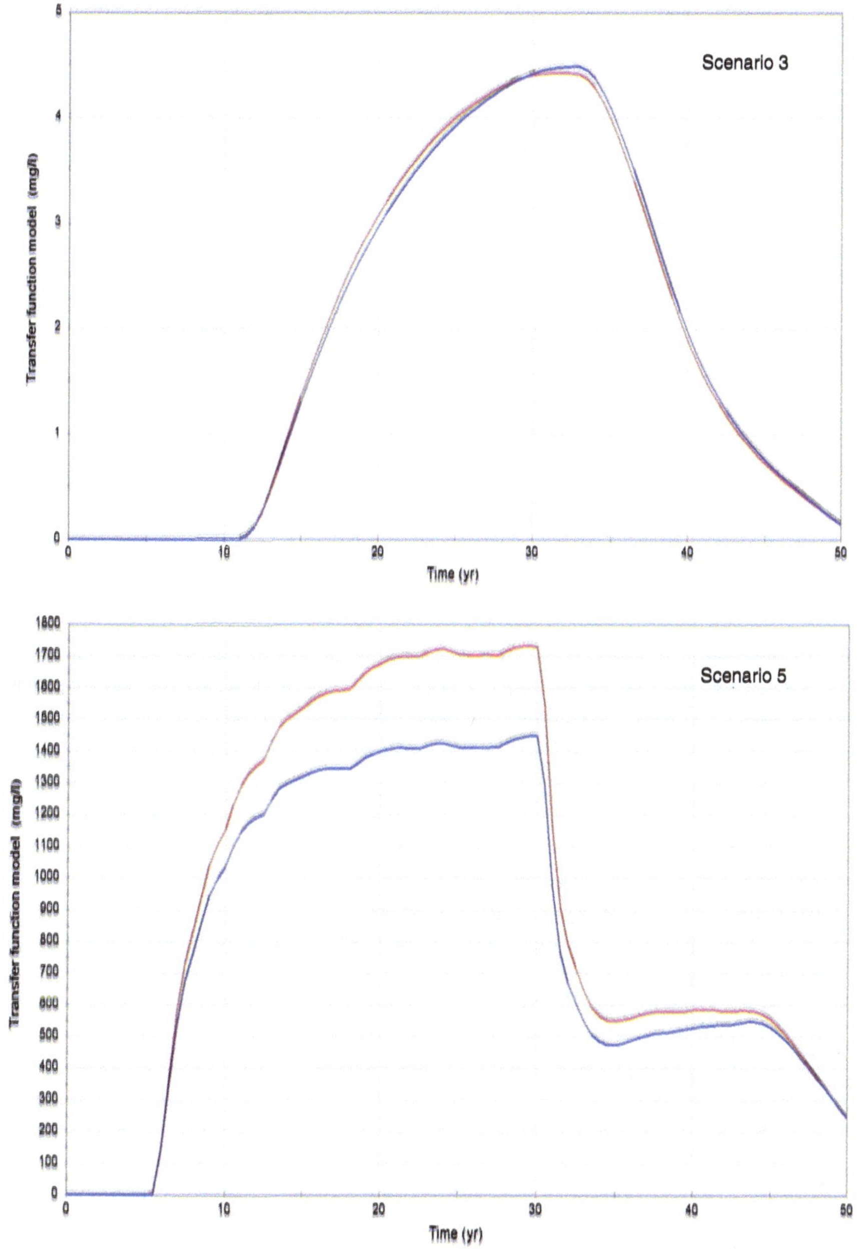

Fig. 18 Leakage through fault and migration in shallow aquifer—comparison between direct implementation in the overall model (*red*) versus an implementation by transfer functions for two different parameter scenarios. *Upper part*—scenario 3 for a low fault permeability and *lower part*—scenario 5 for a high fault permeability. For low concentrations the coupling shows very consistent results, while in scenario 5 with high concentrations there are differences between the two approaches up to about 20 %

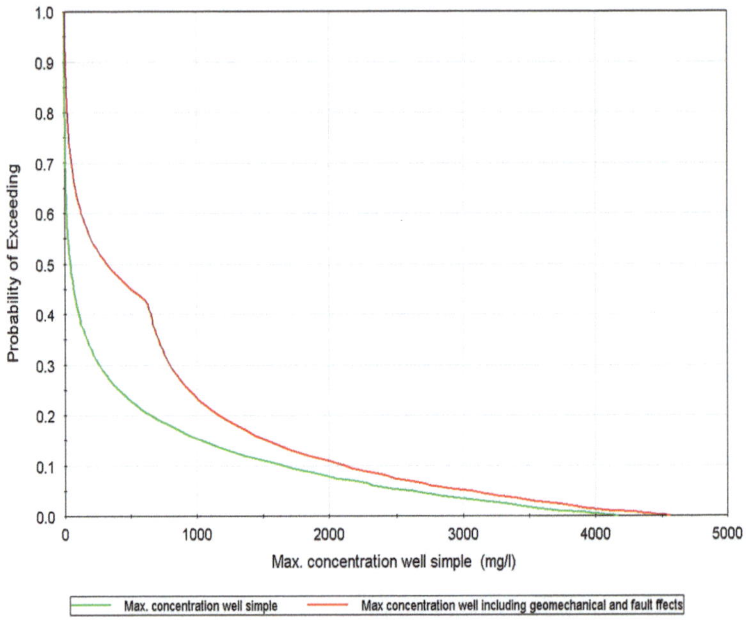

Fig. 19 Probability distribution for CO_2-concentration in a groundwater production well—comparison: *green* reservoir model + aquifer model, *red* reservoir model + fault model + geomechanical model + aquifer model. The difference between the distributions show the additional contributions of a detailed fault model and the incorporation of geomechanical effects

With required adaptations it can be applied to similar application fields in the use of the underground like:

- storage of hydrogen and natural gas
- deposition of brine and final waste deposits
- primary production of geothermal energy, especially petrothermal energy and production of shale gas

Acknowledgments The authors would also like to acknowledge the German Federal Ministry of Education and Research (BMBF—Grant number??).

References

Amann-Hildenbrand A, Bertier P, Busch A, Krooss BM (2013) Experimental investigation of the sealing capacity of generic clay-rich cap rocks. Int J Greenhouse Gas Control 123:20–33

Bielinski A, Kopp A, Schütt H, Class H et al (2008) Monitoring of CO_2 plumes during storage in geological formations using temperature signals: numerical investigation. Int J Greenhouse Gas Control 2(3):319–328

Busch A, Alles S, Gensterblum Y, Prinz D, Dewhurst DN, Raven MD, Stanjek H, Krooss BM (2008) Carbon dioxide storage potential of shales. Int J Greenhouse Gas Control 2:297–308

Busch A, Amann-Hildenbrand A (2013) Predicting capillarity of mudrocks. Mar Pet Geol 45:208–223
Cappa F, Rutquist J (2011) Modeling of coupled deformation and permeability evolution during fault reactivation induced by deep underground injection of CO_2. Int J Greenhouse Gas Control 5:336–346
Celia MA, Bachu S, Nordbotten JM, Gasda SE, Dahle HK (2005) Quantitative estimation of CO_2 leakage from geological storage: analytical models, numerical models and data needs. In: Proceedings of 7th international conference on greenhouse gas control technologies (GHGT-7), Vancouver, Canada, 5–9 Sept 2004, v.I, pp 663–672
Condor J, Unatrakarn D (2010) A comparative analysis of risk assessment methodologies for geological storage of carbon dioxide. Greenhouse gas control technologies (GHGT) conference, Amsterdam
Crank J (1975) The mathematics of diffusion. Clarendon Press, Oxford
CSLF-T (2009) Phase I final report from CSLF risk assessment task force. Carbon sequestration leadership forum, CSLF-T-2009-04, Oct 2009. www.cslforum.org
Cugini A, Guthrie G, DePaolo D, Fox M, Friedmann J, Virden J (2010) US-DOE's National risk assessment program: bridging the gap to provide the science base to ensure successful CO_2 storage. Greenhouse gas control technologies (GHGT) conference, Amsterdam
Deel (2007) Summary of DOE's monitoring, mitigation, and verification program and modeling program, Dawn Marie DeelProject manager carbon sequestration program. National Energy Technology Laboratory
DOE (2009) Monitoring, verification, and accounting of CO_2 stored in deep geologic formations. DOE/NETL-311/081508, Jan 2009. National Energy Technology Laboratory. www.netl.doe.gov
Duan Z, Sun R (2003) An improved model calculating CO_2 solubility in pure water and aqueous NaCl solutions from 273 to 533 K and from 0 to 2000 bar. Chem Geol 193:257–271
Eckart M (2011) BoxModel concept: ReacFlow3D—modelling of the flow of mine water and groundwater, mass and heat transport—program description. DMT GmbH & Co. KG, Essen, Germany
Espie T (2005) The status of risk assessment for geologic storage of CO_2. European CO_2 capture and storage conference—towards zero emission power plants
Faulkner DR, Jackson CAL, Lunn RJ (2010) A review of recent developments concerning the structure, mechanics and fluid flow properties of fault zones. J Struct Geol 32(11):1557–1575
Flemisch B, Fritz J, Helmig R, Niessner J, Wohlmuth B (2007) Dumux: a multi-scale multi-physics toolbox for flow and transport processes in porous media. In: Ibrahimbegovic A, Dias F (eds) ECCOMAS thematic conference on multi-scale computational methods for solids and fluids, Cachan, France, 28–30 Nov 2007
Gudmundsson A, Simmenes TH, Larsen B, Philipp SL (2010) Effects of internal structure and local stresses on fracture propagation, deflection and arrest in fault zones. J Struct Geol 32 (11):1643–1655
Helmig R (1997) Multiphase flow and transport processes in the subsurface—a contribution to the modeling of hydrosystems. Springer, New York
Hildenbrand A, Krooss BM, Schlömer S, Littke R (2003) Dynamic gas leakage through fine-grained seal lithologies. EAGE conference 8–11 September 2003: fault and top seals. What do we know and where do we go? Montpellier, France, pp O-15, 11-10
Hildenbrand A, Schlömer S, Krooss BM, Littke R (2004) Gas breakthrough experiments on pelitic rocks: comparative study with N_2, CO_2 and CH_4. Geofluids 4:61–80
Ho C-H (1992) Risk assessment for the Yucca Mountain high-level nuclear waste repository site: estimation of volcanic disruption. Math Geol 24(4):347–364. doi:10.1007/BF00891268
Jagger M (2009) CCS risk assessment. IEA Summer School, Melbourne
Kahnt R (2006) System simulation mit GoldSim. Delta-h Tagung Dortmund
Kahnt R (2008) Unterstützung bei der Erstellung eines Systemsimulationsmodells für die Systemmodellierung und Risikoanalyse für CO_2-Untergrundspeicher auf der Basis von GoldSim. AITEMIN, Spanien

Kahnt R, Paul M (2008) Integrated methodology for the optimisation of mine closure. Mine closure conference Johannesburg, SA

Kempka T, Kühn M, Class H, Frykman P, Kopp A, Nielsen C, Probst P (2010a) Predictive modelling of Ketzin—CO_2 arrival in the observation well. Int J Greenhouse Gas Control (in press)

Kempka T, Kühn M (2013) Numerical simulations of CO_2 arrival times and reservoir pressure coincide with observations from the Ketzin pilot site, Germany. Environ Earth Sci 70 (8):3675–3685

Kempka T, Kühn M, Class H, Frykman P, Kopp A, Nielsen CM, Probst P (2010b) Modelling of CO_2 arrival time at Ketzin—part I. Int J Greenhouse Gas Control 4(6):1007–1015

Klein E, De Lucia M, Kempka T, Kühn M (2013) Evaluation of long-term mineral trapping at the Ketzin pilot site for CO_2 storage: an integrative approach using geochemical modelling and reservoir simulation. Int J Greenhouse Gas Control 19:720–730

Krooss BM, Leythaeuser D, Schaefer RG (1992a) The quantification of diffusive hydrocarbon losses through cap rocks of natural gas reservoirs-A reevaluation. AAPG Bull 76:403–406

Krooss BM, Leythaeuser D, Schaefer RG (1992b) The quantification of diffusive hydrocarbon losses through cap rocks of natural gas reservoirs-A reevaluation: reply. AAPG Bull 76:1842–1846

LUA (2007) Studien und Tagungsberichte des Landesumweltamtes Band 55. Bericht zur Grundwasserbeschaffenheit im Land Brandenburg für den Zeitraum 2001 bis 2005. Anhang 1. http://www.lugv.brandenburg.de/cms/media.php/lbm1.a.3310.de/wbd55_a1.pdf. Accessed 08 Aug 2014

Maul P, Metcalfe R, Pearce J, Savage D, West J et al (2007) Performance assessments for the geological storage of carbon dioxide: learning from the radioactive wast disposal experience. Int J Greenhouse Gas Control 1(2007):444–455

McDermott CI, Wang WQ, Kolditz O (2010): A hybrid analytical finite element approach for two-phase flow applied to supercritical CO_2 replacing brine in a heterogeneous cap rock, WRR, submitted

Metz et al (2005) IPCC special report on carbon dioxide capture and storage. In: Metz B, Davidson O, de Coninck HC, Loos M, Meyer L (eds) Prepared by working group III of the intergovernmental panel on climate change

Nordbotten JM, Celia MA, Bachu S et al (2005) Injection and storage of CO_2 in deep saline aquifers: analytical solution for CO_2 plume evolution during injection. Transp Porous Media 58:339–360. doi:10.1007/s11242-004-0670-9

North DW (1999) A perspective on nuclear waste. Risk Anal 19:751–758

Nowak T, Kunz H, Dixon D, Wang WQ, Kolditz O (2010) THM-coupled numerical modelling at large scale with high performance computing: application to the Whiteshell underground laboratory. Int. J. Numer. Anal. Meth. Geomech, under revision

Oladyshkin S, Class H, Helmig R, Nowak W (2009) An integrative approach to robust design and probabilistic risk assessment for CO_2 storage in geological formations. Comput Geosci, Springer, 15(3):565–577, 2011. doi:10.1007/s10596-011-9224-8

Oldenburg C (2007) Screening and ranking framework for geologic CO_2 storage site selection on the basis of health, safety, and environmental risk. Environ Geol

Park C-H, Taron J, Görke J-U, Wang W, Kolditz O (2010) Hydromechanical analysis of caprock failure due to CO_2 injection in a deep saline aquifer: Numerical study, WRR (submitted)

Pawar R, Carey J, Chipera S, Fessenden J, Kaszuba J, Keating G, Lichtner P, Olsen S, Stauffer PH, Viswanathan H, Ziock H, Guthrie G (2006) Development of a framework for long-term performance assessment of geologic CO_2 sequestration sites. In: Proceedings of the 8th international conference on greenhouse gas control technologies, Trondheim, NO

Pruess K (2004) The tough codes—a family of simulation tools for multiphase flow and transport processes in permeable media. Vadose Zone J 3:738–746

Pruess K (2005a) ECO_2 N: a TOUGH2 fluid property module for mixtures of water, NaCl, and CO_2, earth sciences division. Lawrence Berkeley National Laboratory University of California, Berkeley, CA 94720, LBNL-57952

Pruess K (2005b) ECO_2 N: a TOUGH2 fluid property module for mixtures of water, NaCl, and CO_2. Report LBNL-57952. Lawrence Berkeley National Laboratory, Berkeley, CA

Pruess K, Bielinski A, Ennis-King J, Fabriol R, Le Gallo Y, Garcia J, Jessen K, Kovscek T, Law DH-S, Lichtner P, Oldenburg C, Pawar R, Rutqvist J, Steefel C, Travis B, Tsang C-F, White S, Xu T (2003) Code intercomparison builds confidence in numerical models for geologic disposal of CO_2. In: Gale J, Kaya Y (eds) GHGT-6 conference proceedings: greenhouse gas control technologies, Kyoto, Japan, pp 463–470

Pruess K, Spycher N (2007) ECO_2n—a fluid property module for the tough2 code for studies of CO_2 storage in saline aquifers, energy conversion and management. Energy Convers Manag 48 (6):1761–1767. doi:10.1016/j.enconman.2007.01.016

Rutqvist J, Tsang C-F (2002) A study of caprock hydromechanical changes associated with CO_2 injection into a brine aquifer. Environ Geol 42:296–305

Rutqvist J, Wu Y-S, Tsang C-F, Bodvarsson GA et al (2002) Modeling approach for analysis of coupled multiphase fluid flow, heat transfer, and deformation in fractured porous rock. Int J Rock Mech Min Sci 39:429–442

Rutqvist J, Birkholzer J, Cappa F, Tsang C-F et al (2007) Estimating maximum sustainable injection pressure during geological sequestration of CO_2 using coupled fluid flow and geomechanical fault-slip analysis. Energy Convers Manage 48:1798–1807

Rutqvist J, Birkholzer JT, Tsang CF et al (2008) Coupled reservoir-geomechanical analysis of the potential for tensile and shear failure associated with CO_2 injection in multilayered reservoir–caprock systems. Int J Rock Mech Min Sci 45:132–143

Rutqvist J, Vasco DW, Myer L (2009a) Coupled reservoir-geomechanical analysis of CO_2 injection and ground deformations at In Salah, Algeria. Int J Greenhouse Gas Control. doi:10.1016/j.ijggc.2009.10.017

Rutqvist J, Barr D, Birkholzer JT, Fujisaki K, Kolditz O, Liu Q-S, Fujita T, Wang W, Zhang C-Y et al (2009b) A comparative simulation study of coupled THM processes and their effect on fractured rock permeability around nuclear waste repositories. Environ Geol 57(6):1347–1360

Saripalli KP, Mahasenan NM, Cook EM (2003) Risk and hazard assessment for projects involving the geological sequestration of CO_2. In: Gale J, Kaya Y (eds) Proceedings of the 6th international conference on greenhouse gas control technologies (GHGT-6), Kyoto, Japan, Pergamon, 1–4 Oct 2002, v.I, pp 511–516

Schlömer 1998) Abdichtungseigenschaften pelitischer Gesteine—Experimentelle Charakterisierung und geologische Relevanz, Institut für Chemie und Dynamik der Geosphäre 4, Jülich, p 212

Seis W, Staub M, Massat L, Grützmacher G, Thomas L, Taute T (2013) Geological CO_2 storage and other emerging subsurface activities—catalogue of potential impacts on drinking water production. Report Cosma-1 D 1. Kompetenz-Zentrum Wasser, Berlin, Germany

Song J, Zhang D (2013) Comprehensive review of cap rock-sealing mechanisms for geologic carbon sequestration. Environ Sci Technol 47(1):9–22

Stauffer PH, Viswanathan H et al (2005) CO_2-PENS: a CO_2 sequestration systems model supporting risk-based decisions. AGU Fall Meeting

Stauffer PH, Viswanathan HS, Pawar RJ, Klasky ML, Guthrie GD (2006) CO_2-PENS a CO_2 sequestration system model supporting risk-based decisions. In: Proceedings of the 16th international conference on computational methods in water resources, Copenhagen, Denmark, 19–22 June 2006

Stauffer P, Viswanathan H, Pawar R, Guthrie G (2007) Predicting engineered natural systems (PENs): applying GoldSim to geological $C0_2$ sequestration. GoldSim Newsletter

Stauffer PH, Viswanathan HS, Pawar RJ, Guthrie GD et al (2009) A system model for geologic sequestration of carbon dioxide. Environ Sci Technol 43:565–570

Viswanathan HS, Stauffer PH et al (2005) The development of a performance assessment framework for geologic CO_2 sequestration. AGU Fall Meeting

Viswanathan HS, Pawar RJ, Stauffer PH, Kaszuba JP, Carey JW, Olsen SC, Keating GH, Kavetski D, Guthrie GD et al (2008) Development of a hybrid process and system model for the assessment of wellbore leakage at a geologic CO_2 sequestration site. Environ Sci Technol 42:7280–7286

Walton FC, Tait JC, LeNeveu D, Sheppard MI (2005) Geological storage of CO_2: a statistical approach to assessing performance and risk. In: Proceedings of the 7th international conference on greenhouse gas control technologies (GHGT-7), Vancouver, Canada, 5–9 Sept 2004, v.I, pp 693–700

Wang WQ, Kosakowski G, Kolditz O et al (2009) A parallel finite element scheme for thermo-hydro-mechanical (THM) coupled problems in porous media. Comput Geosci 35(8):1631–1641

Watanabe N, McDermott C, Wang W, Taniguchi T, Kolditz O et al (2010) Uncertainty analysis of thermo-hydro-mechanical processes in heterogeneous porous media. Comput Mech 45 (4):263–280

Wildenborg AFB, Leijnse AL, Kreft E, Nepveu MN, Obdam ANM, Orlic B, Wipfler EL, van der Grift B, van Kesteren W, Gaus I, Czernichowski-Lauriol I, Torfs P, Wojcik R (2005) Risk assessment methodology for CO_2 sequestration scenario approach, carbon dioxide capture for storage in deep geologic formations—results from the co_2 capture project. In: Benson SM (ed) v. 2: geologic storage of carbon dioxide with monitoring and verification. Elsevier Science, London, pp 1293–1316

Zhang Y, Oldenburg CM, Finsterle S, Bodvarsson GS (2006) System-level modeling for geological storage of CO_2: TOUGH2 users conference, earth sciences division. Lawrence Berkeley National Laboratory, University of California, Berkeley, California

Zhang K, Wu Y-S, Pruess K (2008) User's guide for TOUGH2-MP—a massively parallel version of the TOUGH2 code. Report LBNL-315E. Earth Sciences Division, Lawrence Berkeley National Laboratory, Berkeley, CA

Zhou W, Stenhouse MJ, Arthur R, Whittaker S, Law DH-S, Chalaturnyk R, Jazwari W (2005) The IEA Weyburn CO_2 monitoring and storage project—modeling of the long-term migration of CO_2 from Weyburn. In: Proceedings of the 7th international conference on greenhouse gas control technologies (GHGT-7), 5–9 Sept 2004, Vancouver, Canada, v.I, pp 721–730. Volume 1: peer-reviewed papers and plenary presentations. Elsevier, UK

Zyvoloski GA, Robinson BA, Dash ZV, Trease LL (1997) Use's manual for the FEHM application—a finite-element heat- and mass-transfer code. LA-13306-M. Los Alamos National Laboratory, Los Alamos, New Mexico

Saltwater Monitoring Using Long-Electrode ERT

Thomas Günther, Mathias Ronczka and Thomas Voß

Abstract Saltwater rise is a possible result of CO_2 injection into deep saline aquifers that could threaten aquifer quality. Hence, long-term monitoring of the groundwater system is a required task during and after the injection phase. However, point measurements in boreholes cannot describe large systems with sufficient precision. Therefore cost-efficient monitoring methods are needed, like surface geophysical techniques. Electrical Resistivity Tomography (ERT) is well-suited for saltwater problems and can be deployed using steel-cased boreholes as electrodes, so-called long electrode (LE) ERT. In the project Saltwater Monitoring using Long Electrode Geoelectrics (SaMoLEG) a research institute group focussing on numerical modelling cooperates with a private borehole logging company that has long been active in the region of interest, eastern Brandenburg, where CO_2 injection was planned. A numerical framework for simulation and data analysis is developed and verified using a controlled laboratory experiment. Two differently scaled field sites in the Federal State of Brandenburg, Germany, where natural saltwater rise occurs were repeatedly investigated by LE-ERT. The medium-scale site was permanently wired and allows cost-efficient monitoring. The inversion results agree well with geology and measured in-situ fluid conductivities. The large-scale site, at the area of an existing water works, proves the applicability of the developed method in the catchment scale and gives valuable insight into deeper salinisation. The results show that the LE-ERT method is a suitable and cost-efficient method for saltwater rise and groundwater quality monitoring in the frame of geological CO_2 storage and is ready for application at real storage sites.

T. Günther (✉) · M. Ronczka
Leibniz Institute for Applied Geophysics (LIAG), Stilleweg 2, 30655 Hannover, Germany
e-mail: thomas.guenther@liag-hannover.de

T. Voß
Bohrlochmessungen—Storkow GmbH, Schützenstraße 33, 15859 Storkow, Germany

A. Liebscher and U. Münch (eds.), *Geological Storage of CO_2 – Long Term Security Aspects*, Advanced Technologies in Earth Sciences,
DOI 10.1007/978-3-319-13930-2_8

1 Introduction

From the variety of different CCS technologies the storage of captured CO_2 in saline aquifers is a promising approach due to the abundance of this type of storage formations in the sedimentary basins worldwide. Particularly the North German sedimentary basin (NGSB) features many potential CO_2-storage structures, for example salt pillows overlain by saltwater bearing sandstones, sealed to the surface by thick claystone formations.

While some of the injected CO_2 will be dissolved in the pore water, the major part of the injected CO_2 displaces formation water with an injection pressure exceeding the hydraulic potential. So, even if the storage structure is tight against the buoyant force of the CO_2, the overall increase of formation pressure might cause saltwater to migrate into higher, freshwater-bearing aquifers. Due to natural and anthropogenic causes in many areas of the NGSB the salt-/freshwater boundary is already close to the surface, as it is the case in the central part of the German federal state Brandenburg, which is chosen as the study area of the SaMoLEG project. In some cases, the shallow freshwater/saltwater boundary affects even the production of water works that have to avoid further rise of saltwater by high production rates. Therefore, the further development of efficient saltwater monitoring methods that could be integrated in an early warning system plays an important role in the future for environmentally safe geological CO_2-storage in saline aquifers, not only in the study area, but for all onshore potential storage sites worldwide where groundwater is the main source of drinking water.

The basic and still most common method to measure and monitor groundwater salinity utilizes a network of groundwater measuring wells (GMW) distributed in the survey area (Hannapel et al. 2007). If properly constructed, these wells allow taking depth-selective water samples that can be analysed in laboratories regarding different parameters like ion-sensitive salt contents or even the age of the water by measuring isotope ratios. The major drawback of GMWs is that only point-wise information about the groundwater mineralisation can be retrieved. Another disadvantage is the fact that if the saltwater/freshwater boundary has already reached the filter screen of the GMW, information about further saltwater rise cannot be deduced by sampling anymore.

When salinisation is measured as a total value, GMWs can be also used to conduct inductive electrical well logging inside the casing which allows the monitoring of pore water conductivity of all penetrated aquifers. Borehole logging companies have been applying this method for many years. While the vertical resolution of induction logging is very high ($\sim$10 cm), the information retrieved from one borehole represents only a thin cylinder of the full underground volume. Furthermore induction logging works only in electrically non-conductive casings (e.g., PVC or PE), which are nowadays mostly used for new GMWs.

Surface geophysical methods can be used to close the gap between boreholes by providing two-dimensional or three-dimensional images. Amongst them, electrical and electromagnetic measurements are best suited due to the strong relation of the imaged bulk resistivity to fluid conductivity. Attwa et al. (2011) gave a comparison of methods for imaging saltwater intrusion. As a fast method for long 2D profiles they identified frequency domain electromagnetics (FDEM) on the ground (Kempka et al. 2015), which is however very sensitive to noise and electric conductors such as fences. Airborne FDEM (e.g., Siemon et al. 2009) is a very important tool for characterizing aquifers in the catchment scale, but is limited with regard to investigation depth and cost, particularly if it comes to monitoring. A classical and cost-saving method is electrical resistivity tomography (ERT) on surface 2D profiles, but this method is also restricted with regard to investigation depth. Transient electromagnetics (TEM) can accurately resolve good conductors at great investigation depth but is mainly limited to layered structures.

The ERT method is geometrically scalable due to its static nature. Large penetration depths can be reached if large distances between the electrodes and large signal strengths can be accomplished. This requires sufficiently large electrode surfaces, which is satisfied if steel casings of existing boreholes can be deployed. The so-called Long-Electrode (LE) ERT method was first suggested by Daily et al. (2004) to monitor reservoirs during EOR operations at low cost. Rucker et al. (2010) applied it to near-surface site characterization and used it for monitoring (Rucker et al. 2011), but all in a rather small scale. For the data inversion process they simulated the boreholes by adding conductive cells at the borehole location. A more rigorous approach of simulating extended boreholes was introduced by Rücker and Günther (2011) and is referred to as complete electrode model (CEM). The electrode surface is discretised and coupled to the subsurface using a contact impedance. This approach is followed here.

A special feature of many areas in the eastern part of the NGSB is the exceptionally high density of abandoned GMWs, drilled before 1989 during extensive exploration campaigns for lignite and groundwater with depths ranging from 50 to 300 m. They were mostly equipped with steel casings. The locations of these dense borehole networks coincide in many cases with potential CO_2-storage sites making them a promising instrument for a cost-efficient saltwater monitoring. The concept of the SaMoLEG project is to use these steel-cased wells as long electrodes for LE-ERT measurements. If still accessible, old wells can provide water samples for calibration of the ERT data inversion. Another advantage of the SaMoLEG approach is the availability of the usually well documented drilling reports of these wells (most of them can be found in the archives of the state authorities) which can be used to extract the geological background from the measured ERT data.

2 Numerical Techniques

2.1 Modelling

The geoelectrical boundary value problem (BVP) reads

$$\nabla \cdot (\sigma \nabla u) = -\nabla \cdot \mathbf{j} \text{ in } \Omega$$
$$\frac{\partial u}{\partial n} + \alpha u = 0 \text{ on } \delta\Omega$$

It is solved using the Finite Element (FE) method (Rücker et al. 2006). Incorporating extended electrodes using the complete electrode model (CEM) adds two further equations to the BVP (Rücker and Günther 2011):

$$z_l \sigma \frac{\partial u}{\partial n} + u = U_l \text{ on } \delta\Omega_E$$
$$\int \sigma \frac{\partial u}{\partial n} ds = I$$

They define the coupling of the potentials on the electrode surface and in the subsurface. Furthermore, it is ensured that the total injected current is integrated. Figure 1 shows the current distribution for a Wenner array with 100 m spaced, 100 m deep boreholes in a two-layer case. Current flow is mainly horizontal in the upper layer, but significantly stronger and more vertical in the lower.

Figure 2 shows an exemplary synthetic modelling for a laterally moving saltwater body. Saltwater is already detected far away from the electrodes and the effects are stronger for boreholes compared to surface electrodes (Voss et al. 2013).

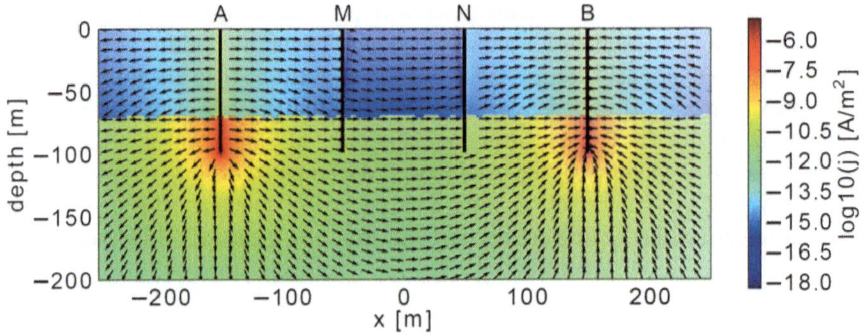

Fig. 1 Distribution of current density amplitude (*colour*) and direction for a Wenner array (current injection into *A* and *B* and potential measurement through *M* and *N*) with 100 m long casings in a two-layer case (100 Ωm over 1 Ωm)

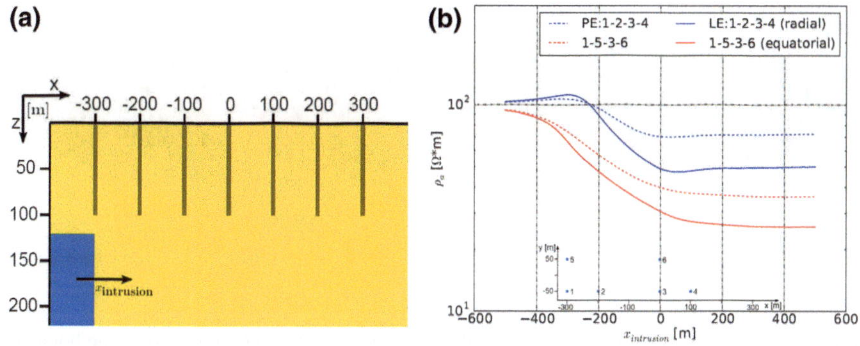

Fig. 2 **a** Model of a saltwater body (1 Ωm; *blue box*) moving horizontally in a homogeneous half-space (100 Ωm); **b** resulting apparent resistivity for radial (*blue*) and equatorial (*red*) dipole-dipole arrays as a function of saltwater position for long electrodes (LE, *solid lines*) and point electrodes (PE, *dashed lines*)

Varying contact impedances along electrodes affect quadrupole measurements only to a limited degree (Ronczka et al. 2013).

2.2 Inversion

We use the open-source software Boundless Electrical Resistivity Tomography (BERT), which is explained in more detail by Günther et al. (2006). A Gauss-Newton scheme is used to minimize a combination of error-weighted data misfit and model roughness that are weighted by a regularization parameter. The CEM is integrated to allow for borehole electrodes using a total potential approach (Rücker et al. 2006) for the forward calculation. We make use of triangular prism elements for both finite element calculations and inversion which allows for improved handling of predominantly layered structures.

For testing the methodology, we created a synthetic model (Fig. 3a) that is similar to the first field case. A conductive (3 Ωm) saltwater body penetrates a sandy layer (200 Ωm) lying on top of an impermeable silt layer (50 Ωm). A grid of 5 × 3 boreholes with 50 m depth was assumed and the synthetic data were contaminated with 2 % plus 100 μV Gaussian noise. The inversion result shown in Fig. 3b demonstrates that both the layer boundaries and the good conductor can be resolved. However, the latter appears only in the middle part, where resolution is good, and at greater depth. Furthermore, it dominates the layer boundary in the whole left part, whereas in the right part the lithology is well reproduced.

Whereas in static inversion the roughness of the model is minimized, in time-lapse inversion the resistivity ratio between subsequent time steps is constrained (LaBrecque and Yang 2001). The so-called difference inversion avoids artifacts in the models and is furthermore able to remove systematic error sources.

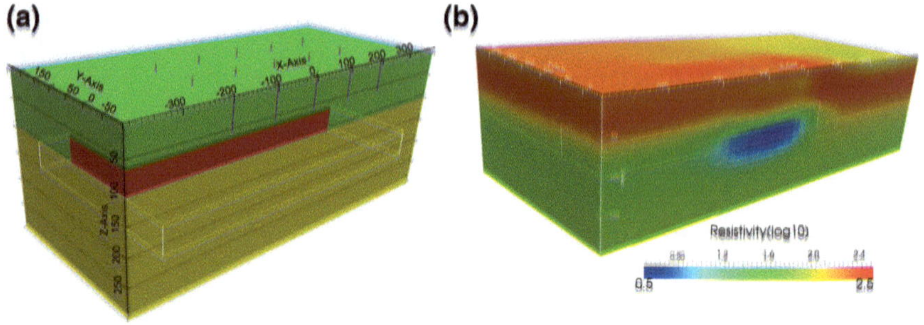

Fig. 3 (**a**) Synthetic model comprising a saltwater block (3 Ωm; *red*) moving over the boundary between a sandy (200 Ωm; *green*) and a silty (50 Ωm; *brown*) layer. A 5 × 3 grid of 100 m long electrodes (*purple*) is considered for simulating data and reconstructing the resistivity distribution (**b**)

2.3 Sensitivity and Resolution

The sensitivity function describes how the measured apparent resistivity changes as a result of a resistivity change in the subsurface. It illustrates the region of detection, but also helps to understand resolution properties. In Fig. 4 two exemplary sensitivity distributions are illustrated for different dipole-dipole configurations at homogeneous conditions. Pure LE measurements with equally long electrodes shift the point dipole-dipole sensitivity down to borehole depth. The constant sensitivity above indicates a loss of vertical resolution as stated by Rucker et al. (2011). Varying borehole lengths or combinations with point electrodes can regain some of the lost vertical resolution (Ronczka et al. 2013).

A more thorough appraisal of resolution properties can be achieved by computing the formal model resolution matrix and deriving a resolution radius (Friedel 2003). Lateral resolution is, as expected, in the range of medium borehole distances, vertical resolution strongly depends on the distribution of the borehole lengths and can significantly be improved by additional surface electrodes.

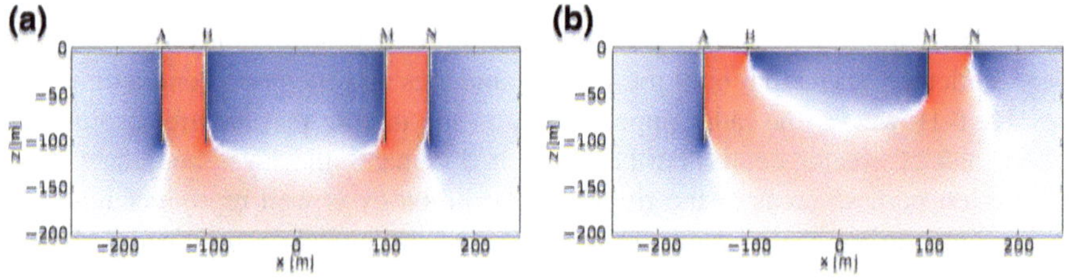

Fig. 4 Sensitivity distribution (positive/*red*, zero/*white*, negative/*blue*) of dipole-dipole array (current injection into *A*–*B* and potential measurement along *M*–*N*) using long electrodes (**a**) and mixed borehole lengths with point electrodes (**b**)

2.4 Optimization of Experimental Design

The optimization of the measuring protocol, i.e. the choice of quadrupole combinations to be used, is of crucial importance for cost and time of experiments, particularly if fast processes are to be monitored. The central objective is maximum resolution for a given number of data or, conversely, the minimum number of data for a desired resolution threshold. Different from other experiments, the effort of a large-scale dipole-dipole survey with independent data loggers is the number of current injections, since all voltages are registered simultaneously. Similar to Stummer et al. (2004), we use the diagonals of the data information matrix for predicting the performance of a measuring sequence. Starting with all logistically possible current injections, we iteratively remove those with the least total contribution of individual information contents until an appropriate trade-off between resolution and effort for the experiment is reached.

3 Laboratory Measurements

Several laboratory experiments on different tank and electrode geometries were conducted of which only one is reported here. A tank of 90 × 75 × 45 cm extension was filled with two layers of sandy material with slightly different grain sizes (both ~1 mm). Twelve electrodes were inserted 20 cm deep in a 4 × 3 grid. An optimized electrode array of 140 measurements was repeatedly measured (in total 150 frames in ~5 h).

First, the tank was fully saturated with tap water (ρ_f = 20 Ωm). A hydraulic gradient of 1 cm/m was established between two semi-permeable walls of the tank. Then, saltwater (ρ_f = 4.2 Ωm) was mixed with a coloured dye tracer and inserted in the source chamber. Due to the hydraulic gradient it moved horizontally through the tank. Repeated photographs at the front-wall were taken to visually verify the reconstructed resistivity distribution. After the tank was completely flooded, tap water was injected in the source chamber and displaced the saltwater again. Using the constant initial state, a formation factor of 4.3 was derived that can be used to derive fluid conductivity from the inversion results. Tap water corresponds to a bulk resistivity of 86 Ωm, whereas saltwater leads to 18.1 Ωm.

Figure 5 shows the comparison between photographs and reconstructed resistivity for three time steps (early and late saltwater flooding, freshwater displacement). Generally, the images exhibit similar behaviour and the chosen iso-surface threshold (30 Ωm) of the smooth resistivity distribution corresponds well with the sharp infiltration front delineated by the dye tracer. The horizontal hydraulic conductivity in the upper layer is slightly higher than in the lower. Faster movements seem to occur at the side walls which could be attributed to boundary effects, i.e. increased porosity and thus hydraulic conductivity at the side walls.

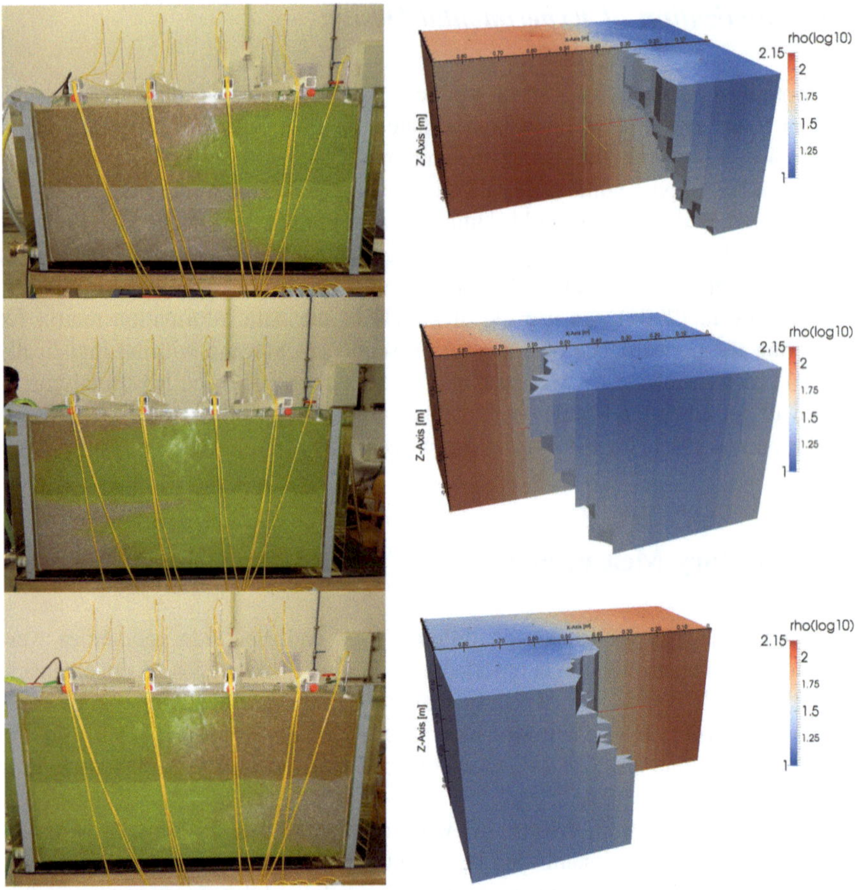

Fig. 5 Photographs (*left*) and reconstructed resistivity (*right*) for three time steps (1 h, 1 h 30 min, 2 h 20 min after tracer injection). First, bromide tracer (*yellow*) was injected from the right before tap water was re-injected after ~1 h 50 min

4 Field Application

4.1 Geology and Description of Survey Areas

Two survey areas were chosen for the medium- and large-scale field experiments according to the criteria accessibility, knowledge of geological settings, anomalously high saltwater-/freshwater boundary as well as a high density of steel-cased boreholes. Both are situated in the central eastern part of the federal state Brandenburg, Germany, only about 15 km separated from each other. From adjacent towns or villages the test sites are named Briesen (large-scale) and Müllrose (medium-scale, see Fig. 6). Geomorphologically the survey areas are part of the

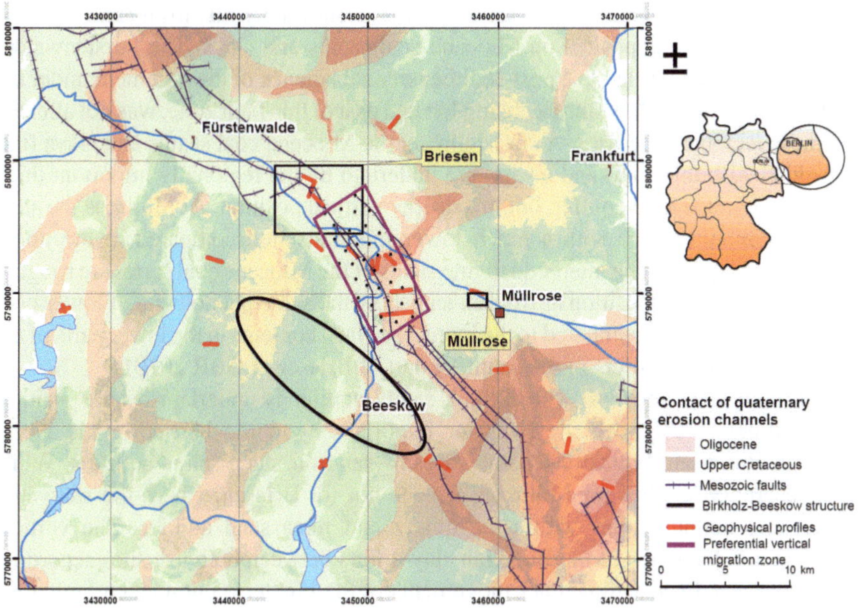

Fig. 6 Geomorphology and geology (NW-SE striking Guben-Fürstenwald Mesozoic fault zone and erosion channels) of eastern Brandenburg with location of the two survey areas. Planned CO_2 injection structure Birkholz-Beeskow (*black ellipse*). *Red lines* show EM profiles in the brine project, the *purple rectangle shows* the zone of assumed vertical migration of saltwater (Kempka et al. 2015)

ancient Berlin-Warsaw glacial valley. Typically fluviatile sandy deposits of the Weichselian glaciation form the uppermost geological layer with thicknesses varying from 20 to 40 m. Due to the morphological condition the water level is comparatively high. Infiltration rates are high due to the good permeability of the loosely packed sands. Because both sites are under forest land use with only low human impact, the upper, unprotected aquifer represents an important regional groundwater resource. At test site Briesen the aquifer has been used since the 1960s as main drinking water source for Frankfurt/Oder (~65,000 inhabitants). The medium-scale test site Müllrose was explored in the 1980s as a possible location for a second water works. During the exploration campaign, NaCl contents above the legal limits were surprisingly found in the upper aquifer and prevented the construction of the water works. Also, some wells of the water works Briesen are currently threatened by high NaCl-contents and are not in operation anymore.

For both test sites numerous well logs and water samples from different depths showed that this shallow salinisation problem is not caused by vertical flow of saltwater from deeper aquifers. At Müllrose the upper aquifer is underlain by ~60–70 m thick undisturbed series of Miocene lignite seams and lignite-bearing clays interbedded with thin fine sands, followed by a Miocene-Oligocene aquifer with a total thickness of about 50 m. At Briesen the upper aquifer is followed down to a depth of ~100 m by Pleistocene silts, sandy silts or occasionally tills from earlier glaciations, which were deposited on Miocene coal seams and clays. The Fürstenwalde-Guben fault zone has led to a post-sedimentary subsidence of the basement under the test site Briesen, while Müllrose, being located east of this fault zone, was not affected by these tectonic processes. As in Müllrose, the Miocene coal seams and clays (interbedded with thin fine sand layers) are underlain by the regionally important thick quartz- and mica sand aquifer, but in a deeper vertical position. At Müllrose the sand aquifer contains water with a much lower mineralisation compared to the upper unconfined aquifer. The interbedded fine sands of the Miocene coal series are roughly in the same vertical position at Briesen and show the same phenomena of lower pore water mineralisation compared to the increased NaCl contents in the lower parts of the upper uncovered aquifers in the southern part of Briesen (LGBR 2010).

On both sites, the ~40–50 m thick sandy aquifer is underlain by the ~50 m thick Rupelian clay, which forms an important hydraulic barrier in the North German Basin and typically separates freshwater from deep saline aquifers. While this separation applies for Müllrose, it does not for Briesen, where the saltwater/freshwater boundary lies within the quartz and mica sand aquifer. The vertical position and thickness of the Rupelian clay differs between both test sites according to the post- and syn-sedimentary movements of the Fürstenwalde-Guben fault zone (FGFZ). In Briesen the FGFZ is fragmented into smaller blocks that moved differently in the past which caused strongly differing vertical positions of the Rupelian clay in the area. Locally this might have lead even to a full offset of the clay which could have opened gaps in the barrier and subsequently formed pathways for an upward migration of deep saline waters into the overlaying quartz sand aquifer. Another suspected reason for the increased salinisation on the Briesen test site is the occurrence of a very small Zechstein salt diapir in the western part of the site, which moved (presumably caused by tectonics of the FGFZ) with its cap up to 150 m depth. It is very likely that the water inside the quartz and mica sand aquifer surrounding the diapir (with a lateral extension of only 300–400 m) can continuously dissolve NaCl from this source, even though the diapir is covered by gypsum caprock. Still the intermediate freshwater-bearing zone, which can be found on both test sites in the depth range of ~70–120 m, indicates that no direct vertical saltwater flow takes place there. Instead the reason for the shallow salinisation of the upper aquifer must be lateral inflow. The regional groundwater salinity mapping was re-analysed using old well-logs and hydrogeological research of the State authorities (Hotzan and Voss 2013). The latter data and investigations in the brine project (Kempka et al. 2015) suggest that vertical upflow of saltwater occurs in the region between both surveys areas (see PVMZ in Fig. 6). Several factors promote the rise of saltwater from deeper saline aquifers: (i) the fragmentation of the basement and the Rupelian by the tectonics inside the FGFZ, (ii) the existence of a Pleistocene erosion channel ("glacial buried valley") which partly or fully eroded the Rupelian clay layer, and (iii) the morphological low position of this spot close to the Spree river even in comparison with the surrounding terraces of the ancient glacial valley. The low position of the valley and the subsequently lesser overload of fresh water act as pump for the deeper aquifers that often show artesian behaviour. As shown by well logs, water with a mineralisation of ~2–5 g/l reaches depths of 20–30 m below the surface in areas with vertical upflow. Due to its origin it can be assumed that the predominant part of this mineralisation is made up of NaCl, i.e. the values are by far beyond the legal limit for the Cl-content of drinking water in Germany (250 mg/l). From this area that behaves like a watershed, mineralised water is carried away in the Spree floodplain towards Briesen and with the groundwater flow towards the draining Oder-Spree-channel near Müllrose.

It can be assumed that particularly the shallow salinisation phenomena on both test sites are strongly influenced by climate, seasonal precipitation, water levels in the river as well as human impact, such that relatively fast changes of the groundwater salinisation over time can be expected. For Müllrose water samples show that the NaCl-concentration varies significantly over small distances. This makes especially the Müllrose site well-suited for monitoring experiments.

4.2 Medium-Scale Test Site Müllrose

At the eastern part of the area, 12 steel-cased boreholes with typical borehole distances of 50–100 m were identified and chosen for LE-ERT measurements. An area of about 500 × 500 m was chosen, since it is small enough to allow a centralized cable layout. Additionally, seven surface electrodes were installed using steel rods. After the first experiment with a high-voltage generator, all electrodes were permanently wired to a central point such that a full measurement can be cost-efficiently deployed in about 30 min with a standard resistivity meter (4-point light 10W by LGM electronics). Only very low power is needed due to the low contact impedances of the boreholes and the small distances between them.

At initial stage (autumn 2012), four classic surface ERT profiles with 5 m electrode distance were spanned across the area for verification of LE-ERT. The robust Wenner-alpha array (see Fig. 1) was measured due to the relatively bad coupling in the dry topsoil. Although almost arbitrary arrays could be measured with LE-ERT, an autarkic bipole-dipole, i.e. smallest possible potential dipole and logistically possible current bipoles, protocol was defined which is the only choice if permanent wiring is not possible anymore. An error model of 5 % plus 300 μV was used for inversion on a triangular prism mesh with 46,126 cells. Figure 7 shows the retrieved resistivity distribution along with the independently obtained 2D resistivity models of the classic ERT profiles.

The 2D and 3D distributions are matching very well. Main contrast is the change from the relatively resistive (50–300 Ωm; orange-yellow colours in Fig. 7) Quaternary sands and silts to the relatively conductive (10–30 Ωm; bluish colours in Fig. 7) Tertiary lignite silts at a depth of about 40 m. However there is also a resistive coarse layer that can be attributed to gravelly sediments.

In order to relate the bulk resistivity to fluid conductivity, point conductivities were extracted from the 3D model and compared to the measured fluid conductivities excepting those from the central monitoring well group. A modified Archie

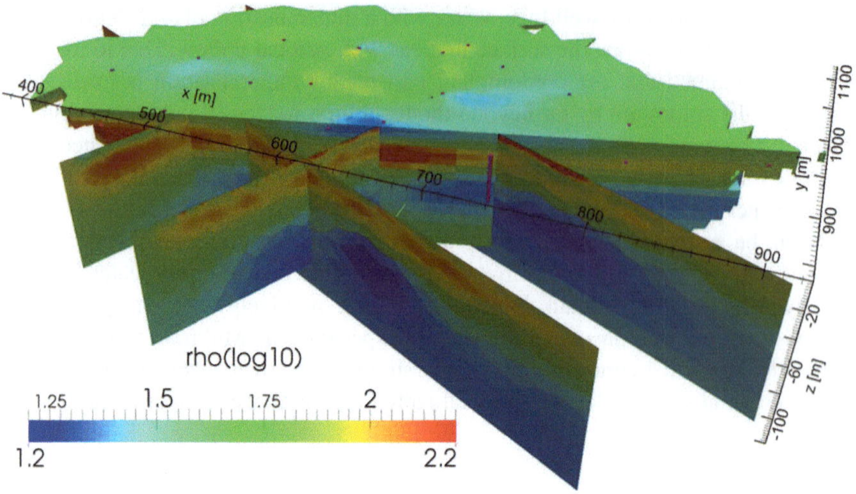

Fig. 7 3D resistivity model (*volume*) retrieved from LE-ERT and 2D resistivity models from classical ERT profiles (*slices*). The electrodes (*purple columns*) mark the position of points or boreholes representing holes in the mesh

equation (Waxman and Smits 1968) was fitted to the points yielding a formation factor of F = 5.22 and a surface conductivity of σ_s = 5.2 mS/m.

In the sequel, repeated measurements were routinely conducted in time intervals of about 3 months and accompanied by fluid sampling in the monitoring borehole. As we do not expect any changes in the Tertiary, we decoupled the second layer from the first and restricted its value to a range of 10–30 Ωm. Figure 8 shows the

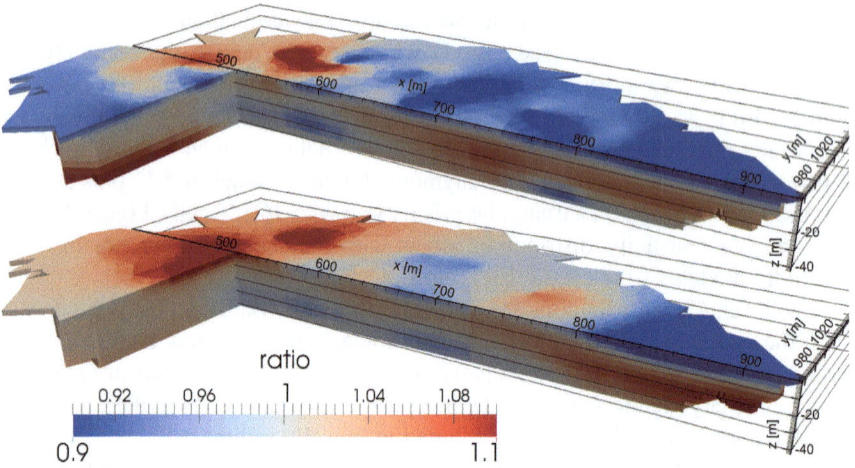

Fig. 8 Resistivity changes (ratio of later by former resistivity) between September and December 2013 (*top*) as well as between December 2013 and March 2014 (*bottom*)

two subsurface resistivity changes as ratios between September 2013, December 2013 and March 2014, retrieved from time-lapse inversion. Due to the decoupling, there are no changes in the Tertiary. The changes in the Quaternary sand are in the range of a few per cent. At the surface, a resistivity increase is observed that can be attributed to both precipitation and temperature effects. At the bottom of the aquifer, the decreasing resistivity can be explained by increased salinisation, which is probably caused by lateral inflow. Other time-steps revealed similar changes, mainly affected by precipitation and water use of the trees. The observed inversion values agree with the trends of the measured fluid samples. However, full understanding of the processes requires longer monitoring periods.

4.3 Large-Scale Test Site Briesen

At the large-scale test site Briesen about 80 steel-cased boreholes are available that can be used for LE-ERT experiments. The number of used boreholes was reduced to 42 by neglecting boreholes that are either too shallow or too close to others. The first two-week experiment was focused in the south-eastern part of the area, i.e. between the artificial recharge ponds of the water works and the river Spree (Fig. 9). Additional to the boreholes, two surface electrodes and two subsurface electrodes in PVC-cased boreholes were placed. Eight newly developed three-channel data loggers were installed and connected to neighbouring points. A set of 30 current injections providing optimum resolution was computed. The high-voltage generator injected square waves with a period of 5 s and a maximum amplitude of 25 A into the ground.

Analysis of the time series involved a software solution of the lock-in method that is usually applied in commercial resistivity meters. Pre-processing excluded data with unreliably low or high apparent resistivities and high geometric factors. Inversion was carried out on a triangular prism mesh with smoothness constraints and a flatness ratio (of horizontal/vertical penalties) of 0.015. Figure 10 shows the

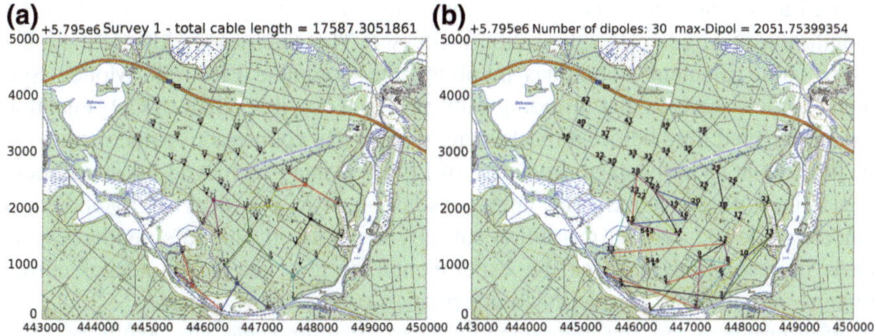

Fig. 9 **a** Experimental design with potential dipoles (colour by data logger marked by *rectangles*); **b** Optimized current bipoles. x/y positions in m (UTM32)

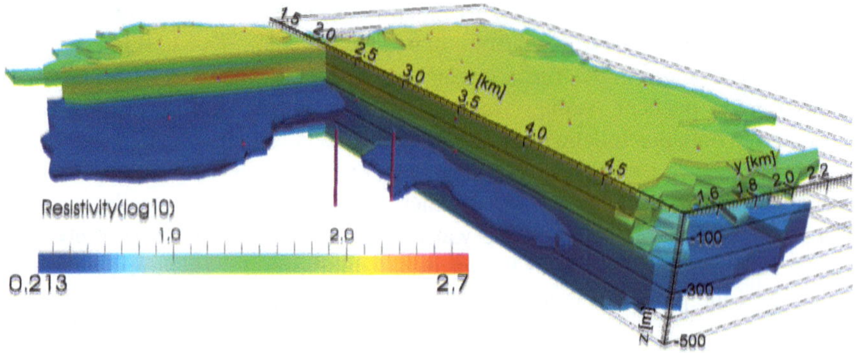

Fig. 10 3D inversion result (with isobodies <3 Ωm) derived from the August 2013 survey at the large-scale test site Briesen (north is along y axis)

resistivity distribution down to the investigation depth of about 500 m. It does not primarily correspond to geology, but to electrical downhole measurements that already showed that a salinisation of the deeper aquifers is present. The conductive iso-body (<3 Ωm) in the south-west of the survey area outlines the saltwater rise along the Fürstenwalde-Guben fault zone and its intrusion into tertiary aquifers. The isolated western conductor was not known before.

A second field campaign was conducted in spring 2014 and used part of the existing potential and current bipoles so that reproducibility could be checked. Further data loggers were installed in the western part and additional current injections were accomplished in order to extend the whole measuring area.

5 Conclusions

Long-electrode ERT is a technique that is able to image three-dimensional resistivity distributions on various scales. If steel-cased boreholes are available, it can be applied up to the size of catchment areas with sufficient spatial resolution for the monitoring of large-scale salinisation processes. The electrical fields are modelled using Finite Element techniques involving the Complete Electrode Model. Synthetic models indicate that intruding saltwater, as a possible result of CO_2 injection, can be detected long before it reaches any well so that counteractive measures could be initiated at an early stage. Inversion of LE-ERT data is achieved by smoothness-constrained minimisation. The lateral resolution is in the order of borehole distances, while the vertical resolution strongly depends on borehole depths. On the basis of resolution matrices the experimental design, i.e. the choice of current bipoles, is optimized to maximize information content at given cost.

Measurements on the permanently wired medium-scale site Müllrose show accordance with geology and classical ERT. Increased salinisation at the bottom of the shallow Quaternary aquifer is supported by fluid measurements. Dedicated time-lapse algorithms applied to repeated measurements are able to image small changes in the subsurface and enable long-term monitoring of groundwater systems. However, in the relatively short period of investigation only small changes, mostly of seasonal origin, could be observed.

The method was successfully applied to characterize the deeper aquifer structure in the production area of the water works Briesen, however with considerable logistic effort. For monitoring purposes only a subset of the measurements would be needed and permanent wiring would further reduce the logistic effort. The developed technology has proved in practice and can be routinely applied to areas where the necessary long-electrode infrastructure is available. The feasibility of the monitoring approach for real CO_2-storage projects likely depends on this limitation due to drilling costs. However, for many potential onshore CCS sites in areas with extensive geological usage in the past (e.g. in the North German sedimentary basin) the existence of steel-cased boreholes can be assumed. Possible gaps in the electrode networks above a CO_2-storage structure can be filled with new installations. The parameters (e.g. positions for new installations or time intervals between single measurements) should be optimised beforehand with the modelling and optimisation know-how acquired with in the SaMoLEG project in accordance with the peculiarities of each individual storage area.

Acknowledgments The project was funded by the Germany Ministry of Education and Research (BMBF) in the framework of the R&D program GEOTECHNOLOGIEN (grants 03G0774A and 03G0774B). We are grateful to Dr. Carsten Rücker (TU Berlin) for his help with the modelling. We like to acknowledge the technical crew (Frank Oppermann, Robert Meyer, Wolfgang Südekum and Dieter Epping) for their help in the field and for laboratory measurements. We thank the water works FWA Frankfurt (Oder), the forestry administration of the state Brandenburg as well as Komturei Lietzen for permission to access test sites Briesen and Müllrose and for logistic support during field surveys. We are grateful to the colleagues from brine project (BTU Cottbus) for cooperating and sharing data on the measuring area.

References

Attwa M, Günther T, Grinat M, Binot F (2011) Evaluation of DC, FDEM and IP resistivity methods for imaging perched saltwater and a shallow channel within coastal tidal flat sediments. J Appl Geophys 75:656–670

Daily W, Ramirez A, Newmark R, Masica K, Livermore L (2004) Low-cost reservoir tomographs of electrical resistivity. Lead Edge 23:472–480

Friedel S (2003) Resolution, stability and efficiency of resistivity tomography estimated from a generalized inverse approach. Geophys J Int 153(2):305–316

Günther T, Rücker C, Spitzer K (2006) Three-dimensional modelling and inversion of dc resistivity data incorporating topography—II: inversion. Geophys J Int 166(2):506–517

Hannapel S, Hermsdorf A, Pohl S, Rietz C, Koseck R (2007) Aufbau von Sondermessnetzen zur Überwachung der geogenen Grundwasserversalzung in Brandenburg (in German). Brandenburg Geowiss Beitr 14(1):5–14

Hotzan G, Voss T (2013) Komplexe hydrogeochemisch-genetische Kartierung zur Einschätzung der Salzwassergefährdung pleistozäner und tertiärer Grundwasserleiter im Raum Storkow-Frankfurt (Oder)-Eisenhüttenstadt (in German). Brandenburg Geowiss Beitr 20(1/2):63–82

Kempka T, Herd R, Huenges E, Endler R, Jahnke C, Janetz S, Jolie E, Kühn M, Magri F, Meinert P, Moeck I, Möller M, Munoz G, Ritter O, Schafrik W, Schmidt-Hattenberger C, Tillner E, Voigt H-J, Zimmermann G (2015) Joint research project brine: carbon dioxide storage in Eastern Brandenburg: implications for synergetic geothermal heat recovery and conceptualization of an early warning system against freshwater salinization. In: Liebscher A, Münch U (eds) Geological storage of CO_2: long term security aspects. Springer, Singapore

LaBrecque DJ, Yang X (2001) Difference inversion of ERT data: a fast inversion method for 3-D in situ monitoring. J Environ Eng Geophys 6:83

Landesamt für Bergbau, Geologie und Rohstoffe Brandenburg (LGBR) (2010) Atlas zur Geologie von Brandenburg, 4.Auflage, Cottbus

Ronczka M, Günther T, Rücker C (2013) Long electrode ERT for salt water monitoring—modelling, sensitivity and resolution Ext Abstr, 19th EEGS near surface meeting, Bochum, Germany

Rücker C, Günther T (2011) The simulation of finite ERT electrodes using the complete electrode model. Geophysics 76(4):F227

Rücker C, Günther T, Spitzer K (2006) Three-dimensional modelling and inversion of DC resistivity data incorporating topography—I: modelling. Geophys J Int 166(2):495–505

Rucker DF, Loke MH, Levitt MT, Noonan GE (2010) Electrical-resistivity characterization of an industrial site using long electrodes. Geophysics 75(4):95–104

Rucker DF, Fink JB, Loke MH (2011) Environmental monitoring of leaks using time-lapsed long electrode electrical resistivity. J Appl Geophys 74(4):242–254

Siemon B, Christiansen AV, Auken E (2009) A review of helicopter-borne electromagnetic methods for groundwater exploration. Near Surf Geophys 7:629–646

Stummer P, Maurer H, Green AG (2004) Experimental design: electrical resistivity data sets that provide optimum subsurface information. Geophysics 69(1):120–139

Voss T, Ronczka M, Günther T (2013) Saltwater monitoring with long-electrode electrical resistivity tomography. Biul Panstw Instytuto Geol 456:621–626

Waxman MH, Smits LJM (1968) Electrical conductivities in oil-bearing shaly sands. Soc Pet Eng J 8:107

Joint Research Project Brine: Carbon Dioxide Storage in Eastern Brandenburg: Implications for Synergetic Geothermal Heat Recovery and Conceptualization of an Early Warning System Against Freshwater Salinization

Thomas Kempka, Rainer Herd, Ernst Huenges, Ricarda Endler, Christoph Jahnke, Silvio Janetz, Egbert Jolie, Michael Kühn, Fabien Magri, Peter Meinert, Inga Moeck, Marcus Möller, Gerard Munoz, Oliver Ritter, Wladislaw Schafrik, Cornelia Schmidt-Hattenberger, Elena Tillner, Hans-Jürgen Voigt and Günter Zimmermann

Abstract Brine was a scientific joint-project implemented to accompany a prospective CO_2 storage site in Eastern Brandenburg, Germany. In this context, we investigated if pore pressure elevation in a CO_2 storage reservoir can result in shallow freshwater salinization involving the conceptual design of a geophysical early warning system. Furthermore, assessments of a potential synergetic geothermal heat recovery from the CO_2 storage reservoir and hydro-mechanical integrity were carried out. The project results demonstrate that potential freshwater salinization is strongly depending on the presence and characteristics of geological weakness zones. The integrated geophysical early warning system allows for reliable monitoring of these potential leakage pathways at different spatial and time scales.

T. Kempka (✉) · E. Huenges · R. Endler · E. Jolie · M. Kühn · F. Magri · P. Meinert · I. Moeck · M. Möller · G. Munoz · O. Ritter · C. Schmidt-Hattenberger · E. Tillner · G. Zimmermann
Helmholtz Centre Potsdam, GFZ German Research Centre for Geosciences Telegrafenberg, 14473 Potsdam, Germany
e-mail: kempka@gfz-potsdam.de

R. Herd · C. Jahnke · S. Janetz · W. Schafrik · H.-J. Voigt
Brandenburg University of Technology Cottbus—Senftenberg, Platz der Deutschen Einheit 1, 03046 Cottbus, Germany

A. Liebscher and U. Münch (eds.), *Geological Storage of CO_2 – Long Term Security Aspects*, Advanced Technologies in Earth Sciences,
DOI 10.1007/978-3-319-13930-2_9

1 Introduction

Geological CO_2 storage can induce far-field pressure elevation in the storage reservoir resulting in formation fluid displacement via geological weakness zones such as faults or geological windows (Schwartz 2014; Tillner et al. 2013a; Walter et al. 2012; Lemieux 2011; Pruess 2011; Oldenburg and Rinaldi 2011; Birkholzer et al. 2009; Nicot 2008). In this context, we addressed the following questions in the present study:

1. Does pore pressure elevation in a CO_2 storage reservoir result in groundwater salinization becoming a potential threat for freshwater resources?
2. How to detect and monitor an upward formation fluid migration process induced by underground pore pressure elevation by means of an early warning system against freshwater salinization?
3. Is a synergetic geothermal heat recovery from a CO_2 storage reservoir feasible by integration of pressure management wells to overcome potential competition with other concepts of underground utilization?

We integrated static and dynamic modeling and simulation as well as the assessment of potentially suitable geophysical monitoring methods to address the aforementioned questions. In a first step, a 3D structural geological model was implemented based on available data and extended with the availability of further data. This 3D static model represented fundamental information for all work packages involved in numerical simulation and geophysical monitoring of the Buntsandstein sandstone units. Subsequently, coupled thermo-hydro-mechanical multiphase flow numerical simulations were employed to elaborate the knowledge required to assess the project of geological CO_2 storage. Further models employed considered the main fault systems and migration of formation fluids along these hydraulic conduits. For the assessment of brine migration scenarios, a coupling of deep reservoir models with shallow hydrogeological models was carried out taking into account different brine migration scenarios. Numerical simulation results were used as input data for geophysical inverse simulations on CO_2 and brine migration detectability aiming at the implementation of a conceptual early warning system against freshwater salinization. Coupled hydro-mechanical simulations were applied to assess the reservoir, caprock and fault integrity in the study area depending on different injection and stress regimes.

The geophysical work packages carried out extensive field and monitoring campaigns in the study area. Besides groundwater sampling and water analyses, electrical resistivity and electro-magnetic campaigns as well as magnetotelluric profile measurements were conducted. Furthermore, a conceptual early warning system was elaborated and verified by numerical simulations. Gathered data supported the calibration but also verification of geophysical forward models, and thus were of substantial importance for the success of the entire project.

2 Study Area Beeskow-Birkholz

The geological structure located in the study area Beeskow-Birkholz (State of Brandenburg, cf. Fig. 1) is an asymmetric saline uplift with an extent of about 5 km × 15 km striking in NW-SE direction parallel to the major fault zones (Lausitzer Abbruch fault system in the South and Guben-Fürstenwalde fault system in the Northeast). The salt structure is limited by significant lineations of Variscan strike interpreted as tectonic zones (ZGI 1990a, b). Accordingly, the Beeskow-Birkholz structure is of salt tectonic origin and potentially influenced by the Guben-Fürstenwalde fault system located 5–10 km to the Northeast. Amplitude of salt accumulation is about 400 m with the top Zechstein salt at about 1,200 m depth (ZGI 1990a). The Mesozoic overburden is uplifted and partially eroded resulting from that salt accumulation. In the centre of the salt structure, Cretaceous and

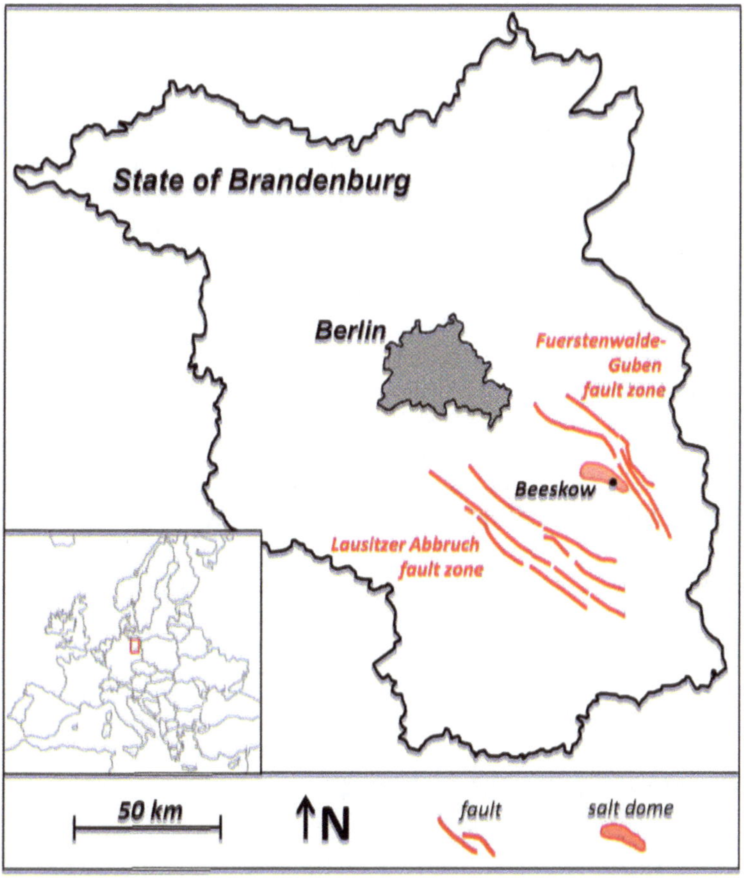

Fig. 1 Study area Beeskow-Birkholz is located approximately 80 km southeast of Berlin close to the small town of same name

Jurassic sediments are missing and the Keuper is in direct contact with the Cenozoic unconsolidated sediments. At the boundary of the structure, Jurassic elements are still present. Nevertheless, Cretaceous is only present in the far-field. The potential storage reservoir is located in the Middle Buntsandstein sandstone at depths from 800 to 1,000 m and is overlain by the Upper Buntsandstein, Muschelkalk and Keuper.

Jahnke (1999) documented the hydrogeochemistry of the freshwater aquifers located in Eastern Brandenburg. Characteristic chemical tendencies were determined in the hydrostratigraphic units of the Cenozoic. As shown in Grube et al. (2002), the concentration of total dissolved solids (TDS) in the region of investigation is comparatively high, even at shallow depths. It ranges from 1 to 10 g/l above and 10 to 50 g/l below the Rupelian clay and from 50 to 350 g/l below the Tertiary base. Isotope hydrogeological investigations in unconsolidated aquifers were carried out by Trettin et al. (1997), providing an overview on the isotope fractions and age of the Cenozoic groundwater. Freshwater salinization was observed close to the study area along the Fuerstenwalde-Guben fault zone by Trettin et al. (1997) and Stackebrandt and Manhenke (2004). In addition to that, local salinization was determined along the Spree River in the West of the Beeskow-Birkholz structure. The origin of the salinization are brine bearing aquifers of the Mesozoic, which are in direct contact with Permian and Mesozoic salts (Müller and Papendieck 1969; Lehmann 1974; Reinsch and Voigt 1975; Tesmer et al. 2007). Brackish groundwater from the Post-Rupelian-Tertiary may also contribute by a lesser amount to the observed salinization effects. The Oligocenic Rupelian clay is the aquiclude complex between the overlying freshwater and the brine bearing aquifers. This unit represents the most important barrier against fluid migration from the deep underground into the freshwater horizons, and is thus responsible for availability of the freshwater aquifers for potable water production. Glacial erosion channels in the Rupelian clay and fault systems can limit the barrier function of the Rupelian clay, so that brine intrusion into freshwater horizons may become possible depending on the hydrogeological setting. Presence of salinization areas in the study area demonstrates that this is partially the case. Dynamics of brine migration may be triggered by different processes: natural hydraulic gradients (e.g. Schirrmeister and Voigt 1976) or by free convection resulting from density differences determined by the temperature field. Natural migration pathways of brine, as e.g. fractures and faults in the Rupelian clay may be reactivated by the pore pressure elevation determined by CO_2 storage.

3 Main Project Results

3.1 Structural Geological Modeling

The main goal of the static modeling activities in the brine project was the implementation of a 3D structural-geological model of the Beeskow-Birkholz anticline. The focus of the model is on the target formations for CO_2 storage, namely Hardegsen, Detfurth and Volpriehausen (Middle Bunter) including relevant structural elements such as the major fault systems. The regional-scale model was mainly developed from available interpreted data gathered during hydrocarbon explorations in the 1970s, available by means of isobaths and isopach maps of the representative geological horizons (Stackebrandt and Manhenke 2010) and log data from four deep boreholes. The extent of the regional model was 40 km × 40 km × 3.5 km (Fig. 2).

Eleven 2D seismic reflection profiles of the Beeskow-Birkholz anticline, derived from hydrocarbon exploration campaigns, were provided by Vattenfall Europe Mining AG and used for a detailed re-interpretation of the fault system structure as indicated in Fig. 3. 100 km of 2D seismic reflection profiles with an areal extent of about 13 km × 18 km were integrated into a core model covering parts of the Beeskow-Birkholz anticline (Fig. 4).

Especially the representation of fault systems in the Mesozoic and Cenozoic not considered in the former hydrocarbon exploration could be improved by the integration of the 2D seismic reflection profiles. The core model differentiates between subsaline and suprasaline faults (Fig. 5). The model development was carried out using the Schlumberger Petrel (Schlumberger 2011) and earthVision^{TM} software packages.

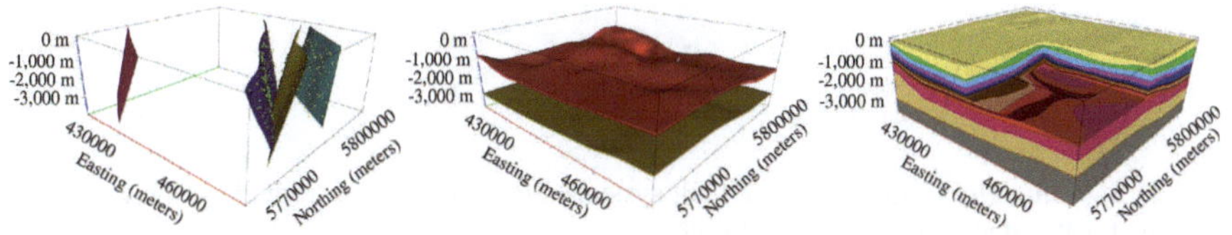

Fig. 2 Schematic workflow of static regional-scale model implementation. Model development proceeded in three steps following digitizing and georeferencing: (1) development of a fault model, (2) implementation of a stratigraphic model and (3) integration of the fault and stratigraphic models

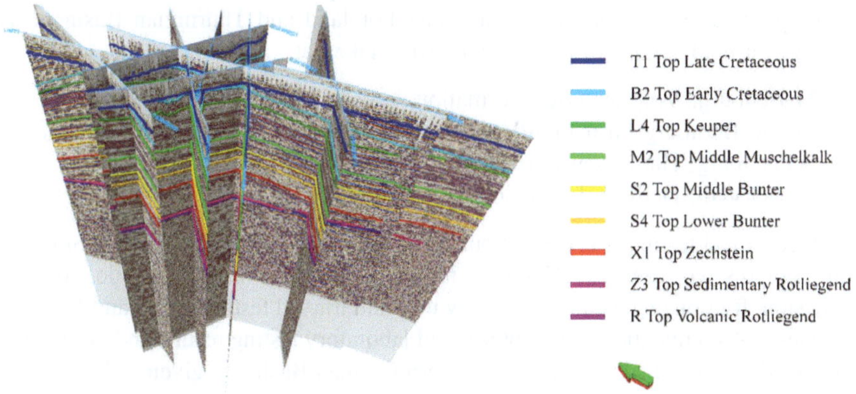

Fig. 3 Digitized 2D seismic profiles in the study area. See Fig. 4 for location of these profiles

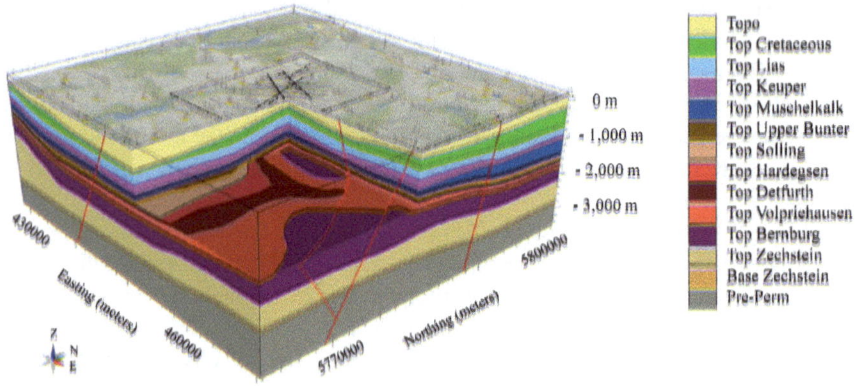

Fig. 4 Regional-scale model of the Beeskow-Birkholz anticline (40 km × 40 km × 3.5 km) including the core model (black bounding box in the center with 13 km × 18 km)

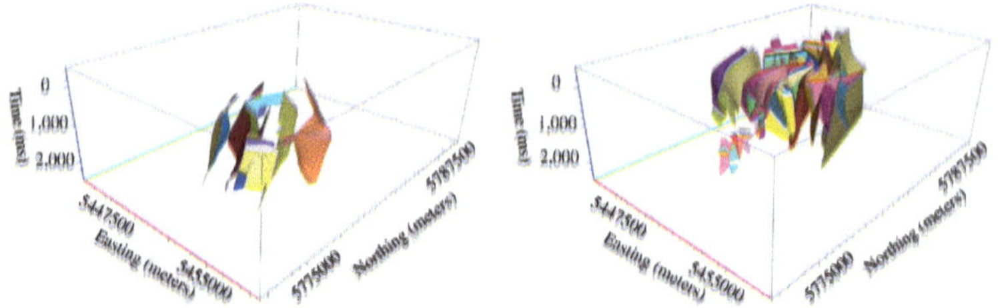

Fig. 5 Subsaline (*left*) and suprasaline fault systems (*right*) interpreted from 2D seismic reflection profiles for the study area Beeskow-Birkholz

Outcrop analogues (as depth equivalent) for mechanical laboratory testing were sampled based on surface geology and paleographic facies maps of the Middle Bunter sandstone to support numerical model parameterization. In total, eight outcrops have been identified in the Harz Foreland and Thuringian Basin in the scope of a field campaign and samples were taken at:

1. Marienburg: Volpriehausen Formation
2. Nebra: Bernburg Formation/Volpriehausen Formation/Hardegsen Formation
3. Schoenburg: Hardegsen Formation
4. Beesenstedt: Detfurth Formation

Mechanical testing was carried out on the acquired specimens to determine data on compressive strength, Young's modulus, Poisson's ratio, tensile strength, cohesion, friction angle and porosity by uni- and triaxial testing, shear and Brazilian disc tests. A compilation of the interpreted laboratory testing results and verification against data published for the Northeastern German Basin are given in Magri et al. (2013) and Röhmann et al. (2013).

3.2 Dynamic Multi-phase Flow Simulations to Assess Brine Displacement

Dynamic multi-phase flow simulations were employed to assess the synergetic potential of production well application for geothermal heat recovery and reservoir pressure management as well as potential freshwater salinization by brine displacement into shallower aquifers via the fault systems present in the study area. Thereto, the static structural geological model was revised and further developed to realize a spatial grid discretization and model parameterization. The resulting model with a size of 42 km × 42 km × 766 m consists of four reservoirs (Detfurth, Hardegsen, Muschelkalk and Stuttgart formations) and four regional faults, representing the Fuerstenwalde-Guben and Lausitzer Abbruch fault zone in the northeast and southwest, respectively. Three cap rocks mainly made up of anhydrite, claystones and marls from the Upper Buntsandstein to the Lower Keuper separate the four reservoirs. The proposed CO_2 injection well is located at the top of the Beeskow-Birkholz anticline structure at about 1,080 m and completed over the entire Detfurth Formation (average formation thickness is 23 m).

Assessment of synergetic reservoir utilization taking into account CO_2 storage and simultaneous geothermal heat recovery was assessed using a numerical model considering the Detfurth Formation only (Fig. 6). The Detfurth Formation has an average porosity of 23 % and permeability of 400 mD, whereby further model parameterization data are discussed in detail in Tillner et al. (2013b).

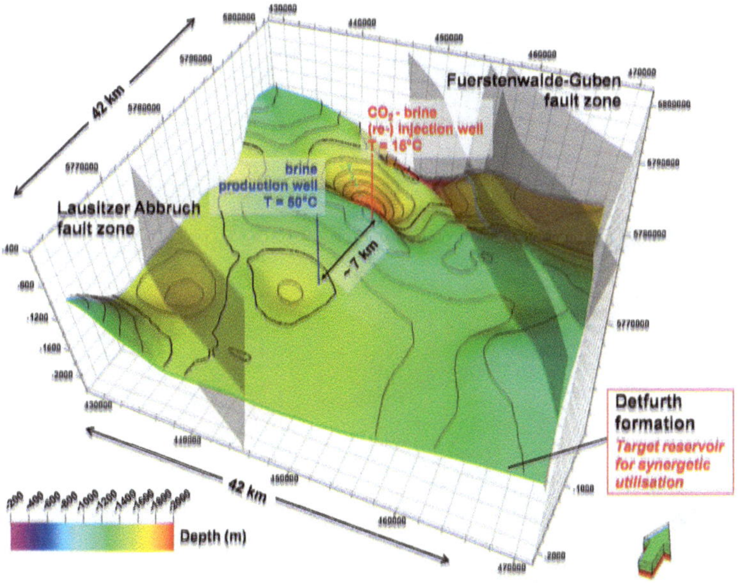

Fig. 6 3D geological model of the Detfurth formation (storage horizon) applied for assessment of synergetic reservoir utilization. Brine is produced using the *blue production* well and then reinjected together with the CO_2 using the *red injection* well (modified from Tillner et al. 2013b)

Different scenarios were assessed including varying permeability anisotropy to account for geological uncertainties. 1.7 Mt CO_2/year and 0.65 Mt brine/year were injected via the injection well at the same time for 20 years. A differential temperature for geothermal heat extraction of 30 K was determined, so that brine re-injection was carried out at 15 °C into the 50 °C (bottomhole temperature) reservoir. Simulation results discussed in detail by Tillner et al. (2013b) show that a geothermal heat recovery is feasible for at least 18 years until CO_2 arrives at the production well for the most unfavorable case of spatial permeability distribution and taking into account hydraulically non-conductive fault systems. However, an isotropic horizontal permeability distribution in the Detfurth Formation would allow for about 67 years of synergetic CO_2 storage and geothermal heat recovery. Radius of temperature decrease in the reservoir is about 2 km after 100 years of operation. Consequently, extracting geothermal heat from an operating CO_2 storage reservoir can significantly contribute to efficient and sustainable reservoir utilization.

Brine migration was investigated using a specific numerical model implementation based on the Virtual Element framework introduced by Nakaten et al. (2013). Virtual Elements allow for an implementation of additional grid elements at discrete fault planes to simulate fluid flow along these faults. The simulation model used is plotted in Fig. 7 and all details regarding model implementation and parameterization presented by Tillner et al. (2013a).

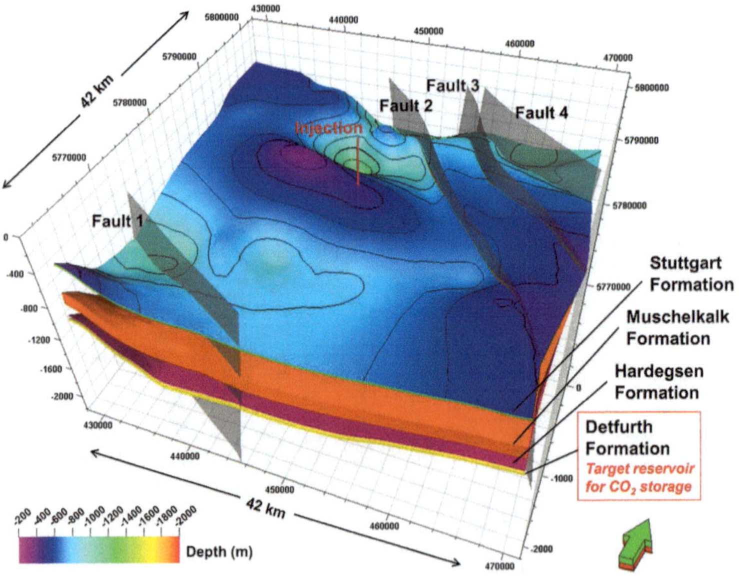

Fig. 7 3D structural geological model applied for investigation of brine migration via hydraulically conductive faults. Four geological units and four vertically conductive faults are present in the simulation model (modified from Tillner et al. 2013a). *Color scale* shows the total depth of the uppermost Stuttgart Formation

Several scenarios considering different fault permeabilities and boundary conditions were assessed by numerical simulations, whereby fault permeabilties ranged from 100 to 10,000 mD. Simulation results indicate that after an injection of 1.7 Mt CO_2/year over 20 years the average pressure in the Detfurth Formation increases by 6–17 % depending on the chosen fault permeability and boundary conditions, whereby a pressure elevation by 3.5–5.5 % is achieved in the uppermost Stuttgart Formation (cf. Fig. 7). About 67.5 Mt brine are displaced out of the Detfurth Formation mainly into Fault 1 and Fault 2 (Fig. 7) and across the model boundaries. The average salinity in the Stuttgart Formation increases by 0.24 % (407 mg NaCl/kg H_2O) after 20 years of injection assuming open model boundaries in the lower Buntsandstein formations (representing an infinite aquifer) and by about 0.17 % (290 mg NaCl/kg H_2O) assuming closed boundaries in the lower Buntsandstein formations (representing a reservoir compartmentalization). Fault permeability does not influence the displaced formation fluid amount, but the mass flow only. Maximum salinity increase is observed in close vicinity of the fault systems (Fig. 8). The data on brine migration was further used in order to evaluate specific geophysical monitoring tools in inverse and forward modeling as well as the applicability of the conceptual early warning system (Sect. 3.8).

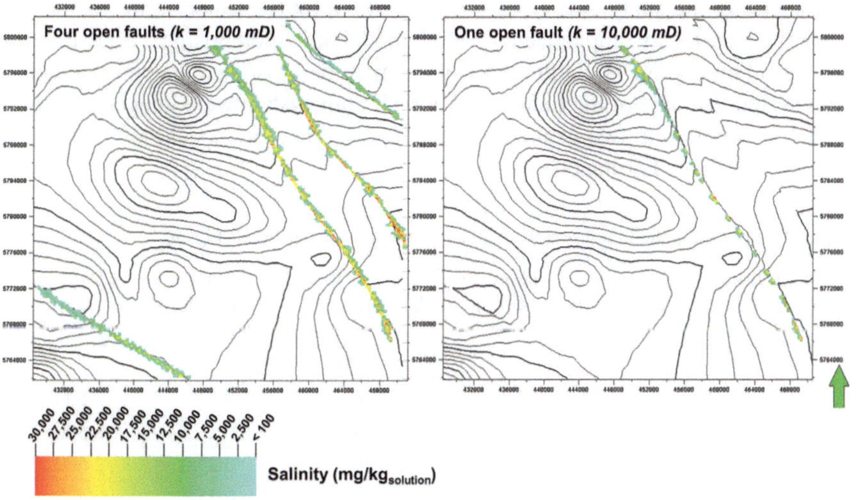

Fig. 8 Local increase in salinity in the Stuttgart Formation for scenarios considering closed boundary conditions in the lower Buntsandstein formations and four open faults (*left*) as well as only one open fault with a higher permeability (*right*) after injecting 1.7 Mt CO_2/year into the Detfurth Formation over 20 years (modified from Tillner et al. 2013a)

3.3 Hydro-mechanical Simulations for Assessment of System Integrity

Pore pressure changes in the storage reservoir, faults and shallow groundwater aquifers determined by dynamic multi-phase flow modeling (cf. Sect. 3.2) were integrated into the hydro-mechanical simulations to account for integrity of reservoir and caprock as well as of the fault systems present in the study area. The open-source scientific numerical simulator OpenGeoSys (OGS) (Wang and Kolditz 2007) was employed to carry out the hydro-mechanical simulations. Thereto, the Petrel-based reservoir model was applied as a common basis for the coupled numerical simulations. This structural model was first converted from Petrel to OGS using a conversion tool especially developed for that purpose. Then, hydro-mechanical parameters (e.g. storage coefficient, porosity, permeability, Young's modulus, Poisson's ratio) were assigned to the numerical model grid as given in Table 1. During the interaction between the TOUGH2-MP/ECO2N (Zhang et al. 2008; Pruess 2005) and the OGS simulator, we benefitted from the 3D interpolation procedure already implemented in the OGS software package by means of transferring corner grid data (OGS) to cell centered data (TOUGH2). A validation of the TOUGH2-MP/ECO2N and OGS coupling was carried out by application of 2D and 3D simulation model benchmarks discussed in Magri et al. (2015).

Regarding the study area Beeskow-Birkholz, a finite element grid was derived from the previously implemented Petrel model grid comprising seven different geological units (Fig. 9). Furthermore, three main faults of the Fürstenwalde-Guben fault system and one of the Lausitzer Abbruch fault system were integrated into the model, in addition to two geological units representing the model base (Basement) and top (Stuttgart Formation).

Table 1 Geomechanical parameters assigned to the geological units (Fig. 9)

Geological unit	Hydraulic permeability k (m^2)	Porosity Φ (−)	Young's modulus E (GPa)			Poisson's coefficient ν (−)	Density ρ_s (kg/m^3)
			Min	Med	Max		
Top	10^{-16}	0.2	4	6	9	0.2	2,600
Stuttgart	10^{-13}	0.25	2	4	5	0.2	2,500
Grabfeld	10^{-20}	0.01	3	6	8	0.27	2,600
Muschelkalk	2×10^{-13}	0.2	27	33	45	0.25	2,400
Upper Buntsand	10^{-20}	0.01	6	8	11	0.27	2,670
Hardegsen	3×10^{-13}	0.16	25	27	30	0.29	2,100
Detfurth claystone	10^{-20}	0.01	5	7	9	0.27	2,670
Detfurth	4×10^{-13}	0.17	20	23	25	0.29	2,200
Basement	10^{-20}	10^{-5}	60	60	60	0.3	2,800
Faults	10^{-12}	0.17	2	23	30	0.29	2,100

Modified from Magri et al. 2013

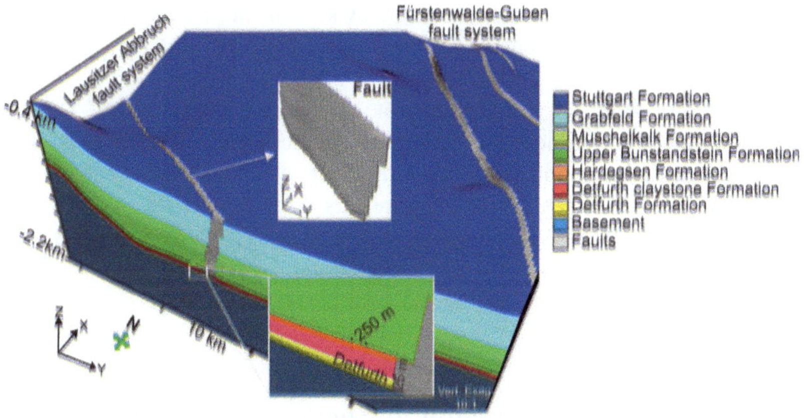

Fig. 9 Hydro-mechanical model representing the relevant geological units related to the brine project activities in the study area Beeskow-Birkholz (modified from Magri et al. 2013)

The applied mesh is made up of about one million hexahedron elements with a spatial discretization of 250 m × 250 m variable in vertical direction. The Detfurth Formation (CO_2 storage reservoir) is discretized by 2.5 m thick elements. Geomechanical parameters applied listed in Table 1 were derived from different sources including NER (2001), Ocak (2008), Alber and Heiland (2001) and Reyer and Philipp (2012) in addition to the data determined by laboratory mechanical testing.

Boundary conditions were set by implementing a constant stress at the model top as well as a constant pore pressure in the entire model. Deformations perpendicular to the respective lateral and bottom boundaries were not allowed. Pore pressure distribution was derived from the dynamic multi-phase modeling discussed in Sect. 3.2. Figure 10 shows the calculated vertical displacements resulting from the

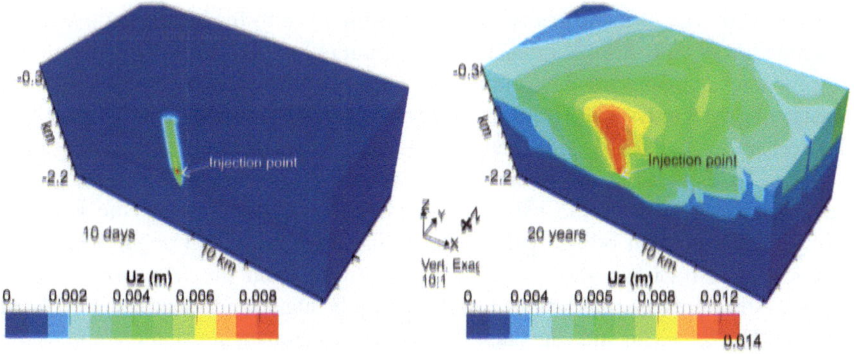

Fig. 10 Vertical displacements calculated by hydro-mechanical simulations for 10 days (*left*) and 20 years of injection (*right*) (modified from Magri et al. 2013)

stress changes induced by pore pressure elevation in the reservoir. A maximum displacement of about 1.4 cm is achieved at a depth of 300 m located at the injection well.

Deformation potentially able to reactivate the faults is not observed. The Mohr-Coulomb failure criterion does not exhibit any model elements being at or close to shear failure. This is supported by supra-regional scale simulations carried out by Röhmann et al. (2013). Our simulation results indicate that the scheduled CO_2 injection amount can be safely stored in the Detfurth Formation at Beeskow-Birkholz, whereby the integrity of the reservoir and caprock as well as of the main fault systems is not compromised by the pore pressure elevation.

3.4 Assessment and Simulation of Shallow Groundwater Salinization

Hydro-chemical investigations of freshwater aquifers in the study area and the implementation of a Cenozoic hydrogeological 3D model comprising model parameterization and dynamic flow and transport simulations were carried out for the assessment of a potential anthropogenic salinization of shallow freshwater aquifers. Thereto, a coupling between the deep reservoir (Sect. 3.2) and the shallow hydrogeological model was realized based on the implementation of a Neumann boundary condition in the shallow hydrogeological model in order to transfer the displaced brine mass flow. Hydro-chemical investigations carried out were based on a review and selection of existing groundwater sampling wells, groundwater sampling and analysis from selected locations and the hydro-chemical assessment as well as data interpretation. Geological and hydrogeological modeling consisted of review of available geological data, the implementation of a geological 3D model and its hydrogeological parameterization based on the study area. In the next step, density-coupled 3D flow and transport modeling was undertaken to analyze the natural salinization processes in the freshwater aquifers followed by the coupled approach for investigation of CO_2 storage impacts on brine migration into the shallow aquifers.

Cenozoic sediments are determined by a cumulative thickness of 100–300 m and consist of five sandy-gravelly main aquifers separated by aquitards (glacial tills, Quarternary and Tertiary silts and clays). The Rupelian clay with a 50–70 m thickness in the study area is the base of the freshwater complex and acting as the major freshwater—brine barrier. The Mesozoic aquifers below the Rupelian clay generally contain brines with a salinization increasing with depth. The hydo-chemical investigations show that brine migration from the Mesozoic aquifers and salt structures is responsible for the partially high salinity in the freshwater aquifers (Fig. 11). This indicates that flow processes connecting shallow and deeper aquifers exist, which induce upward brine and downward freshwater migration. Salinization of shallow aquifers is mainly bound to the flatland areas located between Fürstenwalde and Müllrose (discharge areas with saline water intrusions).

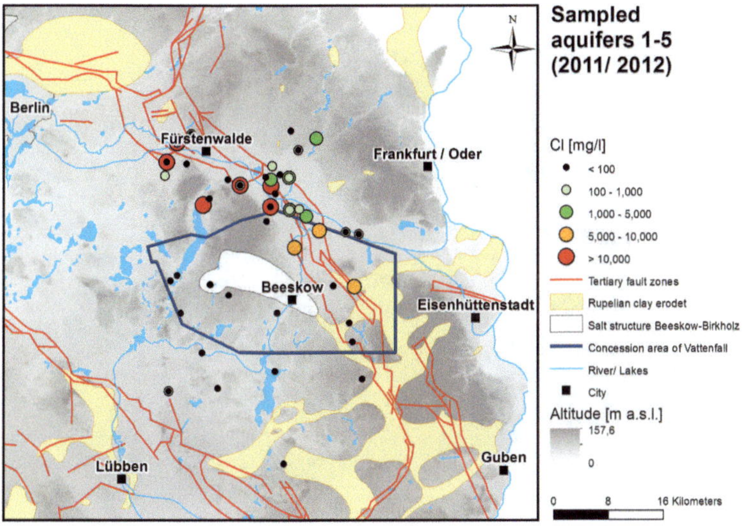

Fig. 11 Chloride concentrations at the sampled groundwater wells in the study area

Depending on the hydrogeological windows in the Rupelian clay, communication between the aquifers as well as the hydraulic gradient may occur. However, in the highland areas (Barnim in the east of Fürstenwalde) infiltration of freshwater into deeper aquifers is observed down to the Rupelian base sands as a result of groundwater recharge. Investigations using chemical and isotopic methods revealed that water composition of the deeper Cenozoic aquifers with salt water influences exhibit a mixed character. Age determination of the water samples with different isotope systems (T/He, $^3He/^4He$, ^{14}C) resulted in ages of up to 23,000 years, and thus can be assigned to Holocene and Pleistocene ages.

The hydrogeological 3D model is based on an extension of the study area considering the hydrological catchments around the Beeskow-Birkholz area as well as regional geological underground structures such as salt diapirs and faults. Hydraulic and chemical impacts of CO_2 storage were not known in advance of the numerical model coupling, and hence a relatively large area was chosen for model implementation. A structural model of the Cenozoic was elaborated containing 25 sedimentary units (Fig. 12). The Mesozoic strata and major fault systems were schematized (with average values) and the model was extended down to the Muschelkalk unit to allow for a direct coupling with the dynamic multiphase flow model (Sect. 3.2). Based on the geological structural model, a hydrogeological 3D model with a lateral size of 74 km × 85 km and a vertical thickness is about 2.4 km was implemented using the finite element simulator FeFlow^{TM}. Details on model parameterization and boundary condition implementation are presented in Kempka et al. (2014).

The dynamic flow and transport model was implemented based on density-driven flow with a constant viscosity. In a first step, long term simulations were

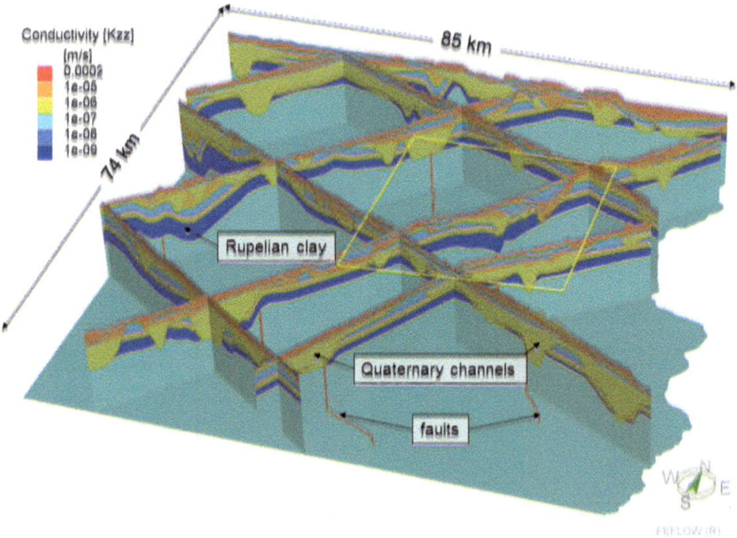

Fig. 12 3D hydrogeological model structure illustrated by eight cross-sections plotting vertical permeabilities (m/s). Vertical exaggeration by 25. *Yellow box* indicates the image section of Fig. 13

used to approximate the current groundwater salinization as an initial state. In a second step, influences of the storage processes were simulated in different scenarios based on the modeled brine displacement in the storage formation (Sect. 3.2).

The results of two scenarios of these coupled simulations are shown in Fig. 13. The first scenario (Scenario 1) can be taken as worst-case scenario, since the calculations carried out with the deep reservoir model (Sect. 3.2) assumed that the storage formation does not have any pore compressibility and that the reservoir is compartmentalized, and thus not allowing any pressure dissipation into the far-field (see Sect. 3.2). However, in Scenario 2 reservoir rock pore compressibility is considered in addition to open boundaries in the reservoir model (infinite aquifer). As a result, mass flow rates are about one third lower compared to Scenario 1. Resulting increase in NaCl concentration in the freshwater aquifers at 50 and 0 m below sea level is plotted for both scenarios before the start and at the end of injection for two different depth slices (Fig. 13).

The simulation results of the different scenarios demonstrate that brine migration into the freshwater aquifers is generally possible as a result of brine displacement out of the CO_2 storage reservoir. Simulated freshwater salinization follows the migration pathways known to be relevant for geogenic salinization already observed in the study area. Increase in NaCl concentration in the brine upward displacement zones depends on the brine volume displaced from the reservoir and on local hydraulic and hydrogeological boundary conditions. A detailed modeling of these processes requires exact hydrogeological data on geological structures as

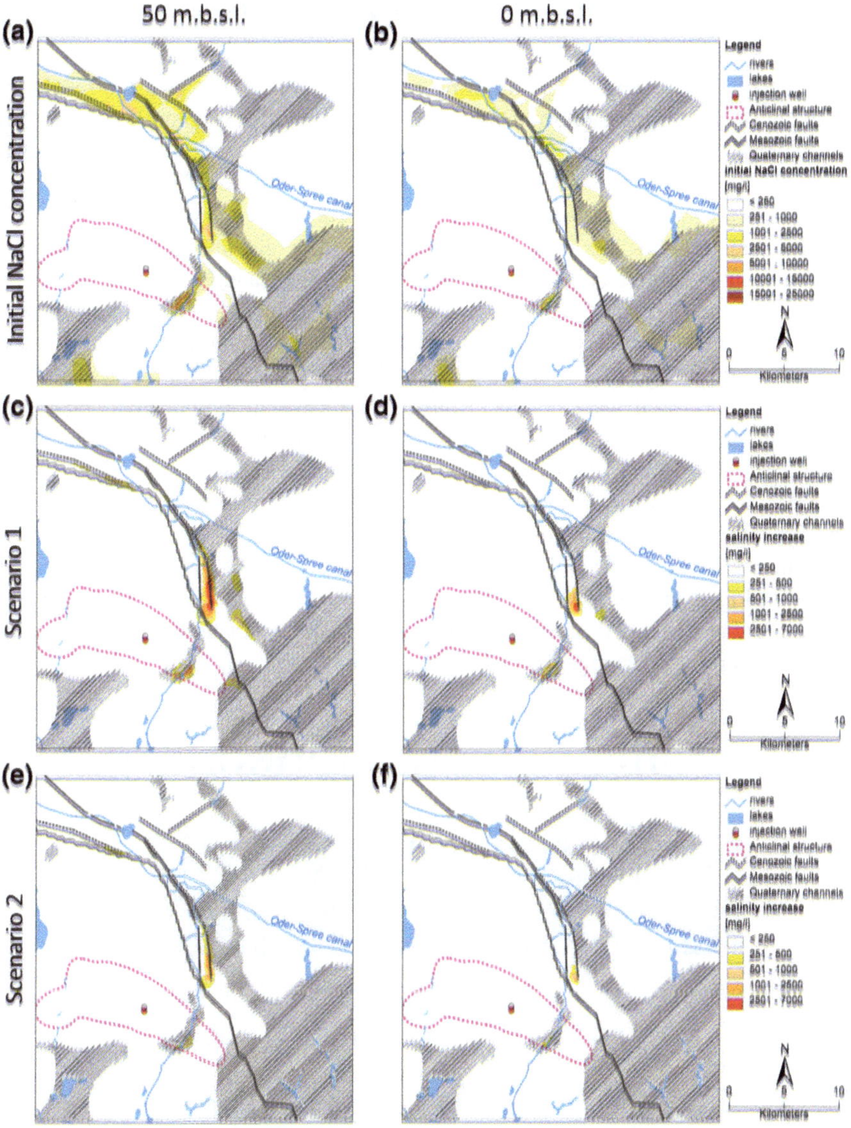

Fig. 13 Increase in NaCl concentration due to brine displacement at −50 m.b.s.l. and at sea level depth based on coupled numerical simulations. Initial NaCl concentrations (**a**, **b**) as well as Scenario 1 (**c**, **d**) considering an incompressible and Scenario 2 (**e**, **f**) considering a compressible storage formation are plotted

well as model parameters that were available only in a limited amount in the present study, mainly determined from point focused studies. Consequently, the presented results are rather of general nature and further research is required.

3.5 Large-Scale Monitoring Using Magnetotellurics

The aim of this geophysical monitoring activity was to investigate the resolution potential of the magnetotelluric (MT) method by natural and controlled sources in the geological environment given in the study area. The work was carried out in four phases comprising synthetic modeling using passive MT data, MT field campaigns for synthetic model calibration, synthetic modeling of controlled source electromagnetic (CSEM) data and conception of a geophysical early warning system against freshwater salinization with the two ERT and EM monitoring tasks.

Focus of the first phase was the simulation using general geological structures in the Northeast German Basin. Thereto, the spatial distribution of resistivity was derived from the geological model (Sect. 3.1) based on sedimentary horizons. Resistivity data used were derived from experience with similar geological conditions in Eastern Brandenburg (Fig. 14). 2D and 3D models were employed for these tasks and discretized according to the geological data.

A MT field campaign was undertaken in March 2012 in the study area based on 26 MT stations to determine the electric and magnetic fields as functions of time in a frequency range between 10 and 1 mHz. Data processing was carried out with high developed geophysical methods (e.g. robust stacking, remote reference processing, delay line filtering, etc.) and provided transfer functions of good quality, despite of strong noise signals present in the study area.

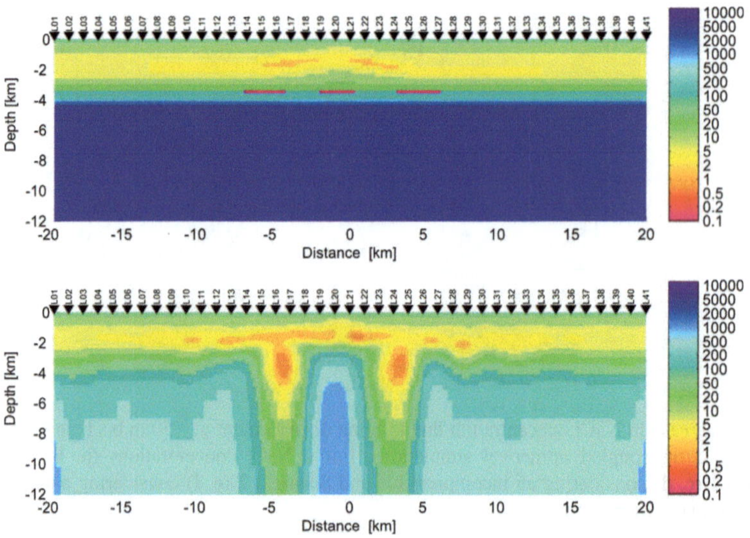

Fig. 14 Investigation of the resolution potential for a 40 km profile with a station distance of 1 km and data in the frequency range of 1 mHz to 1 kHz. The colours represent the apparent resistivity (in Ωm). The synthetic model (*top*) comprises the anticline structure and several low-thickness conductive anomalies (*red lines*). Inversion results (*bottom*) allow for a detection of the about 3.5 km deep anomaly location at the anticline flank

Synthetic 2D and 3D models calibrated using field campaign results demonstrate that the presence of CO_2 is accounted by the transfer functions in a significant way. Hereby, changes amount to about 20 % in the apparent resistivity and about 5° in the phases.

The electric field shows a higher sensitivity for CO_2 compared to the magnetic field. Furthermore, monitoring configurations with horizontal sources at the surface exhibit a relatively low sensitivity to CO_2 migration, whereby those with vertical sources can be improved by positioning the source in about 1 km distance to the CO_2. Configurations using vertical sources in wells show a higher sensitivity, if the source is at a depth of >300 m. Optimum source to receivers configurations consist of a vertical source in a well (500–1,000 m deep) with about 1–2 km distance to the injection well. Hereby, the horizontal distance is almost equal to the target depth. The vertical magnetic field is the most relevant monitoring component using this configuration. However, this field component can be only determined inside the observation wells due to technical reasons. Monitoring configurations with a vertical dipole penetrating the target formation exhibit a very high sensitivity regarding CO_2 resistivity, but a low sensitivity considering the CO_2 migration geometry. CO_2 migration can be detected with a resolution of about 300 m depending on the storage operation and phase distribution involving a resistivity of >5 Ωm. In addition to the CO_2 migration scenarios, brine migration scenarios were investigated. Monitoring of brine migration using deep wells is mainly a task carried out by electrical resistivity tomography (ERT) discussed in Sect. 3.6. Modeling of these scenarios using the CSEM approach is an interface between both monitoring applications. A detailed model of the fault zone was established using multiple dipole sources in one well that penetrates the reservoir and a fault zone. Resistivity was derived from dynamic flow simulations carried out as discussed in Sect. 3.2. Hereby, CSEM modeling results show a good agreement with those carried out for the ERT monitoring technique. Application of higher frequencies allows for a higher sensitivity regarding small-scale geological structures.

3.6 Local-Scale Monitoring Using Electrical Resistivity Tomography (ERT)

The main objective of the geoelectrical monitoring activity was the development of an optimal electrode array design to detect time-lapse effects of the subsurface resistivity distribution. It was shown in recent scientific studies, that surface-based ERT was successfully carried out to assess fluid migration processes at different lateral and vertical scales (Storz et al. 2000; Michot et al. 2003; Daily et al. 2004), and that cased wells permanently equipped with electrode arrays allow for continuous time-lapse monitoring in the near-wellbore area (van Kleef et al. 2001; Bryant et al. 2001). Because ERT provides high spatial resolution on the local scale, it has been intended for borehole and near surface investigations only. Based on these experiences, forward modeling studies were performed for different ERT acquisition geometries (single borehole, crosshole, surface-downhole) and various electrode configurations (pole–pole, dipole–dipole and bipole-bipole), supported by a real-scale three-layer resistivity model. In addition to the field-scale investigations, the ERT feasibility has been validated under defined laboratory conditions.

A cylindrical sandbox model with a diameter and a height of 0.6 m was set up for development and verification of appropriate geoelectrical acquisition geometries as prerequisite of the ERT monitoring task (Fig. 15). First experiments started with homogeneous *Fontainebleau* sand filling with a porosity of 45 % and a permeability of 23 Darcy. Synthetic brine with a concentration of 200 g/l NaCl was injected at the bottom of the sandbox and automated via a LabView™ control interface.

Miniaturized electrode arrays of different geometries were developed and installed in the sandbox. They were modeled as horizontal and vertical strings of different lengths and with a defined number of metallic pins as electrodes (Fig. 16). The geoelectrical measurements were accompanied by a fiber-optic sensor to determine the salinity of the sandbox fluid.

Basic experiments with combinations of 50 surface and 25 borehole electrodes were carried out as standard measurement configurations (Wenner, Schlumberger, dipole–dipole) and accompanied by numerical forward and inverse modeling

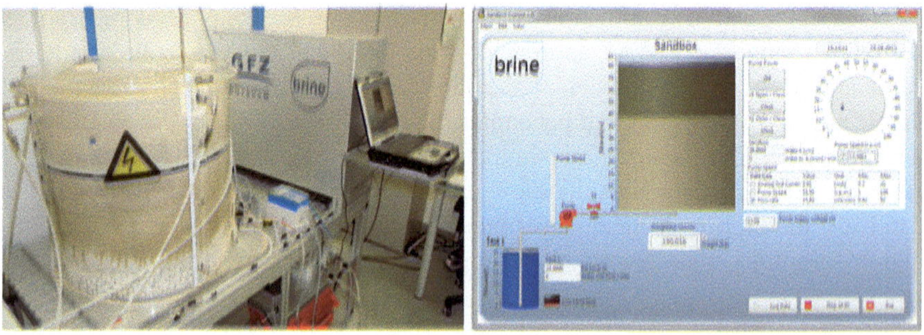

Fig. 15 Cylindrical sandbox with pump and central control unit, mechanical base with load cell and monitor (*left*). Graphical user interface of the control unit during a laboratory test (*right*)

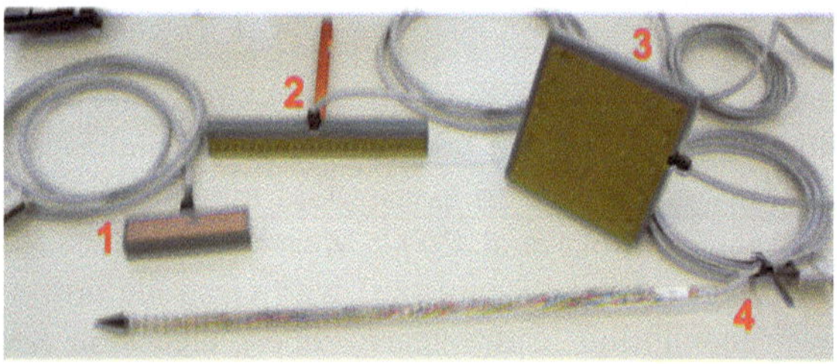

Fig. 16 Realization of different geoelectrical acquisition geometries consisting of surface profiles with electrode distances of 0.5 cm (*1*) and 1 cm (*2*), 2D arrays (*3*) and permanent borehole electrodes (*4*)

(Wagner et al. 2013). In the following experiments, the salinity was increased stepwise from 0 to 250 g/l (NaCl), and aquiclude layers were integrated between the aquifer layers in the sandbox model. The various miniaturized monitoring layouts were tested in terms of their spatio-temporal resolution of brine migration in the sandbox. At laboratory scale, NaCl concentrations of up to 250 mg/l can be monitored, but the ERT resolution decreases towards higher NaCl concentrations in a non-linear manner. In general, resistivity changes can be mapped by well-known inversion techniques as e.g., ratio inversion and difference inversion methods (Hayley et al. 2011). Because many algorithms tend to smooth the results especially for small resistivity changes, the use of directly measured and non-inverted data provides an alternative option (Fig. 17).

From the field-based modeling and underlined by the numerical modeling results of Tillner et al. (2013a), one can conclude that permanent borehole electrodes have to be placed above and below the potential brine intrusion zone in order to reliably detect the migration scenarios in time. An adapted electrode array has been suggested consisting of segments with short electrode distances near the reservoir and near the aquifer zone as well as larger electrode distances in the other regions. The simulation results show that the change in salinity occurs in days, weeks or longer periods of time (Fig. 17). Therefore, ERT readings have to be adjusted in time according to the present salinization situation. One can conclude that the selected acquisition geometries with a limited but properly chosen electrode layout provide meaningful results of the spatio-temporal salinization behavior.

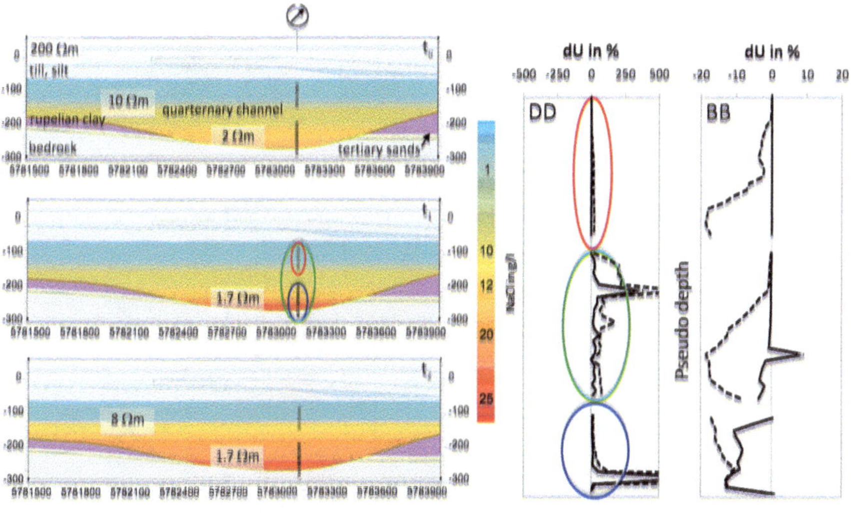

Fig. 17 Detection of brine upward migration through a Rupelian clay defect via shallow guard wells (*left side*) and the corresponding relative change of resistivities derived from Dipole–Dipole (*DD*) and Bipole–Bipole (*BB*) configurations (*right*). The *solid line* corresponds to timestep t_1 (early phase of salinization: 250 days), the *dashed line* to timestep t_2 (advanced phase of salinization: 7,300 days ≡ 20 years)

3.7 Monitoring of Potential Brine Upward Migration Zones Using Electromagnetics

Potential brine upward migration zones in the study area are mainly determined by geological windows as well as hydraulically conductive faults located in the Rupelian clay. For that reason, the aim of the electromagnetics task was the exploration and assessment of existing and potential shallow brine upward migration zones in the area of the Beeskow-Birkholz structure. Electromagnetic multi-frequency (EM) and electrical (ERT) methods were employed, since the expected differences in electric conductivity resulting from potential brine upward migration allow for detection of such processes.

Following data acquisition for selection of monitoring areas, the geophysical monitoring methods discussed above were applied for verification but also in order to achieve a combination of both methods for the same monitoring profiles. EM is generally more sensitive to highly conductive layers (e.g. brine bearing horizons), whereby low conductive layers can be better detected by ERT. In total, 22 locations and about 83 km of profiles were monitored using both geophysical methods (about 63 km by EM and 20 km by ERT) as illustrated in Fig. 18.

An example application of EM and ERT is presented in the following based on the profile Annenhof, where a salinization in the deeper underground (about 90 m.b. s.l.) was investigated. The area is located about 5–7 km from known geological weakness zones like Mesozoic faults and Quaternary erosion channels at about 70 m. a.s.l. (Fig. 19). Thus, elevated groundwater mineralization had not to be expected above the Rupelian clay in this area. Results from EM and ERT monitoring (Fig. 19)

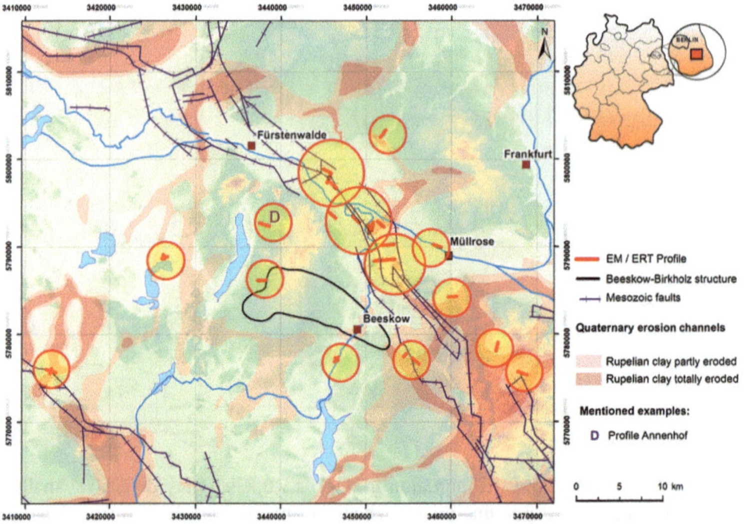

Fig. 18 Study area Beeskow-Birkholz with geophysical profiles (EM/ERT) investigated and main structural characteristics

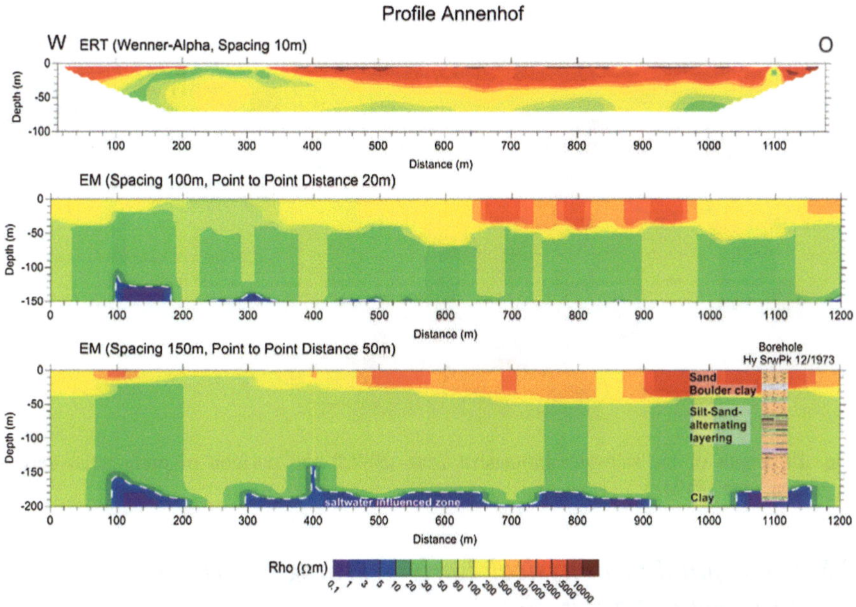

Fig. 19 ERT and EM investigation results for the profile Annenhof. The different colours indicate the resistivity distribution of the underground and the lithology of a nearby groundwater well

show resistivities that correlate well with the geological layering in the area of a close groundwater well. The shallow sand layer is depicted by higher resistivities ($\rho > 100\ \Omega m$), whereby medium values of $\rho < 80\ \Omega m$ were observed for the bolder clay as well as sand and silt intercalations. From a depth of 160 m, a low-resistivity layer ($\rho < 10\ \Omega m$) is detected by EM, which can be interpreted as salinization slightly above the Rupelian clay (horizontal migration of saline water) or the Rupelian clay itself.

The presented example as well as the ones discussed in Kempka et al. (2014) demonstrate the complex hydrogeological and geological situation in the study area. To allow for a comparison between the investigation results, the term saltwater influenced zone (SWIZ) was introduced defining an area with an electric resistivity of the rock below 10 Ωm or a groundwater conductivity >1,000 μS. Consequently, the SWIZ determines the part of the lithological profile with saline water influence. Geological conditions in the study area allow to define lithological units with resistivities below 10 Ωm as being influenced by saline water due to the lack of clay sediments above the Rupelian that may exhibit similar resistivity values. Hence, the top of the saltwater influenced zone can be interpreted as the saltwater-freshwater boundary in the study area (Fig. 20). The area north of Beeskow with elevated SWIZ values is located at the intersection of different zones of weakness (Mesozoic fault and Quaternary erosion channel) and represents a preferential vertical migration zone (PVMZ) for saline water.

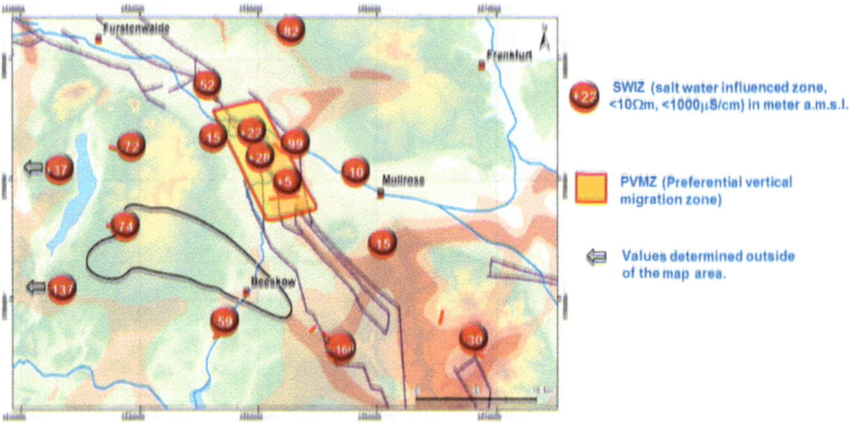

Fig. 20 Depth of the saltwater influenced zone (*SWIZ*) and position of preferential vertical migration zone (*PVMZ*)

3.8 *Conceptual Geophysical Early Warning System Against Freshwater Salinization*

Based on the geophysical monitoring results a concept for an early warning system against freshwater salinization was implemented using MT, ERT and MT monitoring methods (Fig. 21). The monitoring goals for the conceptual design were: monitoring of CO_2 migration and changes in salinity in the caprock as well as monitoring of changes in the freshwater-salt water boundary.

As a result of the physical boundary conditions (resistivity conditions and contrasts in the underground) and the given methodology only MT and CSEM in

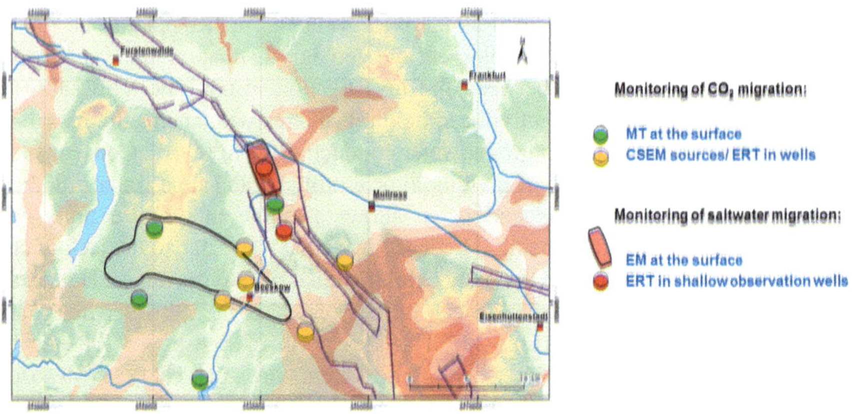

Fig. 21 Geophysical early warning system concept against freshwater salinization based on MT, CSEM, ERT and EM-systems

combination with ERT in wells are relevant for monitoring of CO_2 migration. Especially low resistivity contrast between reservoir sandstones and the saline pore fluid ($\rho \approx 3\ \Omega m$) at monitoring depths of about 1,000 m are a challenge and require a signal to noise ratio that cannot be achieved by MT measurements in anthropogenic determined areas. The concept thus foresees mobile MT station installations at selected locations in the vicinity and at the boundaries of the anticline structure. Furthermore, fixed installations of CSEM sources and ERT electrode arrays in scheduled injection and observation wells are part of the concept. Permanently installed electrodes reaching from the reservoir to the caprock can be applied to determine the initiation of a NaCl concentration increase at an early stage. Hence, potential leakage can be detected and mitigated by selected reservoir management strategies. We expect that a resolution in the ten meter range is feasible with this concept. Investigation of the freshwater-saltwater boundary can be realized by EM and ERT methods. Based on the results of the geophysical exploration carried out in the study area aiming at geological weakness zones and faults as well as those of the hydrogeological modeling providing saline water migration, installation of monitoring instruments at neuralgic points are suggested. These locations are already today determined by vertical brine migration. The focus of the ERT and EM monitoring methods is laid on these preferential vertical migration zones. EM installations are scheduled to be mobile and aligned in line at the surface to extend the point installations of the ERT locations. ERT installations are located in existing shallow wells (observation wells) and ensure a resolution in the meter range. The suggested geophysical monitoring concept may be extended by integration of long electrode ERT using steel casings in existing boreholes as introduced by Günther et al. (2015).

4 Conclusion

The present study aimed at an investigation of formation fluid migration resulting from pore pressure elevation in a storage reservoir induced by geological CO_2 storage: (1) what are the impacts of brine displacement from the CO_2 storage formation along potential migration pathways on freshwater-bearing aquifers, (2) is a synergetic utilization of formation fluids for geothermal heat recovery feasible, and (3) is it possible to employ an integrated geophysical monitoring system to detect upward formation fluid migration?

Conclusions based on the main scientific results of the present study are:

- Implementation of complex 3D structural models based on data from former hydrocarbon explorations is generally feasible and the resulting models are qualitatively sufficient for numerical modeling by means of providing insights into potential fluid migration.
- Synergetic geothermal heat recovery from storage formation fluids is feasible, if a pressure management is intended to be used in the storage formation. Thereto,

production wells can be applied to produce formation fluids and recover geothermal heat from these fluids.

- Hydro-mechanical simulations showed that storing 34 Mt in the anticline structure Beeskow-Birkholz as initially planned by the prospective site operator does not compromise the storage, caprock and fault integrity given the available geological data and models developed in the scope of the present study. Nevertheless, site-specific assessments are required for other study areas of interest.
- Extensive laboratory and field campaigns in the study area were employed to investigate the detectability of brine migration as well as for calibration of our forward models. Each of the investigated monitoring methods can provide a relevant contribution to the early warning system concept. Hence, an integration of the applied monitoring techniques does allow for implementation of an efficient early warning system to detect and monitor formation fluid migration by means of an early warning system against freshwater salinization.
- Based on the investigations carried out and considering specific results from our field, laboratory and numerical modeling activities, we were able to design an integrated geophysical early warning system for monitoring potential freshwater salinization verified against simulated formation fluid migration scenarios.

Acknowledgments We thank the German Federal Ministry of Education and Research (BMBF) for funding of the brine project in the scope of the GEOTECHNOLOGIEN R&D programme and the Projektträger Jülich (PTJ) for their support. Furthermore, we acknowledge data provision by Vattenfall Europe Mining AG and their subcontractors as well as discussions with representatives of the mining authority of the State of Brandenburg (LBGR).

References

Alber M, Heiland J (2001) Investigation of a limestone pillar failure part 2: stress history and application of fracture mechanics approach. Rock Mech Rock Eng 34:187–199

Birkholzer JT, Zhou Q, Tsang C-F (2009) Large-scale impact of CO_2 storage in deepsaline aquifers: a sensitivity study on pressure response in stratified systems. Int J Greenhouse Gas Control 3:181–194

Bryant ID, Chen M-Y, Raghuraman B, Raw I, Delhomme J-P, Chouzenoux C, Manin Y, Rioufol E, Oddie G, Swager D, Smith J (2001) Utility and reliability of cemented resistivity arrays in monitoring waterflood of the Mans-field sandstone, Indiana, USA, SPE 71710

Daily W, Ramirez A, Binley A, LaBrecque D (2004) Electrical resistance tomography. Lead Edge 23(5):438–442

Grube A, Hermsdorf A, Lang M, Rechlin B, Steffens A (2002) Numeric and hydro-chemical modelling of saltwater intrusion into a pleistocene aquifer—case study Großbeuthen (Brandenburg). In: Proceedings of 17th salt water intrusion meeting, Delft, The Netherlands, pp 6–10

Günther T, Ronczka M, Voß T (2015) Saltwater monitoring using long-electrode ERT. In: Liebscher A, Münch U (eds) Geological storage of CO_2: long term security aspects. Springer, Singapore

Hayley K, Pidlisecky A, Bentley LR (2011) Simultaneous time-lapse electrical resistivity inversion. J Appl Geophys 75(2):401–411

Jahnke C (1999) Ein neues Klassifikationssystem für Grundwässer und seine Anwendung in känozoi-schen Porengrundwasserleitern. Grundwasser 2:62–72

Kempka T, Herd R, Huenges E, Jahnke C, Janetz S, Jolie E, Kühn M, Magri F, Möller M, Munoz G, Ritter O, Schafrik W, Schmidt-Hattenberger C, Tillner E, Voigt H-J, Zimmermann G (2014) CO_2-Speicherung in Ostbrandenburg: Implikationen für eine synergetische geothermische Energiegewinnung und Konzeptionierung eines Frühwarnsystems gegen Grundwasserversalzung. Final report BMBF-Projekt brine, grant 03G0758A/B. TIB Hannover, 121 pp

Lehmann HW (1974) Geochemie und Genesis der Tiefenwässer der Nordostdeutschen Senke. Zeit-schrift für angewandte Geologie 20(11):502–509, 12:551–557

Lemieux JM (2011) Review: the potential impact of underground geological storage of carbon dioxide in deep saline aquifers on shallow groundwater resources. Hydrogeol J 19:757–778

Magri F, Tillner E, Wang W, Watanabe N, Zimmermann G, Kempka T (2013) 3D hydro-mechanical scenario analysis to evaluate changes of the recent stress field as a result of geological CO_2 storage. Energy Procedia 40:375–383. doi:10.1016/j.egypro.2013.08.043

Magri F, Tillner E, Kempka T, Watanabe N, Wang W, Zimmermann G (2015) Coupled THM processes. In: Kolditz O, Shao H, Wang W, Bauer S (eds) Hydro-mechanical-chemical processes in fractured porous media: modelling and benchmarking. Springer, Berlin, pp 311

Michot D, Benderitter Y, Dorigny A, Nicoullaud B, King D, TAbbagh A (2003) Spatial and temporal monitoring of soil water content with an irrigated corn crop cover using surface electrical resistivity tomography. Water Resour Res 39(5):1138

Müller P, Papendieck G (1969) Zur Verteilung, Genese und Dynamik von Tiefenwässern unter be-sonderer Berücksichtigung des Zechsteins. Zeitschrift geologische Wissenschaften 3 (2):167–196

Nakaten B, Tillner E, Kempka T (2013) Virtual elements for representation of faults, cracks and hydraulic fractures in dynamic flow simulations. Energy Procedia 40:447–453. doi:10.1016/j.egypro.2013.08.051

NER (2001) Modulus dispersion in sandstone and claystone. New England Research, Inc. /802-296-2401

Nicot JP (2008) Evaluation of large-scale CO_2 storage on fresh-water sections of aquifers: an example from the texas Gulf coast basin. Int J Greenhouse Gas Control 2:582–593

Ocak I (2008) Estimating the modulus of elasticity of the rock material from compressive strength and unit weight. J South Afr Instit Min Metall 108:621–625

Oldenburg C, Rinaldi A (2011) Buoyancy effects on upward brine displacement caused by CO_2 injection. Transp Porous Media 87:525–540

Pruess K (2005) ECO2N: A TOUGH2 fluid property module for mixtures of water, NaCl and CO_2. Lawrence Berkeley National Laboratory Report LBNL-57952, Berkeley, CA

Pruess K (2011) Modeling CO_2 leakage scenarios, including transitions between super- and sub-critical conditions, and phase change between liquid andgaseous CO_2. Energy Procedia 4:3754–3761

Reinsch D, Voigt HJ (1975) Hydrogeologische Übersichtskarte 1: 500.000 (Nordteil der DDR), Hydrochemische Karte des Buntsandsteins, Hydrochemische Karte des Rät-Unterkreide-Komplexes, Zentrales Geologisches Institut, Berlin

Reyer D, Philipp SL (2012) Heterogeneities of mechanical properties in potential geothermal reservoir rocks of the North German Basin. EGU General Assembly (Vienna, Austria) Geophysical Research Abstracts, 14, EGU2012-346

Röhmann L, Tillner E, Magri F, Kühn M, Kempka T (2013) Fault reactivation and ground surface uplift assessment at a prospective German CO_2 storage site. Energy Procedia 40:437–446. doi:10.1016/j.egypro.2013.08.050

Schirrmeister W, Voigt H-J (1976) The boundary between fresh—and saltwater in saliferous sedimentary basins. Hydrogeology of great sedimentary basins—Proceedings of the Budapest conference, IAH-IAHS

Schlumberger (2011) Petrel seismic-to-evaluation software, Version 2011.1

Schwartz MO (2014) Modelling leakage and groundwater pollution in a hypothetical CO_2 sequestration project. Int J Greenhouse Gas Control 23:72–85

Stackebrandt W, Manhenke V (2004) Landesamt für Geowissenschaften und Rohstoffe Brandenburg LGRB. Atlas zur Geologie von Brandenburg. 3. Auflage. ISBN 3-9808157-1-4

Stackebrandt W, Manhenke V (2010) Atlas zur Geologie von Brandenburg: im Maßstab 1: 1,000,000. Landesamt für Bergbau, Geologie und Rohstoffe Brandenburg, Kleinmachnow, pp 1–142

Storz H, Storz W, Jacobs F (2000) Electrical resistivity tomography to investigate geological structures of the earth's upper crust. Geophys Prospect 48:455–471

Tesmer M, Möller P, Wieland S, Jahnke C, Voigt H-J, Pekdeger A (2007) Deep fluid flow in the North East German basin: origin and processes of groundwater salinisation. Hydrogeol J 15:1291–1306

Tillner E, Kempka T, Nakaten B, Kühn M (2013a) Brine migration through fault zones: 3D numerical simulations for a prospective CO_2 storage site in Northeast Germany. Int J Greenhouse Gas Control 19:689–703. doi:10.1016/j.ijggc.2013.03.012

Tillner E, Kempka T, Nakaten B, Kühn M (2013b) Geological CO_2 storage supports geothermal energy exploitation: 3D numerical models emphasize feasibility of synergetic use. Energy Procedia 37:6604–6616. doi:10.1016/j.egypro.2013.06.593

Trettin R, Hiller A, Wolf M, Deibel K, Gläßer W (1997) Isotopenanalytische Charakterisierung tiefliegender Grundwässer im Raum der Fürstenwalde-Gubener-Störungszone. Grundwasser 97(2):65–76

van Kleef R, Hakvoort R, Bushan V, Al-Khodhori A, Boom W, de Bruin C, Babour K, Chouzenoux C, Delhomme JP, Manin D, Pohl E, Rioufol M, Charara M, Harb R (2001) Water flood monitoring in an Oman carbonate reservoir using a downhole permanent electrode array, in SPE Middle East oil show. In: Proceedings society of petroleum engineers, vol. SPE-68078, CD-ROM, 11

Wagner F, Möller M, Schmidt-Hattenberger C, Kempka T, Maurer H (2013) Monitoring freshwater salinization in analog transport models by time-lapse electrical resistivity tomography. J Appl Geophys 89:84–95. doi:10.1016/j.jappgeo.2012.11.013

Walter L, Binning PJ, Oladyshkin S, Flemisch B, Class H (2012) Brine migration resulting from CO_2 injection into saline aquifers—an approach to risk estimation including various levels of uncertainty. Int J Greenhouse Gas Control 9:495–506

Wang W, Kolditz O (2007) Object-oriented finite element analysis of thermo-hydro-mechanical (THM) problems in porous media. Int J Numer Methods Eng 69:162–201

ZGI (1990a) Zentrales Geologisches Institut der DDR (eds) (1990) Geologische Karte der DDR 1:500.000, Tektonische Karte

ZGI (1990b) Zentrales Geologisches Institut der DDR (eds) (1990) Geologische Karte der DDR 1:500.000, Karte der Fotolineationen nach kosmischen Aufnahmen

Zhang K, Wu Y-S, Pruess K (2008) User's guide for TOUGH2-MP—a massively parallel version of the TOUGH2 code. Report LBNL-315E, Earth Sciences Division. Lawrence Berkeley National Laboratory, Berkeley

Combined Natural and Social Science Approach for Regional-Scale Characterisation of CO_2 Storage Formations and Brine Migration Risks (CO_2BRIM)

Holger Class, Alexander Kissinger, Stefan Knopf, Wilfried Konrad, Vera Noack and Dirk Scheer

Abstract The CO_2BRIM project pursues a combined technical-social sciences approach for investigating participatory modelling applied to different stages of a characterisation of potential CO_2 storage formations. From the technical point of view, the project deals with two topics: The first one is concerned with an early stage site screening, where the Gravitational Number (Gr) is used as an indicator for storage efficiency dependent on the conditions found in German Middle Buntsandstein rock units. The second topic addresses the problem of large-scale salt water displacement due to the injection of CO_2. Both topics are accompanied by a method called Participatory Modelling, where third parties, i.e. stakeholders with different background, are involved into the modelling process. This report gives some details about the methodological aspects of participatory modelling for the two

H. Class · A. Kissinger (✉)
Lehrstuhl für Hydromechanik und Hydrosystemmodellierung, Universität Stuttgart, Stuttgart, Germany
e-mail: alexander.kissinger@iws.uni-stuttgart.de

H. Class
e-mail: holger.class@iws.uni-stuttgart.de

S. Knopf · V. Noack
Bundesanstalt für Geowissenschaften und Rohstoffe (BGR), Hannover, Germany
e-mail: stefan.knopf@bgr.de

V. Noack
e-mail: vera.noack@bgr.de

W. Konrad · D. Scheer
Dialogik gGmbH, Stuttgart, Germany
e-mail: konrad@dialogik-expert.de

D. Scheer
e-mail: dirk.scheer@sowi.uni-stuttgart.de

A. Liebscher and U. Münch (eds.), *Geological Storage of CO_2 – Long Term Security Aspects*, Advanced Technologies in Earth Sciences,
DOI 10.1007/978-3-319-13930-2_10

technical topics and discusses the results for the Gr-characterisation in detail; first results and preliminary discussions from work on large-scale brine displacement are added at the end.

1 Introduction

The idea of setting up a joint technical and social sciences research project originated in 2009 when research on geological storage of CO_2 in Germany and other countries reached a stage where the next logical step would have been the implementation of industrial-scale demonstration projects. Such an endeavour would involve a large number of stakeholders for selecting suitable sites, likely to be accompanied by conflicts of different interests. The CO_2BRIM project aimed at integrating stakeholders into various 'virtual' stages of a site characterisation, using geological and dynamical modelling, and introducing participatory modelling to the CO_2 storage research community.

By the time when the project started in 2011, the general trend in Germany indicated already that carbon capture and storage (CCS) will probably not be implemented in the near future. The delay in the German CCS law, i.e. the *Kohlendioxid-Speicherungsgesetz* (*KSpG*), eventually its contents, and the missing acceptance in public opinion have made the CCS option unattractive for the industry, since the predictability of legal decisions is not given and therefore a reliable planning is not possible. Thus, the whole CCS debate lost the drive. For the CO_2BRIM project this was a disadvantageous situation since the idea of participatory modelling works best when stakeholders show a strong interest. The recruiting of sufficient stakeholders to represent the different parties, like industry, Non-Governmental Organizations (NGOs), governmental agencies, water suppliers, etc. has been more difficult than expected. Nevertheless, the project succeeded in involving expert groups in the technical topics while most of the participants of the social sciences study did not show up on their own initiative but upon personal requests.

At an early planning stage of a CO_2 storage project we assume the screening of different sites to be a task which, for two reasons, does not allow comprehensive numerical modelling. First, there is typically a lack of data before the sites are explored. Second, 3D geological and dynamical modelling is very time-consuming. Therefore, we attempted in a computationally extremely cheap approach to tag different regions in the North German Basin with respect to their expected storage efficiency in the Middle Buntsandstein rock unit. This is realised by evaluating the Gravitational Number (Gr) on the basis of depth, salinity, and geothermal gradient found at different locations which have already fulfilled the criteria to be categorised as worthy of further investigation in the Storage Catalogue of Germany (Müller and Reinhold 2011). The Gravitational Number can be interpreted as a qualitative indicator for storage efficiency. It relates gravitational forces to viscous forces which influence the shape of the CO_2 plume and therefore also the space occupied by the plume. Section 2 explains the method technically and provides a discussion of how stakeholders influenced the results and conclusions from this study in the framework of a Delphi workshop.

The second major (technical) topic addresses brine displacement and migration after CO_2 injection, eventually aiming at evaluating the hazards associated with displaced saline waters. We are aware of the difficulties involved in setting up generic scenarios that are representative for geologic situations encountered in real projects. Nevertheless, we believe that it is helpful to think about potential migration pathways in a generic manner, to include expert opinion in order to prioritise scenarios, and to develop and test the methods to apply for simulating the scenarios. This allows for an improved understanding of the uncertainties and ranges of predicted brine migration when comparing models of different complexity (and thus: accuracy). Section 3 provides the technical descriptions and the participatory approach, here in the form of face-to-face interviews with ten experts. Further participatory methods are being planned, therefore the results presented here are preliminary.

One must be aware that within the CO_2BRIM project only very selected tasks are addressed. For example, scenarios related to the leakage of CO_2 are not part of the investigation, although CO_2 leakage is likely to be the most important issue for a real project. The major aim of CO_2BRIM is to have two topics as described in Sects. 2 and 3 as showcases where participatory modelling can be practiced and the experiences and lessons learnt (see Sect. 4) can hopefully guide future application of this engineering/social science integration method.

2 Storage Efficiency and Gravitational Number as a Screening Indicator

The Gravitational Number (Gr) as an indicator for storage efficiency in a given reservoir is based on its ability to predict the influence of viscous forces versus buoyancy forces during CO_2 migration in a reservoir. Below, we summarise the theoretical background of the study and the availability of data for the North German Basin. Thereafter, the results of the study are briefly presented, thereby discussing how the participatory modelling approach in the frame of a group Delphi workshop took influence. Detailed explanations on this topic are found in Kissinger et al. (2014).

2.1 Background and Available Data

We use here the following definition of Gr:

$$Gr = \frac{(\rho_B - \rho_{CO_2})gk}{\mu_{CO_2}\frac{\dot{m}_{CO_2}}{\rho_{CO_2}}} \tag{1}$$

$\rho_B - \rho_{CO_2}$ is the density difference between the fluids brine and CO_2, k the permeability in the formation, μ_{CO_2} the dynamic viscosity of CO_2, and $\dot{m}_{CO_2}$ the specific mass injection rate. The density differences are the driving force for vertical flow of CO_2. The larger the gravitational forces are in comparison to the horizontal spreading of the CO_2, the more gravity segregation will dominate the spreading of the plume in the reservoir, where the lighter CO_2 spreads along the top of the reservoir. Strong gravity segregation prevents an efficient usage of available pore space. Thus, one is interested in having small Gr numbers for good storage efficiency.

The question emerging for practical use is: Which factors are relevant for the Gr number and how big is the range of Gr within a region to screen for a storage project (the focus is here on the Middle Buntsandstein unit in the North German Basin). In fact, the fluid properties density and viscosity depend on the local thermodynamical conditions, i.e. pressure and temperature, and on the salinity. Therefore, the depth of a formation, the geothermal gradient, and the salinity distribution provide relevant input data for estimating Gr. Furthermore, the permeability appears in the definition of Gr. However, regionalized permeability data is not available for the North German Basin. Thus, it was decided for this study to restrict the influences on Gr to depth, geothermal gradient, and salinity. The obtained regional distribution of Gr can be interpreted as an indicator for storage efficiency with respect to the encountered initial fluid properties.

The area for this study is selected from the project *Storage Catalogue of Germany*, where areas worthy of further investigation for the Middle Buntsandstein were tagged with the criteria that their depth at the top of the rock unit exceeds 800 m and that they have a net thickness of more than 10 m (Reinhold et al. 2011). Depth, temperature, and salinity were determined on a raster with a resolution of 1 km × 1 km. Information on depth was gathered from an unpublished depth grid for the base of the Upper Buntsandstein (corresponding to the top of the Middle Buntsandstein), see Fig. 1. Temperature information was taken from a map showing the lateral variation of the mean geothermal gradient published by the Leibniz Institute for Applied Geophysics (LIAG 2012). Combining the interpolated geothermal gradients with the depth information required the following equation:

$$T_{\text{depth}} = T_0 + (\nabla T * \text{depth}) \tag{2}$$

T_0 is the average ground temperature. According to Schulz (2009) the average value in Germany is 8.2 °C. An interpolated distribution of the mean geothermal

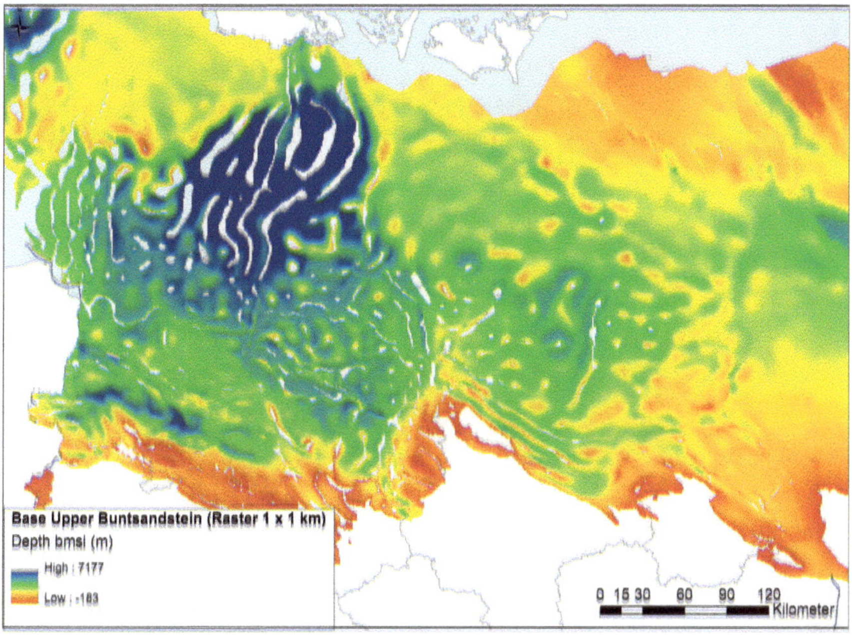

Fig. 1 Depth grid for the base of the Upper Buntsandstein (equivalent with the top of the Middle Buntsandstein), depth information in meter below mean sea level (*bmsl*). Figure modified after Kissinger et al. (2014)

gradient is shown in Fig. 2. Finally, concerning the salinity, we followed the simplified assumption of an average salinity gradient of about 15 g/l per 100 m increasing with depth (Wolfgramm and Seibt 2008), applied throughout the area of investigation. This yields depth-dependent salinity as

$$S_{\mathrm{depth}} = \nabla S * \mathrm{depth} \tag{3}$$

The calculated salinity values were capped by the maximum salinity of 360 g/l (Brasser et al. 2011). Based on the values of depth (pressure), temperature, and salinity, as explained, Gr was calculated for each grid cell (Kissinger et al. 2014).

2.2 Results Before and After Participation

The Gr mapping on the basis of fluid properties dependent on depth, geothermal gradient, and salinity is shown in Fig. 3. A range between 4×10^{10} and 10×10^{10} is observed. We note again that this does not account for the variability of permeability and porosity. Also, the injection rate is assumed to be always the same. Thus, we need to investigate the relevance of this mapping of Gr. Is it useful—as hypothesised—to

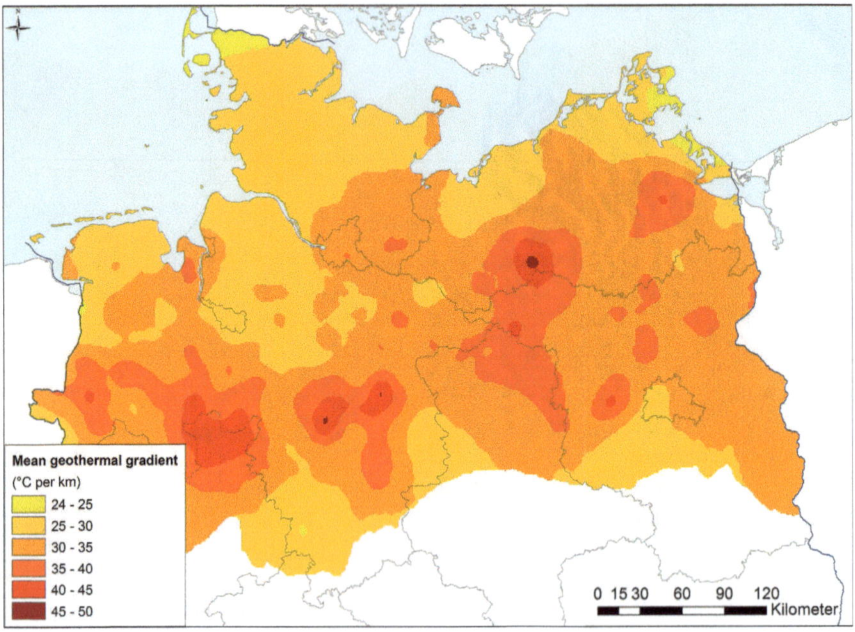

Fig. 2 Interpolated distribution of the geothermal gradient for the area of interest (LIAG 2012). Figure modified after Kissinger et al. (2014)

show a tendency towards better storage efficiency at lower Gr numbers? Or will variations in permeability and porosity dominate the storage efficiency?

A series of numerical studies with the open source simulator DuMu^x (Flemisch et al. 2011) are employed to show the influence of Gr on storage efficiency and on the residual trapping; the latter is considered an important mechanism to increase safety of storage. As a measure for storage efficiency, we choose the mass of CO_2 injected into the homogeneous and radially symmetric reservoir until a "spill point" is reached by the plume (see Fig. 4), divided by the area up to the spill point (lateral footprint):

$$\text{Storage Efficiency}\,(\text{kg/m}^2) = \frac{\text{Mass of CO}_2\text{ in place}}{\pi \cdot (\text{Radius of Spill Point})^2}. \tag{4}$$

The numerical studies are supplemented in Kissinger et al. (2014) by a comparison with an analytical estimate of storage efficiency as proposed by Okwen et al. (2010). The reservoir setup is kept as simple as possible. It consists of a radially symmetric domain as illustrated in Fig. 4 with a CO_2 injection well in the middle. The reservoir has a thickness of 25 m. The reservoir is homogenous and has an effective porosity of 0.2 and a permeability of 10^{-13} m^2. For more details on the reservoir setup we refer to Kissinger et al. (2014). Different representative points from the Gr grid are chosen, covering various depths and the range of Gr values.

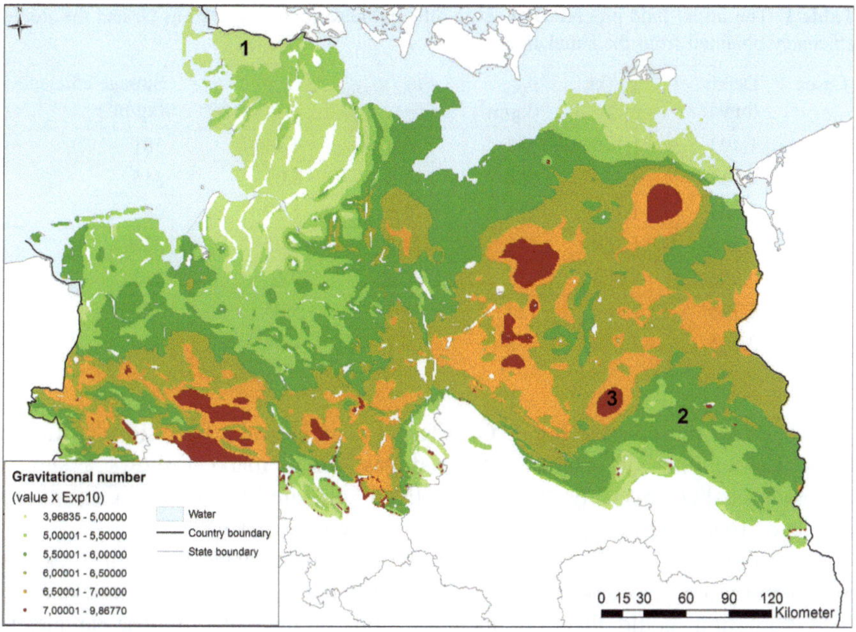

Fig. 3 Gr distribution for the Middle Buntsandstein unit in the North German Basin. Numbers refer to areas selected for investigating the relevance of the Gr-based ranking (see below and details in Kissinger et al. 2014; *Area* 1 low Gr, *Area* 2 medium Gr, *Area* 3 high Gr). Figure modified after Kissinger et al. (2014)

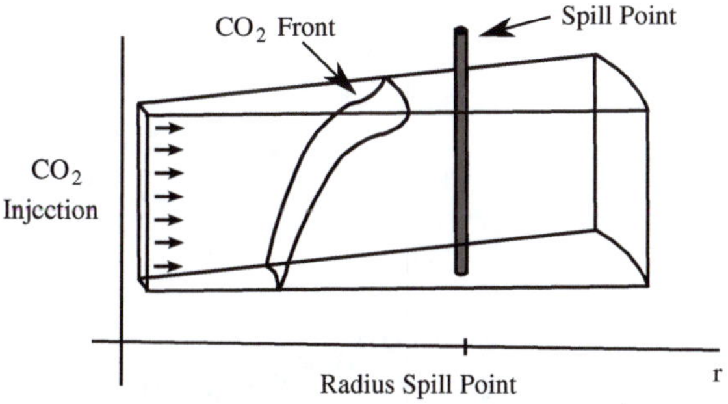

Fig. 4 Schematic setup of the model domain for the numerical simulations. Figure modified after Kissinger et al. (2014)

The varying properties of 7 of these grid cells which are used for the different simulations are listed in Table 1. The spill point in Eq. 4 can be interpreted as the maximum tolerable plume extent (here: 1 km at a constant injection rate of 0.1 Mt/year). Land's model (1968) is used to determine the residually trapped CO_2.

Table 1 The initial fluid properties of seven different raster points as well as Gr and the storage efficiency obtained from the simulations

Cases	Depth (bmsl)	ρ_B (kg/m^3)	ρ_{CO_2} (kg/m^3)	μ_B/μ_{CO_2} (–)	Gr (–) × 10^{10}	Storage efficiency (kg/m^2)
1	1,492	1,140	757	10.4	4.44	381
2	1,735	1,154	668	15.2	6.01	315
3	2,304	1,187	611	11.8	7.19	299
4	1,865	1,150	521	14.5	8.05	230
5	906	1,083	371	32.3	9.60	165
6	1,776	1,161	734	15.2	5.05	362
7	3,305	1,200	758	9.1	4.98	450

Results are presented in Fig. 5. Gr is tested as an indicator for residually trapped CO_2. The amount of residually trapped gas (circle marker) shows a similar, although weaker, inversely proportional trend compared to the total CO_2 storage efficiency (residually trapped plus mobile CO_2, cross marker).

As outlined before, the influences of permeability, variable reservoir thickness, and porosity are not considered here. Analysing the significance and relevance of the Gr criterion should, therefore, include a study on the effect of these parameters. Therefore, we select three characteristic areas of high, medium, and low Gr, where additional data on reservoir thickness and porosity was available (see also Fig. 3). Their characteristic parameters are listed in Table 2.

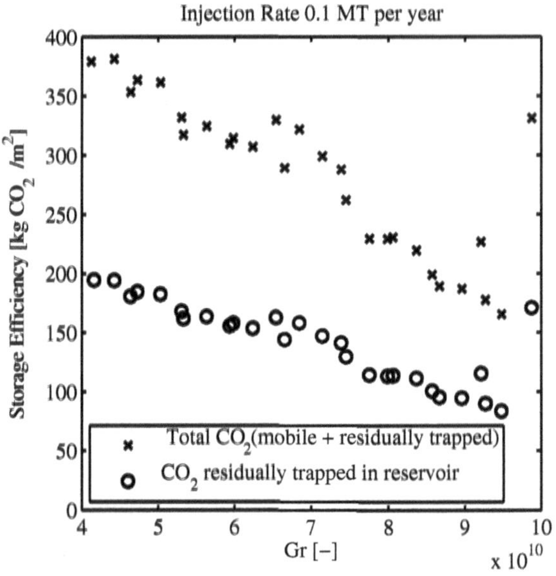

Fig. 5 Correlation of Storage efficiency and the residually trapped CO_2 mass (in the same unit kg/m^2) with Gr. Figure modified after Kissinger et al. (2014)

Table 2 The characteristics of the three different areas shown in Fig. 3

Data type	Area 1	Area 2	Area 3
	Range	9 data points	Range
Effective porosity (−)	0.11–0.30	0.06–0.25	0.04–0.08
Thickness (m)	6–20	5–20	10
Gr (−) × 10^{10}	4.85	5.73	7.27
Pressure (MPa)	20.14	10.89	10.90
Temperature (K)	331	335	351
Salinity $\left(\frac{kgNaCl}{kgBrine}\right)$	0.25	0.23	0.23

For Area 1 and 3, there are only ranges of a minimum and maximum value for effective porosity and thickness. Values for pressure, temperature, salinity and Gr are mean values for each area and used as initial conditions for the simulations

For all areas, the calculated storage efficiency includes now the "full" two-phase flow physics. We performed the same simulations as above with additionally varying porosity and reservoir thickness. The results of the simulations are displayed in Fig. 6. On the left, there is mean storage efficiency and its standard deviation for each area over the mean Gr value of this area. A clear trend of decreasing storage efficiency with increasing Gr is visible. The standard deviations of Area 1 (low) and Area 2 (medium) are higher than for Area 3, due to the small range of porosities (4–8 %) and constant thickness (10 m) in Area 3. The calculated

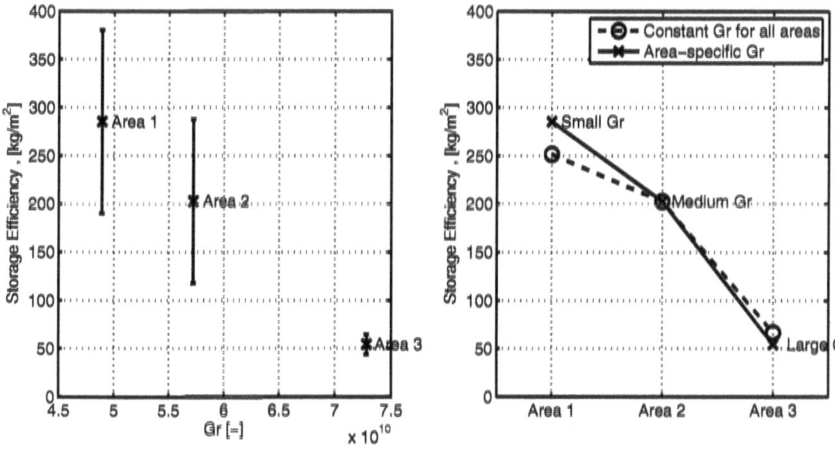

Fig. 6 *Left* Storage efficiency over Gr. The *cross marker* represents the mean storage efficiency; *error bars* indicate the standard deviation. *Right* Mean storage efficiency for the three areas with constant Gr for all areas (*dashed line*) and area-specific Gr (*solid line*). *Note* the three means for the area-specific case (*cross marker*) on the *right* plot are identical to the three means on the *left* plot (*cross marker*). Figure modified after Kissinger et al. (2014)

storage efficiency depends now on Gr and on the joint probability functions of porosity and thickness for each area. It is not obvious which of the two has more influence on storage efficiency. Thus, additional simulations were performed with the same Gr value (here we choose the Gr value of Area 2) for all three areas (constant Gr case). Figure 6 (right) compares the simulated mean storage efficiencies for the constant Gr case (dashed line) with the previously performed simulations with the area-specific Gr values (solid line). This gives evidence that mainly the different distributions of effective porosity and thickness of the three areas determine the trend. It is only increased by the area-specific value of Gr, i.e. a small value of Gr leads to an increase of storage efficiency compared to the constant Gr case and a high value of Gr leads to a decrease of storage efficiency compared to the constant Gr case. For Area 1 the "contribution" of Gr to the storage efficiency is much higher than for Area 3.

The expected qualitative influence of the Gr criterion on storage efficiency in the three areas is supported by the simulation results. However, the trend shown in the left plot of Fig. 6 is mainly caused by the different distributions of porosity and thickness. It is only between Area 1 and Area 2 where the significant increase in storage efficiency can be attributed to the different Gr values.

From a technical/engineering point of view there are different conclusions to be drawn from this study on the Gr number as a criterion for storage efficiency. Depth, temperature, and pressure (depth) of a formation have a clearly noticeable influence. However, their influence is less than that of variable porosity and reservoir thickness, but can still be in the same order of magnitude. Overall, Gr indicates where fluid properties are favourable for an efficient usage of available pore space.

How did German stakeholders evaluate Gr? Did their views and input influence the study on the Gr number? In order to discuss Gr principles and draft findings with stakeholders, a Group Delphi workshop has been carried out (Hannover, November 2012). A Group Delphi is a participatory method to elicit consensus and disagreement as to experts evaluations of controversial issues (Schulz and Renn 2009). 14 experts attended the workshop representing business and industry, public authorities, and the science communities in the fields of geosciences, carbon capture and storage, and water resources.

On the one hand, workshop participants were rather sceptical about Gr as a relevant criterion for reservoir characterisation. Concerns were raised that the Gr approach is difficult to communicate beyond expert communities and Gr results might be misunderstood by stakeholders and the public. A definite need is to contextualise the approach by geo-scientists and experts in order to prevent misunderstandings. On the other hand, experts conceded that Gr may be used as a screening instrument at early assessment stages in order to back and support other methods for characterising storage regions. Thereby, attention should be given that results are not overrated and not used as sole basis for decision making. The Group Delphi led not only to an overall assessment of the Gr approach. Moreover, the workshop delivered a couple of highly useful insights in terms of consolidating the basis of reasoning. One example refers to the data uncertainty in the Gr calculations. For some parameters there is only an incomplete data base available.

However, it was not before the Group Delphi discussions that the need to comment on that issue with more emphasis was fully recognised by the researchers. Notably, the fact that permeability is not varied raised doubts among some experts as to the significance of Gr numbers calculated without real permeability data. Furthermore, the asserted lack of regional permeability data was challenged. Further literature inquiries have been performed by the researchers which confirmed the expected result that permeability data for the Middle Buntsandstein rock unit of the North German Basin is not available. Part of this data search was the futile screening of research reports that according to some workshop participants should contain sufficient permeability data to represent a regional data distribution.

Another example of the Group Delphis influence is the investigation of two instead of just one CO_2 injection strategy. A comparison of a rather low (0.1 Mt/year) with a rather high (1.0 Mt/year) injection rate confirmed the assumed trend decreasing Gr numbers lead to increasing storage efficiency even under different conditions.

In addition, the workshop deliberations resulted in a revaluation of the relevance of the Gr criterion for distinguishing storage regions. Before the workshop, Gr was considered a quite sound screening criterion comparable to others like minimum depth. Thereafter, the assessment was revised, assigning Gr the status of a lower-ranked criterion for the differentiation of regions. This shift of decision-making assignment reflects the stakeholders sceptical view on the validity of a criterion based on calculations leaving out a key parameter (permeability) due to the lack of data. Overall, participants feedback has emphasized that Gr should only be one among many other indicators in the context of a criteria catalogue. Gr should be treated as a further screening tool among others. Hence, rankings based on the Gr score should be handled with care and should not be taken as exclusion criterion in early decision phases. The reconsideration of Gr as mere supporting tool has further been emphasized by additional simulations which revealed that both porosity and reservoir thickness often have a higher influence on storage efficiency as Gr itself.

3 Large-Scale Brine Migration Scenarios

The displacement of resident brine is inevitable during injection of CO_2 into a saline aquifer. Supposably hazardous situations arise when these brines reach drinking water in caenozoic aquifers. For vast areas of the North German Basin for example, this requires that brines penetrate from below through the Rupelian clay barrier. Imaginable scenarios to overcome the Rupelian are manifold. Brine migration through fault zones into the quarternary was investigated by the Geotechnologien project BRINE (see Chap. 9 Kempka et al.). A further possibility is salt water flowing upwards along the flanks of salt diapirs which exist over wide areas of the North German Basin and which may have broken the Rupelian barrier. We put the focus of our investigations on the latter type of scenario, since there is still a poor understanding of the role that salt diapirs would play in CO_2 storage projects. The participation of experts aims at gathering knowledge and inspires an intensified discussion while the geological and numerical modelling serves as a showcase for this type of scenario at realistic temporal and spatial scales.

3.1 Geology and Reference Setup

To construct a geological model for large-scale brine migration scenarios we used subsurface data from a region in the southwestern German North Sea. This area is characterised by an anticlinal structure on top of a salt pillow (Permian Zechstein salt) with overlying Mesozoic and Caenozoic sediments, a configuration representing a typical structural setting of the North German Basin (NGB). The anticlinal structure descends gently into a structural low, bordered on an elongated steeply rising salt wall (diapir) (Fig. 7).

The data base for the 3D structural model consists of digitized depth converted seismic reflector horizons data (Baldschuhn et al. 2001; Bombien et al. 2012) and new interpretations (Kaufmann et al. 2014). These data were used to construct different stratigraphical surfaces by interpolating and gridding the horizons data using the convergent interpolation technique (software Petrel 2012.1). To account for the evaluation of risks of pressure increase and accompanied salt water transport to the near-surface drinking water horizons we added virtual surfaces of important geological layers which are representative for storage units in the NGB to the final data set. This modification of the model concerns predominantly the Middle Buntsandstein unit in which we added a reservoir horizon and the Tertiary. For the latter we modified the surfaces for the Rupelian-clay to be penetrated by the rising salt wall. Subsequently, the resultant modified 2D grids of the geological units were merged into a consistent 3D structural model. Thus, the structural model does not represent a real location but a realistic spatial configuration. The final 3D structural

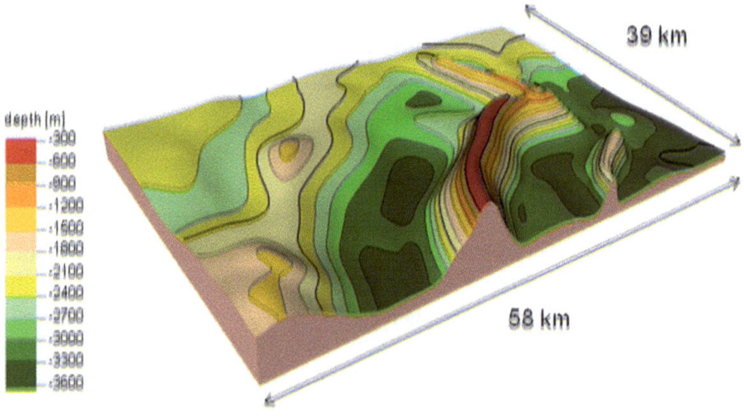

Fig. 7 3D view on top of the salt surface. Vertical exaggeration: 2:1

model covers an area of 58 km × 39 km with a horizontal resolution of 300 m, and a vertical resolution corresponding to the number of 11 stratigraphical units resolved in the model. These layers are from top to bottom: Quaternary, Tertiary post-Rupelian, Tertiary Rupelian, Tertiary pre-Rupelian, Cretaceous, Upper Buntsandstein, upper Middle Buntsandstein, the Solling unit of the Middle Buntsandstein (reservoir horizon), lower Middle Buntsandstein, Lower Buntsandstein and Permian Zechstein salt. The base of the Zechstein salt varies smoothly over the model with depth between 3,300 and 4,000 m. In contrast, the depth position of the Zechstein salt surface is highly differentiated (Fig. 7). This structural pattern affects the configuration of the overlying Mesozoic and Caenozoic layers and is important for the temperature and pressure conditions in the Solling reservoir horizon.

To perform brine migration scenarios a 3D volume mesh was developed and physical parameters depending on the lithology of the respective geological unit were assigned for each layer. Lithological composition and corresponding geophysical parameters of the geological units for the layers are derived from regional literature data and numerical simulation studies (Larue 2010; Reutter 2011; Schäfer et al. 2011; Noack et al. 2013). Table 3 shows the main lithological compositions, the average thickness of the layers, porosity and permeability data assigned to the model layers. According to the structural configuration of the Zechstein salt layer the average thicknesses of the geological layers above increase from top of the salt pillow structure towards the structural low.

Table 3 Properties of the model layers according to Larue (2010), Reutter (2011), Schäfer et al. (2011), and Noack et al. (2013)

Layer	Lithology	Thickness (m)	Porosity (%)	Permeability (mD)
Quarternary	Sand, gravel	300	20	1,000
Tertiary post-Rupelian	Sand, silt	400	15	100
Tertiary Rupelian	Clay	80	10	0.001
Tertiary pre-Rupelian	Sand, sandstone	350	10	100
Cretaceous	Chalk, claystone	900	7	10
Upper Buntsandstein	Salt, anhydrite, claystone	50	4	0.001
Upper Middle Buntsandstein	Siltstone	20	4	0.1
Solling	Sandstone	20	20	110
Lower Middle Buntsandstein	Siltstone	110	4	0.1
Lower Buntsandstein	Clay and siltstone	350	4	0.1
Permian Zechstein	Rock salt	1,200[a]	0	0

[a] The high thickness of the Zechstein salt layer results from the configuration of salt structures and varies considerably within the model domain

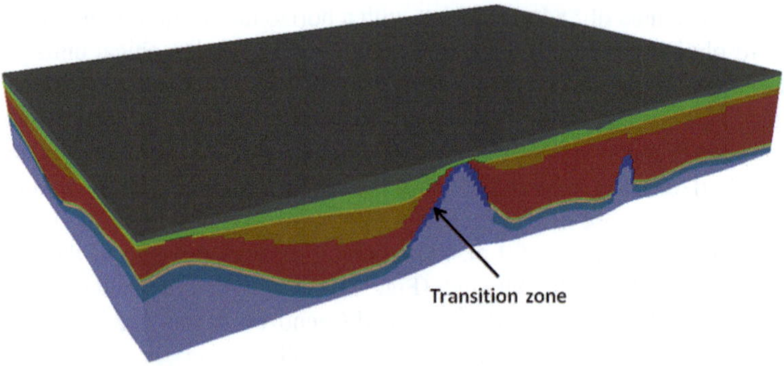

Fig. 8 Perspective view of the 3D model with illustrated trend of the transition zone along the salt wall. Vertical exaggeration: 2:1

For the scenarios we include discontinuities in the regionally important Rupelian aquitard (Tertiary) and a transition zone (Fig. 8) along the salt flank as such discontinuities are supposed to provide permeable pathways for brines which could reach shallow drinking water horizons. Thereby, the transition zone along the salt wall represents a higher porous and permeable interface between rock salt and adjoining rocks.

Based on this model we develop scenarios in which we vary for example hydrogeological parameters of the geological discontinuities, the injection rate and the initial state of the system in terms of the salinity distribution. Furthermore we compare different levels of model complexity with regard to the physical processes considered and their effects on our results. The aim is to identify the level of model complexity which is sufficient for this kind of setting with regard to the limited data availability at hand for the far field.

3.2 Expert Interviews and Evaluation

Stakeholder integration into the process of developing and selecting scenarios was realised in this context through guideline-based expert interviews. The main objective to carry out expert interviews focused on eliciting geological structures and mechanisms that in stakeholders views are most important factors for CO_2 induced brine migration. In order to focus interviewees reflections on the geological model a model sketch has been used to provide the opportunity to fill in drawings of scenario ideas. As a side effect of sketch use, the expert knowledge should lead to critical remarks about the model, thus making valuable contributions towards model construction. In the light of two pre-test interviews carried out the original model sketch was revised by researchers since interviewees had observed some graphical inconsistencies. In the following, we used an updated version for carrying out the interviews. A total of ten face-to-face interviews were conducted in Spring and

Summer 2013 with interviewees representing public authorities, business and industry, the science community, and independent experts. These talks referred to the following key topics:

- Risk evaluation and scenarios for brine migration
- Key parameters and processes influencing brine migration
- Prioritisation of brine migration scenarios
- Specification of brine migration scenarios with the help of the geological model sketch

A first conclusion resulting from the evaluation of the interviews relates to the conceptualisation of damage in case brine would come into contact with drinking water. This issue is assessed from two different angles. Some stakeholders advocate an 'absolute' understanding stressing that as soon as any salt water penetrates drinking water aquifers, this must be called damage. The others hold the opinion that the issue of salinisation of groundwater needs to be considered in 'relative' terms. For proponents of this perspective damage is not a question of whether there is, or is not, a contact of brine with groundwater. To their opinion, damage is rather defined to overcome specific threshold values, i.e. one needs to assess the brine quantity and its salinity to do risk judgements. This issue remained largely unsolved during the interviews and hints to differing risk framing among interviewees. Framing comprises the selection and interpretation what counts as a risk which may vary among different actors groups due to underlying values influencing their interests. What is needed is to properly analyse differing framing concepts and start deliberative processes including all relevant stakeholders in order to identify a common ground for risk management strategies.

Looking at the suggestions for potential brine migration paths, it became obvious that the experts clearly differentiate between man-made and geology-induced risks. The former comprises facilities such as old and new boreholes, drinking water wells, or open-cut-mining sites. Examples of the latter are cracks and faults, salt diapirism/dome, thin and non-continuous seal, or non-continuous Rupelian clay. The distinction between potential migration paths caused by technical installations as well as by geological structures comes along with a clear-cut risk prioritisation. Stakeholders unanimously agreed in estimating geology-based risks as far more relevant while man-made risks were ranked of minor importance. They argued, for instance, that faulty abandoned drill holes open the way only for minor brine flow quantities which can be handled easily with technical measures.

Besides reflecting about the risks of brine migration, the evaluation of the stakeholder interviews revealed a broad range of key elements for scenario building. The most important suggestions are outlined in the following:

- Variable boundary conditions: scenarios should consider different boundary conditions since this has considerable impact on brine displacement and pressure increase mechanisms
- Variable geology: use different geological structures since brine displacement and pressure increase is highly dependent on the geological structure

- Variable space dimensions: investigate scenarios with different spatial dimensions (e.g. a large-scale scenario with 100 km)
- Man-made migration paths: integrate drill holes in order to validate expected impacts such as low displacement quantities and minor increase in pressure
- Variable parameter values: use heterogeneous poro/perm values (rocks) and pressure/density values (fluids)
- Injection strategy: consider different injection points and volumes
- Pressure management: simulate different volumes of brine production
- Variable discretisation: detailed discretisation of geological weak points against rough discretisation of huge spatial structures

3.3 Scenario Setup and Preliminary Results

In an internal project workshop in the Spring of 2014 the information obtained from the expert interviews was condensed to develop scenarios for numerical simulations. The scenarios contain:

- Variable transition zone transmissibility
- Variable injection rates
- Closed and open lateral boundary conditions
- Variable initial salinity distributions
- Pressure management: Additional extraction wells to control the pressure in the reservoir

The scenario setup and the results will be further reviewed in a stakeholder workshop. Figure 9 presents a preliminary simulation result that shows the CO_2 plume extent after 50 years of injection (injection rate 0.5 Mt per year) for the target formation (where the CO_2 is injected). The model used was an isothermal,

Fig. 9 Top view of the target formation (Solling) showing the CO_2 saturation at the end of the injection period (50 years) as well as two iso-lines of pressure increase. The *grey* area depicts the salt wall. The injection rate is 0.5 Mt per year

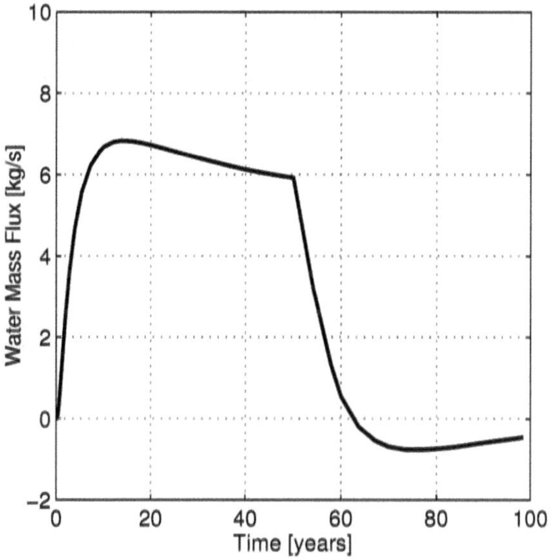

Fig. 10 Water flux into Tertiary post-Rupelian across the transition zone. The injection ends after 50 years. The transition zone is highly permeable with a permeability of 10^{-12} m^2

immiscible 2-phase flow model already presented in Sect. 2.2. The CO_2 plume extent is small compared to the extent of the pressure increase. Therefore, the study area needs to be sufficiently large in order to avoid boundary effects. The CO_2 is injected at the flank of an anticlinal structure. Therefore, the CO_2 is moving up the anticlinal structure (away from the salt wall) due to buoyancy forces and the CO_2 plume does not gather symmetrically around the injection point. Figure 10 shows the water mass flux across the transition zone into the Tertiary post-Rupelian during and after the injection for a scenario where the transition zone along the salt wall is highly permeable (10^{-12} m^2). After rising to a flux of more than 6 kg/s the flux gently declines until the end of the injection. This behaviour can be explained by the movement of the CO_2 away from the salt wall. After the injection stops the flux rapidly drops and even becomes negative as the CO_2 is still moving away from the salt structure with water filling the pore space behind the CO_2 plume. The fluxes slowly return to their initial state as the CO_2 reaches the highest point below the cap rock. It is planned to increase the complexity of the model by including salt transport along with variable initial salinity distributions.

4 Lessons Learnt

We can already summarise some important lessons that were learnt in the first part of the project both from the technical and from the socio-scientific point of view.

The Gr number has an undisputed scientific value. However, it is not intuitive to non-experts and can be used only as a supplementary criterion by geo-scientists and experts. Gr indicates in the absence of permeability data only where the encountered fluid properties are favourable for an efficient usage of pore space. The participatory approach led to a revision of the relevance of the Gr criterion and also to a sharpening of the scientific information.

Based on our experiences so far we conclude, that the transfer of scientific concepts to the practical application can benefit from an early-stage expert evaluation. The Group Delphi appears to be a reasonable method and might be applied to other fields of CCS research.

Acknowledgments The authors gratefully acknowledge the funding for the CO_2BRIM project (03G0802A) provided by the German Federal Ministry of Education and Research (BMBF) and the German Research Foundation (DFG) within the geoscientific research and development program Geotechnologien.

References

Baldschuhn R, Binot F, Fleig S, Kockel F (2001) Geotektonischer Atlas von Nordwest-Deutschland und dem deutschen Nordsee-Sektor. In: Geologisches Jahrbuch, Reihe A 153, p 88

Bombien H, Hoffers B, Breuckmann S, Helms M, Lademann K, Lange M, Oelrich A, Reimann R, Rienäcker J, Schmidt K (2012) Der Geotektonische Atlas von Niedersachsen und dem deutschen Nordseesektor als geologisches 3D-Modell. Technical report

Brasser T, Fischer-Appelt K, Larue J, Moeing J (2011) Hydrochemischer Charakter von Tiefenwässern in ausgewählten Regionen Deutschlands. Technical report, Gesellschaft für Anlagen- und Reaktorsicherheit (GRS) GmbH

Flemisch B, Darcis M, Erbertseder K, Faigle B, Mosthaf K, Lauser A, Müthing S, Nuske P, Tatomir A, Wolf M, Helmig R (2011) DUMUX: DUNE for multi-{phase, component, scale, physics, …} flow and transport in porous media. Adv Water Resour 34(9):1102–1112

Kaufmann D, Heim S, Jähne F, Steuer S, Bebiolka A, Wolf M, Kuhlmann G (2014) GSN Generalisiertes, erweitertes Strukturmodell des zentralen deutschen Nordsee-Sektors Konzept zur Erstellung einer konsistenten Datengrundlage für weiterführende Modellierungen im Bereich des zentralen deutschen Nordsee-Sektors. Technical report, Bundesanstalt für Geowissenschaften und Rohstoffe, Hannover

Kissinger A, Noack V, Knopf S, Scheer D, Konrad W, Class H (2014) Characterization of reservoir conditions for CO_2 storage using a dimensionless gravitational number applied to the North German Basin. Sustain Energy Technol Assess 7:209–220

Land CS (1968) Calculation of imbibition relative permeability for two and three-phase flow from rock properties. Soc Petrol Eng J 8(2):149–156

Larue J (2010) Endlagerung im Tonstein, Entwicklung eines synthetischen Tonsteinstandortes, Teil 2: Standortcharakterisierung. Abschlussberichte zum Vorhaben 3607R02538 planerische Grundsatzfragen. Technical report, GRS-A-3535, Köln

LIAG (2012) Deutschlandkarte des geothermischen Gradienten

Müller C, Reinhold K (2011) Informationssystem Speichergesteine für den Standort Deutschland - eine Grundlage zur klimafreundlichen geotechnischen und energetischen Nutzung des tieferen Untergrundes (Speicher-Kataster Deutschland). Technical report, Bundesanstalt für Geowissenschaften und Rohstoffe (BGR), Berlin/Hannover

Noack V, Scheck-Wenderoth M, Cacace M, Schneider M (2013) Influence of fluid flow on the regional thermal field: results from 3D numerical modelling for the area of Brandenburg (North German Basin). Environ Earth Sci 70(8):3523–3544

Okwen RT, Stewart MT, Cunningham JA (2010) Analytical solution for estimating storage efficiency of geologic sequestration of CO_2. Int J Greenhouse Gas Control 4(1):102–107

Reinhold K, Müller C, Riesenberg C (2011) Informationssystem Speichergesteine für den Standort Deutschland - Synthese -. Technical report, Bundesanstalt für Geowissenschaften und Rohstoffe (BGR), Hannover

Reutter E (2011) Hydrostratigrafische Gliederung Niedersachsens. Geofakten 21

Schäfer F, Walter L, Class H, Müller C (2011) The regional pressure impact of CO_2 storage: a showcase study from the North German Basin. Environ Earth Sci 65(7):2037–2049

Schulz R (2009) Aufbau eines geothermischen Informationssystems für Deutschland. Endbericht. Technical report, LIAG, Hannover

Schulz M, Renn O (2009) Das Gruppendelphi. VS Verlag für Sozialwissenschaften, Wiesbaden

Wolfgramm M, Seibt A (2008) Zusammensetzung von Tiefenwässern in Deutschland und ihre Relevanz für geothermische Anlagenteile. In: GTV-Tagung in Karlsruhe 2008, pp 503–516

Chances for and Limitations of Acceptance for CCS in Germany

Elisabeth Dütschke, Diana Schumann and Katja Pietzner

Abstract This chapter presents two studies on the perception and acceptance of CCS in Germany: the first one is a qualitative case study analysis which examined four German projects which were initiated for CO_2-storage. These include two commercial projects driven by industry (one in North Frisia, the other one in Eastern Brandenburg), a joint research and industry project in the Altmark focusing on Enhanced Gas Recovery (EGR) and a joint research project at Ketzin. Only one of the four projects, the Ketzin project, was successful in proceeding to CO_2 injection and did not elicit local protest. The comparison of the four cases points to differences in project scale and scope, in the perceived risks and benefits and in the communication processes, all of which have possibly influenced project acceptance. The second study investigated and compared the public perception of CO_2 offshore storage, CO_2 onshore storage and CO_2 transport via pipeline based on a national and two regional surveys. It shows that CCS is not unknown amongst the German public; however, the acceptance of CO_2 storage is low independent of the place of storage. Perceived risks and benefits are identified as the main influence factors on attitudes towards CO_2 storage and CO_2 transport via pipeline.

1 Introduction

CO_2 capture and storage (CCS) is perceived worldwide and in the European Union (EU) as an important technology which contributes to greenhouse gas (GHG) emissions mitigation (European Commission 2013; IEA 2013). However, up to now only few large-scale demonstration projects comprising the complete CCS

E. Dütschke (✉)
Fraunhofer ISI, Breslauer Str. 48, 76139 Karlsruhe, Germany
e-mail: elisabeth.duetschke@isi.fraunhofer.de

D. Schumann
Forschungszentrum Jülich, 52425 Jülich, Germany

K. Pietzner
Wuppertal Institut, Döppersberg 19, 42103 Wuppertal, Germany

A. Liebscher and U. Münch (eds.), *Geological Storage of CO_2 – Long Term Security Aspects*, Advanced Technologies in Earth Sciences,
DOI 10.1007/978-3-319-13930-2_11

process chain (capture, transport and storage) exist worldwide; none of these projects includes large fossil fuel power plants yet and none of them has been implemented within the EU (European Commission 2013; Global CCS Institute 2014). Experience with storing CO_2 in Europe has been gained at the storage sites Sleipner and Snøvhit as part of natural gas extraction. Overall the pace of development and uptake of CCS is slow globally and has even stopped in some countries like Germany where the future of CCS is uncertain at present, despite the enactment of the CCS law in August 2012 (Fischer 2015). Among other reasons (cp. Stigson et al. 2012) low levels of acceptance for CCS projects in quite a number of countries (Brunsting et al. 2010; Dütschke 2010; Pietzner et al. 2010; Bradbury 2012) have been a crucial factor for slowing down CCS deployment. The two studies presented in this chapter therefore aim to explore the status quo of public acceptance of CCS in Germany by a combination of methodological approaches and to extend the knowledge of factors which influence CCS acceptance.

Earlier research on CCS acceptance includes two main streams of analysis: on the one hand case study research investigates the development of acceptance for specific CCS projects, on the other hand survey-based research has been undertaken, usually in order to ascertain awareness and knowledge of CCS among the general public, and attitudes towards it. In this paper recent work adding to both streams is included and results are integrated into the discussion section.

Earlier studies about the acceptance of specific CCS projects came to the conclusion, that in regions where activities to implement CCS have been started, the knowledge of the technology is usually higher than among the public in general while acceptance of CCS is often lower (Reiner et al. 2011, recently also Schumann 2015). These lower levels of acceptance have also been proven by public and stakeholder protests against the CCS projects. Earlier case study work has come up with a range of factors that influence the acceptance of a specific project (Oltra et al. 2012). The first study presented in this paper draws on these factors and analyzes them specifically for all CO_2 storage projects in Germany, adding to the knowledge which aspects are likely to be relevant for the acceptance of CCS projects in this country.

Earlier research from the second stream focusing on CCS perception has shown that, although the level of awareness of CCS has been increasing in the past years, the general public has only limited knowledge of CCS as a technology (special Eurobarometer 2011; Pietzner et al. 2010; Schumann 2015). This leads to the fact that public acceptance of CCS can only be reliably assessed if the awareness and knowledge of the technology is also measured (cp. research on so-called pseudo-opinions: de Best-Waldhober et al. 2006).

Furthermore, prior research on CCS acceptance has mainly focused on investigating the capture of CO_2 from fossil energy sources (e.g. de Best-Waldhober et al. 2012; Reiner et al. 2011; Desbarats et al. 2010), with energy production from coal being predominant. So far acceptance research has hardly differentiated between different storage options and if so, the focus was on onshore storage (Pietzner et al. 2010; Pietzner and Schumann 2012; Schumann 2015). Only a few studies on CCS acceptance have investigated the transport step (see Gough et al. 2014, for a recent exception). In general, the research on public acceptance of CCS has been able to point out that the issues around CO_2 storage are more likely to evoke opposition by the public than capture and transport (Hammond and Shackley 2010; Mander et al. 2010; Upham and Roberts 2011; Pietzner et al. 2010; Schumann 2015). This paper tries to extend the current state of knowledge by integrating the transport step as well as offshore storage. Furthermore, the analysis of CCS perception among the general public is compared to the perception of CCS in two German regions in order to relate it to the project-specific research.

For the two studies presented in the following sections, two different methodological approaches were used (qualitative case studies and quantitative analyses of the data from the representative surveys)—a combination which supports painting a differentiated picture of CCS acceptance. More extensive documentation of the results can be found in published project reports (Dütschke et al. 2014; Schumann 2014) and further publications (Pietzner et al. 2014; Schumann et al. 2014).

2 Case Study Analysis: Public Acceptance of CO_2 Storage Projects

In Germany some pilot installations for carbon capture have been implemented (e.g. by Vattenfall in Spremberg, by EnBW in Heilbronn). Demonstration projects have not been realized. For storing CO_2 four projects have been started of which three were finished without injecting CO_2 (see Table 1 for an overview). These three projects include two commercial projects driven by industry, one in North Frisia (led by RWE), the other one in Eastern Brandenburg (led by Vattenfall), as well as a joint research and industry project in the Altmark focusing on Enhanced Gas Recovery (EGR). This last project was based on cooperation between a research consortium led by Deutsches GeoForschungsZentrum (GFZ) and Gas de France SUEZ (GDF SUEZ) leading a cooperation of industry partners. The only project proceeding to store CO_2 was the joint research project at Ketzin. All four projects formed the basis for the case study analysis whose results will be summarized in this section.

The results from the case study analysis presented in this paper follow three levels of analysis:

- *Project characteristics*: Comparing project descriptions regarding project aims, duration, (intended) storage size, project team, comparison of initial plans and actual implementation etc.
- *Communication activities*: Documenting communication strategies and actual activities of the project teams as well as their perception and evaluation by local stakeholders and citizens.
- *Local project acceptance*: Qualitative assessment of the local acceptance for the project by the local public and by local stakeholders. Analysis of the locally perceived risks and benefits of the projects.

After providing more details about the data base used for the analysis, the findings will be outlined.

Table 1 Overview on project characteristics of the four storage projects

Project	Project characteristics
Commercial project North Frisia	• Led by RWE, a major energy company; to be combined with joint research project "COAST" • Long-term aim was to store CO_2 from RWE's coal-fired power plants • Active project preparations from 2008 to 2010, officially given up in 2011 • No field activity at all, research project not started • Accompanied by local protests
Commercial project Eastern Brandenburg	• Led by Vattenfall, a major energy company • Long-term aim was to store CO_2 from Vattenfall's coal-fired power plants • Active project preparations from 2009 to 2012 • No field activity at all • Accompanied by local protests
Research and industry project Altmark	• Close cooperation of industry project by Gas de France SUEZ (GDF SUEZ) and Vattenfall with joint research project "CLEAN", led by Deutsches GeoForschungsZentrum GFZ, Potsdam • Cooperation aimed at combining Enhanced Gas Recovery (EGR) with CO_2 storage in nearly depleted natural gas field • Project run time from 2009 to 2011 • Necessary technical installations for storage were put into place, however, no permission issued for CO_2 injection (goal: 100,000 t). Geological potential for further storage • Finished without CO_2 injection, but extensive scientific research was conducted • Accompanied by local protest since 2010
Joint research project Ketzin	• Cooperation of research institutes and industry, led by Deutsches GeoForschungsZentrum GFZ, Potsdam • Testing all steps of CO_2 storage including the injection of a limited amount of CO_2 (max. 100,000 t CO_2, injected slightly more than 67,000 t) • Site and geological structure has been formerly used for storage of natural gas albeit in much shallower reservoir horizons • Runtime of several research projects from 2004 until today. Currently closing down activities. No plans for further storage • No local protests

2.1 Data Base and Analysis

In order to conduct the case study analysis, a broad range of data was collected (see Dütschke et al. 2014 for more details). The core of the collected data consists of 50 partly standardized interviews that were conducted with individuals from three major groups: (i) members of the respective project teams; (ii) local representatives from the four regions, e.g. local opponents, representatives from the churches, media and local associations, local citizens; (iii) further stakeholders, e.g. from the permitting authority, funding agency, politicians from the Federal State level. The interviews were complemented by a broad document analysis which included, among others, articles from the press, publications on the projects, information material provided by the project teams or opponents, protest notes e.g. from local councils, hearings in local, Federal State or national parliaments.

The interviews were fully transcribed and coded along the levels of analysis. The additionally collected documents were used to validate interview statements. Further results from the case study analysis are described in Dütschke et al. (2014) and Pietzner et al. (2014) and also include an analysis of local media.

2.2 *Overall Results*

This section summarizes the results of the case study analysis. They will be presented in an integrated way, i.e. already combining the contributions from the different levels of analysis. Table 1 gives an overview of the main project characteristics integrating some information about local reactions to the projects as an indicator of their acceptance.

From the four cases analyzed only the project at Ketzin did not meet public protest and this is also the only project that succeeded in progressing to actually injecting CO_2. Thus, while public opposition has been dominant in three out of four projects, a contrasting example exists as well.

The patterns of project characteristics give some idea of hypotheses on the influence of project characteristics on public acceptance: First of all, the size of the storage may play a role for public acceptance. The commercial projects North Frisia and Eastern Brandenburg that aimed for large-scale storage evoked public protest. Both other projects were limited in storage size for the project duration, however, while in Ketzin no plans for further storage ever emerged and the focus was always on a scientific field trial, in the Altmark region commercially sized storage was an option for the long term.

With regard to communication processes different patterns and reactions were observed for the four projects: In North Frisia the first contact about the project with the region took place during official hearings during the permission process. The project developers were surprised by harsh public reactions which followed the announcement of the project and had difficulties in responding with a professional communication strategy. Thus, communication about the project was perceived to be delayed, that decisions were already taken and that local concerns were not taken seriously.

In the Altmark, some early contacts with local politicians and information events about the project took place; however, they were followed by delays in communication activities and disagreement in the project consortium over the communication strategy when difficulties in the permission process arose. This led to a decrease in communication activities which made project opponents the dominant party informing the public about the project in later stages.

In Eastern Brandenburg the project developer sought contact with local stakeholders in the early phases of the project and came up with information for the public across various channels including an information office which was installed in the main town of the discussed storage area shortly after the project was announced. Still, local stakeholders perceived this information as coming too late because documents for the permission processes around geological exploration had been submitted in parallel. Furthermore, information was perceived to be too positive about CCS.

Finally, in Ketzin, prior to specific planning, the project developer was in contact with the local community who welcomed the activities. Information campaigns for the local public emerged in parallel to the actual project implementation and received varying levels of interests and were generally perceived positively.

Overall, these findings indicate, that professional information strategies do not necessarily lead to positive effects. Possibly, the affected public and its representatives seek a participatory role in the siting decision and the further project development process.

A closer look at local acceptance in the four regions also showed the risks and benefits discussed on a local level that came up in the interviews in relation to the projects (see Table 2).

An interview based analysis is always limited as increasing the number of interviewees might always increase the number of risks and benefits mentioned,

Table 2 Overview on perceived risks and benefits associated with the four storage projects

	North Frisia	Eastern Brandenburg	Altmark	Ketzin
Perceived benefits	None	None	Contribution to mitigation of climate change	Contribution to mitigation of climate change
			Positive effects on local economy and infrastructure	Positive publicity for the region
Perceived risks	Risks for humans, animals and wildlife	Risks for humans, animals and wildlife	Risks for humans, animals and wildlife	Risks due to high complexity of technology
	Negative competition for renewables	Negative competition for renewables	Negative competition for renewables	
	Risks for drinking water	Bad publicity for the region	Risks for drinking water	
	Bad publicity for the region		Bad publicity for the region	
	Local economic disadvantages			

i.e. it is not possible to estimate in how far this list is complete or to derive conclusions about which arguments were most relevant in the local debate. However, the results show that in the two regions with commercial projects, North Frisia and Eastern Brandenburg, no benefits through the projects were mentioned in the interviews. In the research and industry project Altmark some benefits were seen, however, they are contrasted by a longer list of perceived risks, while the results for Ketzin are the other way round: On the one hand possible risks are perceived due to the complexity of the technology in general, however, they are not very specific, on the other hand benefits are perceived as well. Thus, this finding could lead to the hypothesis that the uneven distribution of the perception of risks and benefits may have played a role in different local reactions to the CCS projects.

3 Surveys: Public Perception of CO_2 Offshore Storage, CO_2 Onshore Storage and CO_2 Transport Via Pipeline

Previous research by the authors had shown that the acceptance of CO_2 storage amongst the German public is generally lower than the acceptance of CO_2 capture or CO_2 transport (Pietzner et al. 2010; Schumann 2015). However, the previous studies focused on investigating public acceptance of CCS as a process chain by which the CO_2 would be stored onshore. The question how the German public would perceive CCS if the CO_2 was to be stored under the seabed (CO_2 offshore storage) has not been investigated so far.

Therefore, the aim of this study was to examine the perception of CO_2 offshore storage amongst the German public in comparison to the perception of CO_2 onshore storage and CO_2 transport via pipeline. Since a data base which allows investigating and systematically comparing the public perception of CO_2 offshore storage, CO_2 onshore storage and CO_2 transport via pipeline was not available, three representative surveys (a nationwide survey and two regional surveys) of the German public were carried out in 2013.

The survey data was used to analyze and compare the public perception on two dimensions: firstly between the three regions and secondly across the two storage options and CO_2 pipelines. Furthermore, the factors influencing public attitudes towards the two CO_2 storage options and CO_2 transport via pipeline were identified.

In the following the data base and the approaches for the data analysis are described. Afterwards the results of our analyses are summarized.

3.1 *Data Base and Approaches for Data Analysis*

In order to generate a sufficient number of cases for statistical analyses, standardized, representative surveys based on random samples were conducted. Since previous studies have shown that public perception of CO_2 onshore storage differ regionally

(Schumann et al. 2010), two regional surveys were carried out, in addition to a nationwide survey. For the regional surveys two regions located on the coast of the German North Sea were chosen: (1) district of North Frisia (in the following referred to as "North Frisia") and (2) district of Aurich plus the islands of Borkum, Langeoog, Spiekeroog, and Wangerooge (in the following referred to as "Aurich plus islands").

Coastal regions as study areas were mainly selected because of their proximity to possible CO_2 offshore storage areas. From the coastal regions two regions were chosen in which activities against CO_2 storage already existed, since it was of interest whether the perception of CO_2 offshore storage will differ in regions where protest against CO_2 storage in general is already present. However, an important difference between the two regions is that in North Frisia the energy company RWE aimed at exploring the area for the suitability of CO_2 storage and failed due to public resistance (cf. the sections above) while in Aurich plus islands no such project has been started.

The public perception of CO_2 offshore storage, CO_2 onshore storage and CO_2 transport via pipeline was measured in the surveys by using the indicators self-reported awareness, factual knowledge, risk perceptions and attitudes (cf. for the details Schumann 2014). In addition, benefit perceptions of CCS, general values and socio-demographic characteristics of the respondents were surveyed, because it was assumed that they can be relevant factors influencing the attitudes towards CO_2 offshore storage, CO_2 onshore storage and CO_2 transport via pipeline.

The questionnaire for the surveys was developed by the authors and is documented in Schumann (2014). The surveys were conducted by a professional polling firm with computer-aided telephone interviews (CATI) from mid-March 2013 to mid-April 2013.

The participants were recruited using multilevel random sampling. 1,000 interviews were realized nationwide, 503 interviews in North Frisia and 500 interviews in Aurich plus islands. The nationwide survey (in the following referred to as "rest of Germany") does not include respondents from the two coastal regions. The representativeness of the samples for the nationwide and regional populations has been proven by comparing the distributions of the criteria gender, age, professional qualification and household size with the data of official statistics. Overall, the comparisons showed that the samples were very representative.

For the analysis of the survey data a two-dimensional comparative approach was applied: firstly the results according to the three regions were compared, secondly the perceptions of CO_2 offshore storage were compared to the perception of CO_2 onshore storage and CO_2 transport via pipeline.

The comparisons were conducted along the measured indicators self-reported awareness, factual knowledge, risk perceptions and attitudes. For this purpose methods of descriptive statistics (frequencies, means, standard deviations, and correlations) were used. The statistical significance of differences in the results was tested with non-parametrical tests. Therefore, differences in the results, explained in the following section, are statistically significant, unless otherwise stated. In order to identify the factors which determine the attitudes towards CO_2 offshore storage, CO_2 onshore storage and CO_2 pipelines, ordinal regression analyses were carried out.

3.2 *Overall Results*

3.2.1 Self-Reported Awareness of CO_2 Offshore Storage, CO_2 Onshore Storage and CCS

The results of our descriptive statistical analyses showed that CO_2 storage is not unknown amongst the German public: nationwide half of the respondents has heard of "storage of CO_2 in onshore repositories" and "storage of CO_2 under the seabed", respectively (cf. Table 3). In the coastal regions the awareness of the terms "storage of CO_2 in onshore repositories/under the seabed" was perceptibly higher than in the rest of Germany. Furthermore, the percentages of respondents who answered that they know "quite a bit or a lot" about CO_2 storage are approximately twice as high in the two coastal regions compared to the nationwide average.

In Aurich plus islands the self-reported awareness of the term "storage of CO_2 in onshore repositories" was higher than the self-reported awareness of the term "storage of CO_2 under the seabed". Nationwide the statistically significant differences in the awareness of the two terms were only very small, whereas in North Frisia no statistically significant differences were found.

Table 3 Self-reported awareness of CO_2 storage and CCS according to region

Topic	Question: "Have you heard about the following topics?"				
	Region	No, never heard of it (%)	Yes, heard of it, but know nothing/just a little bit about it (%)	Yes, heard of it and know quite a bit/a lot about it (%)	Total (%)
Storage of CO_2 in onshore repositories	Nationwide	49.7	40.6	9.7	100
	North Frisia	32.8	51.1	16.1	100
	Aurich plus islands	38.8	43.4	17.8	100
Storage of CO_2 under the seabed	Nationwide	50.2	43.2	6.6	100
	North Frisia	32.4	51.7	15.9	100
	Aurich plus islands	42.0	44.0	14.0	100
CCS	Nationwide	57.3	38.2	4.5	100
	North Frisia	42.5	45.3	12.1	100
	Aurich plus islands	49.4	38.6	12.0	100

The self-reported awareness of the term "carbon capture and storage or CCS" was lower than the awareness of "storage of CO_2 in onshore repositories/under the seabed". Again, the self-reported awareness was higher in the coastal regions than in rest of Germany and in North Frisia higher than in Aurich plus islands. The percentages of respondents who answered that they know "quite a bit or a lot" about CCS were three times higher in the coastal regions compared to the nationwide average. The regional differences in the self-reported awareness of CCS are still statistically significant, when the influence of socio-demographic characteristics is controlled.

3.2.2 Factual Knowledge About CO_2, CO_2 Storage and Pipelines

In order to assess the factual knowledge of German citizens, the survey respondents were asked an open-ended question about CO_2. Furthermore they were presented with five statements about CO_2, CO_2 storage and pipelines respectively and then asked to say whether these statements are true or false.

The open-ended question "What does the abbreviation CO_2 mean?" was correctly answered by 63 % of the respondents nationwide. In Aurich plus islands 60 % and in North Frisia 53 % knew that the abbreviation "CO_2" stands for carbon dioxide. An incorrect answer to this question was given by 31 % nationwide, in North Frisia by 24 % and Aurich plus islands by 26 %. Accordingly, only 6 % of the respondents nationwide responded that they do not know what the abbreviation "CO_2" means, whereas in North Frisia 23 % and in Aurich plus islands 14 % gave this answer.

Further analyses also revealed that the majority of the survey participants knew that CO_2 is a greenhouse gas. However, in the coastal regions the shares of the respondents who gave this correct answer was five percentage points higher in comparison to the rest of Germany (cf. Table 4). At the same time, the respondents in the coastal regions more often assigned "CO_2" with incorrect negative attributes than the respondents in the nationwide survey. For example, 77 % of the respondents in North Frisia said that CO_2 is poisonous, 38 % answered that CO_2 is a water pollutant and 29 % stated that CO_2 is flammable.

The average of correct answers to the knowledge questions on CO_2 storage was considerably higher in the coastal regions than in the rest of Germany. At the same time the average of incorrect answers was a little higher in the coastal regions compared to the nationwide survey.

The average of correct answers to the knowledge questions on pipelines was perceptibly higher in Aurich plus islands compared to North Frisia or the rest of Germany. The average of incorrect answers was markedly higher in North Frisia than in the other two regions. The highest average of the answer "don't know" was found in the nationwide sample.

The regional differences in the knowledge about CO_2, CO_2 storage, pipelines are still statistically significant, when the influence of socio-demographic characteristics is controlled.

Table 4 Knowledge of attributes of CO_2 according to region (percentage of respondents who answered "true")

Region	Question: Please tell me to the best of your knowledge whether each statement is true or false. CO_2 is...				
	Flammable (%)	A greenhouse gas (%)	Poisonous (%)	Explosive (%)	A water pollutant (%)
Nationwide	20.3	75.1	52.6	18.9	20.0
North Frisia	29.4	79.7	76.9	27.4	38.0
Aurich plus islands	35.6	79.8	68.0	27.2	29.8

3.2.3 Risk Perceptions of CO_2 Offshore Storage, CO_2 Onshore Storage and CO_2 Transport Via Pipeline

The results of further statistical analyses showed that the personal and societal risks of CO_2 offshore storage, CO_2 onshore storage and CO_2 transport via pipeline were perceived markedly higher in North Frisia than in Aurich plus islands and in the rest of Germany. The personal risk of CO_2 offshore storage was assessed slightly higher in Aurich plus islands than in the nationwide average. The personal and societal risk of CO_2 transport via pipeline was perceived somewhat lower in Aurich plus islands than in the rest of Germany.

Concerning the two storage options, our results revealed that nationwide the personal and societal risk of CO_2 onshore storage was assessed higher than the personal and societal risk of CO_2 offshore storage. In North Frisia no statistical significant differences in the perceptions of the personal and societal risk of the two storage options were found. In Aurich plus islands the societal risk of onshore storage was perceived slightly higher than the societal risk of offshore storage, whereas no statistical significant differences were found with regard to the perceptions of the personal risks of the two storage options.

Compared to CO_2 transport via pipeline the personal and societal risks of CO_2 offshore/onshore storage were perceived visibly higher in all regions. Furthermore, the societal risks of CO_2 pipelines were assessed higher than the personal risks.

3.2.4 Initial Preference and General Attitudes Regarding CO_2 Storage

Regarding CO_2 storage, firstly the initial preference of the citizens for CO_2 offshore storage or CO_2 onshore storage was surveyed. For this purpose the respondents received some information about CCS and CO_2 storage and were then asked which storage option they would prefer. Afterwards the respondents were given a second piece of information and were then asked to assess the risks of CO_2 offshore storage, CO_2 onshore storage and CO_2 transport via pipeline (cf. section above) as well as to assess in general the idea to store CO_2 under the seabed, to store CO_2 in onshore repositories and to transport CO_2 via pipeline, respectively.

Concerning the question which storage option would be preferred, the survey results showed that the majority of the German public would spontaneously prefer that CO_2 be stored nowhere at all (cf. Table 5). This general rejection of CO_2 storage, which is reflected in the answer "nowhere", was highest in North Frisia, and nationwide higher than in Aurich plus islands.

This result is all the more remarkable because the interviewers in the surveys only read the predefined answers "under the seabed of the North Sea", "in onshore repositories, nearby the emission source" and "in onshore repositories, only in sparsely populated areas" to the respondents. The answers "I don't care", "nowhere" and "elsewhere" were only written down by the interviewers if they were spontaneously given by the respondents.

The preferences of those respondents who made a choice between the storage options given also differ regionally: nationwide offshore storage would be preferred while respondents from the coastal regions would prefer onshore storage close to the emission source.

The general attitudes of the German public regarding CO_2 offshore storage and CO_2 onshore storage were rather negative (cf. Table 6). In the coastal regions both storage options have been assessed markedly more negatively compared to the nationwide average. The general attitudes are furthermore visibly more negative in North Frisia than in Aurich plus islands.

Nationwide the CO_2 transport via pipeline was on average generally assessed neutral (cf. Table 6). In Aurich plus islands the general attitude towards CO_2 pipelines was slightly more negative and in North Frisia visibly more negative than in the rest of Germany.

CO_2 onshore storage was evaluated more negatively than CO_2 offshore storage in the nationwide survey. In the regional surveys no statistical differences in the general attitudes towards the two storage options could be found. In comparison to CO_2 transport via pipeline the attitudes towards CO_2 offshore storage/CO_2 onshore storage were perceptibly more negative in all regions.

Table 5 Initial preferences regarding CO_2 storage according to region

Question: "Which option for CO_2 storage would you prefer?"			
Initial preference	Region		
	Aurich plus islands (%)	North Frisia (%)	Nationwide (%)
Under the seabed of the North Sea	18.2	15.3	25.9
In onshore repositories, nearby the emission source	24.4	21.3	19.9
In onshore repositories, only in sparsely populated areas	20.0	17.1	14.2
I don't care	9.0	4.8	7.0
Nowhere	25.6	38.6	31.4
Elsewhere	2.8	3.0	1.6
Total	100	100	100

Table 6 General attitudes towards CO_2 offshore storage, CO_2 onshore storage and CO_2 transport via pipeline according to regions

Question: "Overall, how do you assess the idea of CO_2 offshore storage/CO_2 onshore storage/CO_2 transport via pipeline?"						
Region	CO_2 offshore storage		CO_2 onshore storage		CO_2 transport via pipeline	
	Mean[a]	SD[b]	Mean[a]	SD[b]	Mean[a]	SD[b]
Nationwide	3.6	1.8	3.3	1.7	3.9	1.6
North Frisia	2.4	1.8	2.3	1.7	2.9	1.9
Aurich plus islands	2.8	1.9	2.9	1.8	3.6	1.9
Total	3.1	1.9	2.9	1.7	3.6	1.8

[a] Scale from 1 (=very negative) to 7 (=very positive). The higher the mean the more positive CO_2 offshore/onshore storage/transport via pipeline was assessed

[b] *SD* standard deviation

3.2.5 Determinants of General Attitudes Towards CO_2 Offshore Storage, CO_2 Onshore and CO_2 Transport Via Pipeline

In order to investigate the question which factors determine the general attitudes regarding CO_2 offshore, CO_2 onshore storage and CO_2 transport via pipeline three ordinal regressions were performed (cf. for the details Schumann 2014; Schumann et al. 2014).

The results of the regression analyses revealed that the perceptions of the personal and societal risk of CO_2 transport via pipeline/CO_2 offshore storage/CO_2 onshore storage as well as the perceptions of the personal and societal benefits of CCS are the most important determinants of general attitudes towards CO_2 transport via pipeline, CO_2 offshore storage and CO_2 onshore storage. The directions of the correlations between the risk and benefit perceptions and the general attitudes were the same in every region and regression model: the lower the personal or societal risk was perceived by the respondents, the more positive were the general attitudes regarding CO_2 pipelines or CO_2 offshore storage/CO_2 onshore storage. The lower the personal or societal benefit of CCS was assessed, the more negative were the general attitudes regarding CO_2 transport via pipeline or CO_2 offshore/onshore storage.

4 Summary of Findings and Discussion

This chapter outlines findings from two studies on CCS perception and acceptance. The first one, a comparative case study analysis, analysed the four projects North Frisia, Eastern Brandenburg, Altmark and Ketzin for CO_2 storage in Germany of which only Ketzin proceeded to inject CO_2. The Ketzin project is the only one among the four for which no local protests have been documented. The case study analysis developed assumptions about influence factors on local acceptance on three levels of analysis: project characteristics, regional context and local project acceptance. The comparison of the four projects shows that Ketzin was always clearly limited in scope and scale, i.e. being a research project that stores a limited amount of CO_2. The analysis of communication processes around the projects clearly indicates that a professional information strategy does not necessarily lead to an increase of local support for the project. The data suggests that the public and its representatives may be more responsive to a participatory approach where they can get involved before decisions are taken as it was perceived by the local stakeholders at Ketzin. A closer look at the perceptions of the projects themselves additionally pointed to an uneven distribution of risks and benefits in the four regions with the most balanced distribution found for the successful Ketzin project. Thus, while the study points out that storing CO_2 may be acceptable to a local community under certain conditions, it also points out that it is not an easy challenge.

To generalize findings from a post hoc case study analysis means deriving conclusions under high levels of uncertainty as numerous factors have influenced the respective outcome of the single case, i.e. determining which single factor has been decisive for the outcome is not possible. Nevertheless, the findings from the case study analysis can be interpreted in such a way that finding acceptance for industry-driven projects does not seem likely in Germany at the moment—independent of accompanying measures such as professional and early communication. Two possible explanations which, however, can not be proven based on this study are that first trust towards industry is too low and second that higher risks are ascribed to bigger sized storages. Besides this, balancing perceived risks and benefits is a further challenge for which so far no effective instruments have been developed and tested.

The second study investigated the acceptance of CCS by analysing survey data by descriptive statistics and ordinal regressions. The results of the descriptive statistical analyses revealed that CO_2 storage is hardly accepted by the German public. Without being provided with this predefined answer, the majority of the respondents answered spontaneously "none at all" to the question which option of CO_2 storage they would prefer. In the refusal of CO_2 storage no major differences could be discovered between the two storage options offshore and onshore. The citizens of the coastal regions refused both storage options in equal measure. In the rest of Germany CO_2 onshore storage was assessed slightly more negatively than CO_2 offshore storage. In principle the rejection of CO_2 storage was higher in the two coastal regions than in the rest of Germany and highest in North Frisia. This was illustrated by higher risk perceptions and more negative attitudes towards both storage options.

CO_2 transport via pipeline was nationwide generally assessed neutrally on average. In Aurich plus islands the general attitude towards CO_2 pipelines was slightly more negative and in North Frisia visibly more negative than in the rest of Germany. In comparison to CO_2 offshore storage/CO_2 onshore storage the attitudes towards CO_2 transport via pipelines were perceptibly more positive in all regions. Regression analyses showed that the perceptions of the personal and societal risk of

CO_2 transport via pipeline/CO_2 offshore storage/CO_2 onshore storage as well as the perceptions of the personal and societal benefits of CCS are the most important direct determinants of general attitudes towards CO_2 transport via pipeline, CO_2 offshore storage and CO_2 onshore storage.

Taken together both studies point to low levels of acceptance for CO_2 storage, in general in the surveys and, at least for potentially large scale storage, in the case study analysis. In both studies perceived risks and benefits turned out to be relevant factors: in the case study analysis they were found to be unbalanced for projects unsuccessful in storing CO_2. In the surveys the perceived risks and benefits turned out to be the most relevant predictors of attitudes towards CO_2 storage and pipelines.

With regard to the role of information, the case studies suggest that providing extensive and professional information on CCS does not necessarily lead to a positive public response. However, as pointed out above, participatory approaches instead of information focused approaches may be of higher value. The surveys showed that many people have some knowledge of CO_2, CO_2 storage and pipelines while at the same time they have misconceptions about it. This leads to the assumptions that some information is possibly not accepted by the public or that information itself can contain misconceptions.

Thus, while this study points out that research on CO_2 storage might be acceptable to the public as demonstrated by the successful project in Ketzin, to find acceptance for a large scale deployment seems to be a challenge taking into account the public perception of CCS mirrored in both studies.

Acknowledgments This research is part of the project "Chances for and limitations of public acceptance of CCS in Germany (CCS-Chancen)" for which we gratefully acknowledge funding by the Federal Ministry of Education and Research (Grant No. 03G0831A).

References

Bradbury J (2012) Public understanding of and engagement with CCS. In: Markusson N, Shackley S, Evar B (eds) The social dynamics of carbon capture and storage. Routledge, London, pp 45–73

Brunsting S, de Best-Waldhober M, Feenstra C, Mikunda T (2010) Stakeholder participation practices and onshore CCS: lessons from the Dutch CCS case Barendrecht. Energy Procedia 4:6376–6383

de Best-Waldhober M, Daamen D, Faaji A (2006) Public perceptions and preferences regarding large-scale implementation of six CO_2 capture and storage technologies: well-informed and well-considered opinions versus uninformed pseudo-opinions of the Dutch public. Leiden University, Leiden

de Best-Waldhober M, Daamen D, Ramirez A, Faaij A, Hendriks C, de Visser E (2012) Informed public opinion in the Netherlands: evaluation of CO_2 capture and storage technologies in comparison with other CO_2 mitigation options. Int J Greenhouse Gas Control 10(2012): 169–180

Desbarats J, Upham P, Riesch H, Reiner D, Brunsting S, de Best-Waldhober M, Dütschke E, Oltra C, Sala R, McLachlan C (2010) Review of the public participation practices for CCS and non-CCS projects in Europe. Deliverable 1.2: NEARCO2—new participation and communication strategies for neighbors of CO_2 capture and storage operations. IEEP, London

Dütschke E (2010) What drives local public acceptance—comparing two cases from Germany. Energy Procedia 4:6234–6240

Dütschke E, Wohlfarth K, Schmidt A, Pietzner A, Schwarz A, Carpantier R, Schumann D (2014) Akzeptanz von CO_2-Speicherprojekten in Deutschland—Eine Tiefenanalyse basierend auf Fallstudien. Project report CCS-Chancen. Fraunhofer ISI, Karlsruhe, Germany

European Commission (2013) Communication from the Commission to the European Parliament, the Council, the European Economic and Social Committee and the Committee of the regions on the Future of Carbon Capture and Storage in Europe

Fischer W (2015) No CCS in Germany despite the CCS act? In: Kuckshinrichs W, Hake J-F (eds) Carbon capture, storage and use—technical, economic, environmental and societal perspectives. Springer International Publishing Switzerland, Switzerland

Global CCS Institute (2014) The global status of CCS. http://cdn.globalccsinstitute.com/sites/default/files/publications/121016/global-status-ccs-february-2014.pdf. Accessed Feb 2014

Gough C, O'Keefe L, Mander S (2014) Public perceptions of CO_2 transportation in pipelines. Energy Policy 70:106–114

Hammond J, Shackley S (2010) Towards a public communication and engagement strategy for carbon dioxide capture and storage projects in Scotland: Scottish centre for carbon capture. Working paper 2010-08. SCCS, Edinburgh

IEA (2013) Technology roadmap. Carbon capture and storage. http://www.iea.org/publications/freepublications/publication/CCS_Roadmap.pdf. Accessed Sept 2014

Mander S, Polson D, Roberts T, Curtis A (2010) Risk from CO_2 storage in saline aquifers: a comparison of lay and expert perceptions of risk. Energy Procedia 4:6360–6367

Oltra C, Upham P, Riesch H, Boso À, Brunsting S, Dütschke E, Lis A (2012) Public responses to CO_2 storage sites: lessons from five European cases. Energy Environ 23(2–3):227–248

Pietzner K, Schumann D (eds) (2012) Akzeptanzforschung zu CCS in Deutschland. Aktuelle Ergebnisse, Praxisrelevanz, Perspektiven. Oekonom Verlag, München

Pietzner K, Schumann D, Esken A (2010) CO_2-Abscheidung und -Speicherung aus gesellschaftlicher Sicht. ÖkologischesWirtschaften 4:39–42

Pietzner K, Schwarz A, Dütschke E, Schumann D (2014) Media coverage of four carbon capture and storage (CCS) projects in Germany: analysis of 1115 regional newspaper articles. Energy Procedia, pp 7141–7148

Reiner D, Riesch H, Chyong CK, Brunsting S, de Best-Waldhober M, Dütschke E, Oltra C, Lis A, Desbarats J, Pol M, Breukers S, Upham P, Mander S (2011) Near CO_2 WP 2. Opinion shaping factors towards CCS and local CCS projects: public and stakeholder survey and focus groups. University of Cambridge, Cambridge

Schumann D (2014) Akzeptanz von CO_2-Offshore-Speicherung. CO_2-Onshore-Speicherung und CO_2-transport per Pipeline in der deutschen Bevölkerung. Bericht zu Arbeitspaket 2 im Projekt "Chancen für und Grenzen der Akzeptanz von CCS in Deutschland" (CCS-Chancen). Forschungszentrum Jülich, Jülich

Schumann D (2015) Public acceptance. In: Kuckshinrichs W, Hake J-F (eds) Carbon capture, storage and use—technical, economic, environmental and societal perspectives. Cham, Springer International Publishing Switzerland, Switzerland (in press)

Schumann D, Dütschke E, Pietzner K (2014) Public perception of CO_2 offshore storage in Germany: regional differences and determinants. Energy Procedia, pp 7096–7112

Schumann D, Pietzner K, Esken A (2010) Umwelt, Energiequellen und CCS: regionale Unterschiede und Veränderungen von Einstellungen der deutschen Bevölkerung. Energiewirtschaf-tliche Tagesfragen 60(5):52–56

Special Eurobarometer (2011) Public awareness and acceptance of CO_2 capture and storage. http://ec.europa.eu/public_opinion/index_en.htm. Accessed Sept 2014

Stigson P, Hansson A, Lind M (2012) Obstacles for CCS deployment: an analysis of discrepancies of perceptions. Mitig Adapt Strat Glob Change 17:601–619

Upham P, Roberts T (2011) Public perceptions of CCS: emergent themes in pan-European focus groups and implications for communications. Int J Greenhouse Gas Control 5:1359–1367